Inside OrCAD Capture for Windows

The EDN Series for Design Engineers

N. Kularatna *Power Electronics Design Handbook: Low-Power Components and Applications*
J. Lenk *Simplified Design of Microprocessor-Supervisory Circuits*
C. Maxfield *Designus Maximus Unleashed!*
EDN Design Ideas (CD-ROM)
C. Schroeder *Printed Circuit Board Design Using AutoCAD*
J. Lenk *Simplified Design of Voltage-Frequency Converters*
J. Lenk *Simplified Design of Data Converters*
F. Imdad-Haque *Inside PC Card: CardBus and PCMCIA Design*
C. Schroeder *Inside OrCAD*
J. Lenk *Simplified Design of IC Amplifiers*
J. Lenk *Simplified Design of Micropower and Battery Circuits*
J. Williams *The Art and Science of Analog Circuit Design*
J. Lenk *Simplified Design of Switching Power Supplies*
V. Lakshminarayanan *Electronic Circuit Design Ideas*
J. Lenk *Simplified Design of Linear Power Supplies*
M. Brown *Power Supply Cookbook*
B. Travis and I. Hickman *EDN Designer's Companion*
J. Dostal *Operational Amplifiers, Second Edition*
T. Williams *Circuit Designer's Companion*
R. Marston *Electronic Circuits Pocket Book: Passive and Discrete Circuits (Vol. 2)*
N. Dye and H. Granberg *Radio Frequency Transistors: Principles and Practical Applications*
Gates Energy Products *Rechargeable Batteries: Applications Handbook*
T. Williams *EMC for Product Designers*
J. Williams *Analog Circuit Design: Art, Science, and Personalities*
R. Pease *Troubleshooting Analog Circuits*
I. Hickman *Electronic Circuits, Systems and Standards*
R. Marston *Electronic Circuits Pocket Book: Linear ICs (Vol. 1)*
R. Marston *Integrated Circuit and Waveform Generator Handbook*
I. Sinclair *Passive Components: A User's Guide*

Inside OrCAD Capture for Windows

Chris Schroeder

Boston Oxford Johannesburg Melbourne New Delhi Singapore

Newnes is an imprint of Butterworth–Heinemann.

A member of the Reed Elsevier group

Recognizing the importance of preserving what has been written, Butterworth–Heinemann prints its books on acid-free paper whenever possible.

Butterworth–Heinemann supports the efforts of American Forests and the Global ReLeaf program in its campaign for the betterment of trees, forests, and our environment.

Library of Congress Cataloging-in-Publication Data
Schroeder, Chris, 1954–
Inside OrCAD capture for Windows / Chris Schroeder.
p. cm. -- (EDN series for design engineers)
Includes index.
ISBN 0-7506-7063-0 (alk. paper)
1. Electronics—Charts, diagrams, etc.—Data processing. 2. OrCAD SDT. 3. Electric circuit analysis—Data processing. 4. Printed circuits—Design and construction—Data processing. I. Title. II. Series.
TK7866.S3723 1998
621.3815'0285'5369—dc21 98-8023
CIP

British Library Cataloguing-in-Publication Data
A catalogue record for this book is available from the British Library.

The publisher offers special discounts on bulk orders of this book.
For information, please contact:
Manager of Special Sales
Butterworth-Heinemann
225 Wildwood Avenue
Woburn, MA 01801-2041
Tel: 781-904-2500
Fax: 781-904-2620

For information on all Butterworth–Heinemann publications available, contact our World Wide Web home page at: http://www.bh.com

10 9 8 7 6 5 4 3 2 1

Printed in the United States of America

Contents

Preface

This book is about OrCAD Capture, the industry leader in schematic capture software. Unlike the hundreds of books written about mechanical CAD (computer-aided design) software such as AutoCAD, few books have been written about EDA (electronic design automation), and this is the first one that covers Capture. *Inside OrCAD Capture for Windows* was written in recognition of the needs of both new and experienced users. The book is organized into a series of tutorial exercises, preceded by two chapters giving an overview of schematic drafting and Capture installation. Most of the tutorial exercises are divided into two or three sessions. Each session can be completed in less than two hours and contains a mix of theory and practice. The disk that accompanies the book contains sample files for the tutorial exercises and a special utility for sorting bills of materials.

New users can read Chapters 1 and 2 and then complete the first four tutorial exercises in Chapters 3 through 6 to get off to a fast start using Capture. Experienced users can hone their skills with the advanced tutorials starting in Chapter 7. Experienced users can also refer to particular chapters for information about specific subjects, such as creating netlists for PCB design or SPICE circuit simulation.

Certain prerequisites exist for the use and understanding of the material presented in this book. The reader is assumed to have some knowledge of electronics and schematic drafting. Some degree of PC literacy including familiarity with Windows 95 also is required. In order to complete the tutorial exercises, the reader will require a PC running Windows 95 with attached laser or inkjet printer and access to OrCAD Capture software.

Acknowledgments

This book is dedicated to the memory of my grandmother, Emmy Schaefer. During the postwar years in Germany, she provided a loving home environment while my parents pursued careers in medicine. After moving to the United States with my parents in 1960, I spent many wonderful summers visiting my grandparents in the old country, and they continued to influence my life. Some would argue about whether or not it takes a village to raise a child, but I have no doubts about the importance of a grandmother's love.

I would like to thank the staff at Crane Cams for encouraging me in this project. Special credit goes to my boss, Bill Gaterman, the vice-president of engineering, for giving me the freedom to pursue creative activities that do not always show an immediate return on investment. I am also grateful to Rob Chamblin for his suggestions after trying some of the techniques presented in the book.

On a final note, I would also like to thank the staff at Butterworth-Heinemann, especially Jo Gilmore.

About the Author

Chris Schroeder received his B.S. in Engineering from the University of Michigan in 1976. He is currently chief engineer for the automotive electronics business unit of Crane Cams in Daytona Beach, Florida, and has previously been involved with sales and marketing of computer graphics equipment and design of industrial electronics. He enjoys flying small airplanes, playing with electronics, and writing. Chris lives with his wife, Tina, and young son, Garrett, in Ormond Beach, Florida.

1

Introduction to Schematic Capture

Schematic diagrams are used to graphically represent the components and interconnections of electrical circuits. In the past, schematics were drafted with manual drawing techniques. Up until the late 1970s, the only schematic drafting aids were plastic drawing templates. The author still has a large collection of schematic, flow chart, and PCB (printed circuit board) design templates from Bishop Graphics and Berol dating back to those days. Other than the use of templates and new symbols for solid state devices, little had changed for almost fifty years. Back in the late 1970s, CAD (computer-aided design) systems based on mainframe and minicomputer technology were just starting to see use in larger companies, but these systems were very expensive. At over $100,000 per seat, simple tasks like routine schematic drafting were difficult to cost justify. That situation quickly changed with the advent of the IBM PC and low cost CAD software such as AutoCAD.

OrCAD SDT schematic capture software was originally introduced in 1985 and quickly became accepted, largely because of ease of use, speedy performance on PC workstations, and low cost. Today, with over 100,000 copies sold, the Windows version, OrCAD Capture, has become the most widely used schematic software. While OrCAD was not the first company to offer schematic capture on the PC, it popularized the concept and is now the undisputed industry leader.

When the author first started working in the electronics industry back in the mid-1970s, CAD systems were rare, and engineers usually drew up a rough schematic by hand. The circuit was then prototyped on a wire-wrap board. Once the circuit was debugged, the drafting department redrew the schematic and started the PCB layout. Schematic capture is now one of the first steps in the design process, and the term EDA (electronic design automation) is used in place of CAD. In today's environment of competitive pressures to reduce time to market and with the widespread use of SMT (surface mount technology), the engineer usually creates the schematic using an EDA tool such as OrCAD Capture as the first step toward generating a prototype PCB.

Schematic capture creates a database of components and interconnections along with the graphical schematic.

Via postprocessing steps, this database can automatically generate netlists for PCB and programmable logic device design, bill of materials listings, and various output formats used for timing and circuit analysis. This database capability results in significant cost and time savings in the overall design process.

While the focus of this book is familiarizing the reader with OrCAD Capture, a thorough understanding of modern schematic drafting practices is a prerequisite for effectively utilizing the software. OrCAD Capture (henceforth referred to as Capture) enforces a degree of discipline through concepts such as the organization of multiple-sheet schematics into a structured hierarchy and the use of standardized libraries of parts. In addition, Capture has certain unique constructs, including invisible power and ground pins and hierarchical ports (internal connections between sheets) that are not commonly found in traditional manually drafted schematics. These concepts and constructs must be considered and adhered to in order to obtain all the benefits of schematic capture.

The assumption is made that the reader has some knowledge of schematic drafting and the use of Microsoft Windows on PC systems. The orientation of this book is towards automotive, computer, and industrial control electronics. Most of the examples in this and subsequent chapters are taken from real-world applications.

Using Electronic Symbols

Electronic schematics consist of symbols that represent the individual electronic parts used in the circuit. These symbols are interconnected with lines that represent the actual electrical connections. Figure 1-1 shows symbols for the most common of all electronic parts, the resistor. On a typical schematic, each symbol represents an individual part. The symbols are annotated with text. The basic schematic symbols are highly standardized, since the most common parts such as resistors and capacitors have been in use for almost a century.

During the last two decades there has been an ever-accelerating proliferation of complex integrated circuit devices. An early attempt was made by the IEEE (Institute of Electrical and Electronics Engineers) to standardize the representation of these parts with a complex new symbology. This approach was feasible with decoders, counters, and bus-oriented devices such as latches and drivers. The IEEE symbols did not keep pace with the advent of VLSI (very large scale integrated circuit) devices such as communications controllers, microcontrollers, programmable logic, and other devices that sometimes have hundreds of pins.

Capture provides support for IEEE symbols, but few companies still use them. For the most part, they have been forgotten.

Reference Designators

Each symbol is annotated with text that includes a reference designator, for example R1 or R2, and a description of the part. In the case of R1 on Figure 1-1, the description consists of the value (1.0K or 1000 ohms) and wattage rating (.25 watts). Other descriptive text might include the tolerance (1% for R2), a voltage rating, or a manufacturer's part number.

It is very important to clearly understand the importance of the reference designator and the rules for assigning reference designators. An alphanumeric reference designator is used to uniquely identify each part. A given circuit might have ten 1.0K resistors used in different locations. Each of these resistors is given a unique reference designator, for example, R1, R5, and R7. In addition to the schematic, the reference designators also appear on the PCB legend silkscreen, assembly drawing, and bill of materials. Manufacturing uses the reference designators to determine where to stuff parts on the board. Field service uses them to identify and replace failed parts.

Standards have evolved for assigning reference designators. A reference designator consists of an alphabetic prefix and a numeric suffix. Each class of electronic parts has a one or two letter prefix. Most companies use the ANSI (American National Standards Institute) reference designator prefixes with minor modifications as given in Table 1-1. The numeric suffix is numbered starting from one for each class of part, for example, C1, C2, C3, R1, R2, U1, and U2. With manual drafting, the convention is to number each class starting at the upper left-hand corner and then going from left to right in rows from top to bottom. Capture can automatically annotate (assign and number) reference designators. The prefix is predefined in the part library, which is described in more detail later in this chapter. Capture will number parts in each class in the order they are placed into the design. If the design is created with a reasonable flow, automatic annotation produces acceptable results.

Some parts, such as logic ICs (integrated circuits), consist of multiple subparts or gates. In this case, common practice is to add an additional alphabetic suffix to the reference designator, starting with the letter A. For example, the four individual gates of a CMOS 4001 quad NOR gate might be designated U5A, U5B, U5C, and U5D.

Table 1-1 Reference Designator Prefixes

A	assembly, subassembly, device, or function block that is separable and/or repairable
AT	attenuator, isolator (RF devices)
B	fan, motor
BT	battery, photocell
CB	circuit breaker
CP	coupler, junction (RF devices)
D	diode, any two-terminal semiconductor device
DC	directional coupler (RF device)
DL	delay line
DS	alarm, buzzer, visual, or audible signaling device
E	antenna, any miscellaneous electrical device
F	fuse
FL	filter
G	generator (rotating machine)
H	hardware
HY	circulator (RF device), hybrid circuit
J	receptacle (stationary connector)
JP	jumper plug (common usage on computer boards)
K	contactor, relay (**CR** often used in industrial electronics)
L	inductor, coil (single winding that may have multiple taps)
LS	loudspeaker, horn, any audio/ultrasonic output transducer
M	meter, clock, strain gauge, any miscellaneous instrument
MG	motor generator
MK	microphone, any audio/ultrasonic input transducer
MP	mechanical device, any without electrical connections
P	plug (removable connector)
PS	power supply
Q	transistor, MOSFET, SCR, any three-terminal semiconductor
R	resistor, any fixed or variable (Capture uses **RN** for resistor network)
RT	thermistor
RV	varistor
S	switch, thermostat, thermal cutout
T	transformer, including autotransformer with single winding
TB	terminal board (obsolete)
TC	thermocouple
TP	test point
U	integrated circuit (use of IC is obsolete), nonrepairable assembly
V	electron tube, vacuum/ion device including high power RF

Table 1-1 Reference Designator Prefixes (Cont'd)

VR	obsolete usage for zener diode or voltage regulator
W	waveguide, transmission line (RF device)
X	socket for lamp or fuse
Y	crystal, ceramic resonator, tuning fork device
Z	tuned cavity or circuit, other miscellaneous RF networks

Table 1-1 is by no means all-inclusive, some companies and industries use varying practices, and the evolution of prefixes is ongoing. For example, the prefix CR has largely become obsolete, and D is now used for most two-terminal semiconductor devices, including LEDs. Likewise the prefixes IC and VR are obsolete, with U now being used for all integrated circuits including voltage regulators. In some areas of industrial controls, CR is used to refer to relays and contactors and PL is used for plugs.

On a final note, remember that the reference designator gives information only about the class of part (resistor, diode, and integrated circuit) and the location of the part on the schematic. The reference designator does not give any information about the electrical parameters of the part.

Part Descriptions

The part description must give concise information about all relevant electrical properties. An appropriate part description depends on the type of part. At first glance, it might appear that many passive components such as resistors and capacitors have been highly standardized and only the part value and tolerance would be required to specify the part. For example, one might assume that all 1000 ohm .25 watt 5% tolerance resistors are readily interchangeable, and therefore the part value and tolerance should be a sufficient description. In fact, most schematics are still drawn with such assumptions. Unfortunately, matters have been complicated by two recent trends: schematic capture as the starting point for the design process and the proliferation of SMT.

When schematic capture is used as the starting point, the bill of materials is generated from the schematic database. To the extent feasible, all relevant information available about a given part should be entered into this database. While this will entail more effort up front, updated bills of materials can be efficiently generated as the design evolves through successive engineering changes. This concept of the schematic as database is in sharp contrast to previous industry practice whereby bill of materials preparation and schematic drafting were distinct activities, and the schematic contained only a subset of the descriptive information contained in the bill of materials.

SMT is the other factor driving the need for more detailed parts descriptions. Today, a 1000 ohm .25 watt 5% resistor could be axial lead through hole, cylindrical MELF (metal electrode face), or EIA (Electronic Industry Association) 1210 size rectangular chip. In some cases, due to considerations of voltage standoff requirements or pulse power handling capability, a design could have a mix of both standard and SMT devices with the same electrical parameters.

Capture makes provision for multiple fields in the part description. Efficient and rigorous use of this capability during the early stages of the design process will greatly reduce errors and manual editing of the bill of materials. Detailed guidelines as to what information should accompany individual classes of parts are given in the following sections.

First, let's review the units and associated symbols used to describe the values of electrical circuits and parts commonly found on schematics. There is an immediate problem that needs to be addressed. ANSI/IEEE standards call for a mix of upper case and lower case letters and some Greek letters, such as Ω for ohms, which is the unit for resistance. This is in direct conflict with drafting convention that only upper case letters appear in drawings and with the limitations of ASCII keyboards and output devices. Capture supports lower case letters; however, problems can occur in later postprocessing operations. The use of mixed capitalization is not recommended. Table 1-2 gives multiplier prefixes for use with engineering units, and Table 1-3 gives the most common electrical units and associated symbols used by convention.

Table 1-2 Multiplier Prefixes

UNIT	ANSI SYMBOL	MULTIPLIER
femto	f	10^{-15}
pico	p	10^{-12}
nano	n	10^{-9}
micro	μ (use u)	10^{-6}
milli	m	10^{-3}
kilo	k	10^{3}
mega	M	10^{6}
giga	G	10^{9}

Micro and milli prefixes appear to be a problem if only upper case letters are used. The situation is not as bad as it appears because there are no common devices in which both these prefixes are likely to occur. Resistors are usually in the range of .01 ohm to 22 megohm. By convention, when M is used with resistance values, it always stands for megohm (for example, 10M is 10 megohm).

Inductors are usually in the .1 microhenry to 10 henry range, with millihenry values quite common. By convention, MH stands for millihenry.

Table 1-3 Engineering Units

UNIT	CONVENTIONAL SYMBOL
Capacitance	
picofarad	PF
nanofarad	NF
microfarad	UF or no symbol (MFD is archaic)
Inductance	
microhenry	UH
millihenry	MH
henry	H
Resistance	
milliohm	write out in decimal, for example .001
ohm	no unit symbol (R used in Europe)
kilohm	K
megohm	M
Electrical Units	
microampere	UA
milliampere	MA
ampere	A
microvolt	UV
millivolt	MV
volt	V
kilovolt	KV
milliwatt	MW
watt	W
kilowatt	KW
Mechanical Units	
microsecond	US or USEC
millisecond	MS or MSEC
second	SEC
minute	MIN

Table 1-3 Engineering Units (Cont'd)

UNIT	CONVENTIONAL SYMBOL
Mechanical Units (Cont'd)	
hour	HR
mil (.001 inch)	MIL
inch	IN
foot	FT
centimeter (.01 meter)	CM
meter	M
ounce	OZ
pound	LB
gram	GM
kilogram	KG

While on the subject of units, let's briefly discuss numbering. Preferred practice with schematic capture is to use decimal values rather than fractions. Use .25 watts rather than 1/4 watt. Unlike practice on some mechanical drawings, leading zeros are not used in front of decimal points. Postprocessing routines may encounter difficulties unless these considerations are observed.

Dropping the units symbol is accepted practice with resistors. The appearance of *ohms* or the proper Greek symbol (Ω) is now rare. In Europe, the letter R is sometimes used both as a unit symbol and decimal point placeholder. A 4R7 resistor is 4.7 ohms. Good practice is to add a text note to the schematic indicating that all resistance values are in ohms unless otherwise specified.

Common film and electrolytic capacitors are in the microfarad range. The use of the symbol UF is common. The use of MFD has become archaic. The trend is to entirely drop the units symbol for microfarad range parts. Again, good practice would dictate a text note to this effect.

OrCAD Capture Symbols for Electronic Parts

This section provides an overview of some of the most widely used types of schematic symbols available in the Capture libraries and includes background information on appropriate part descriptions. Figure 1-1 shows Capture symbols for the ubiquitous resistor — the most common of all electronic parts. Two styles are used for fixed resistors. R1 is the style used in Europe and industrial controls in North America. R2 is the more traditional style.

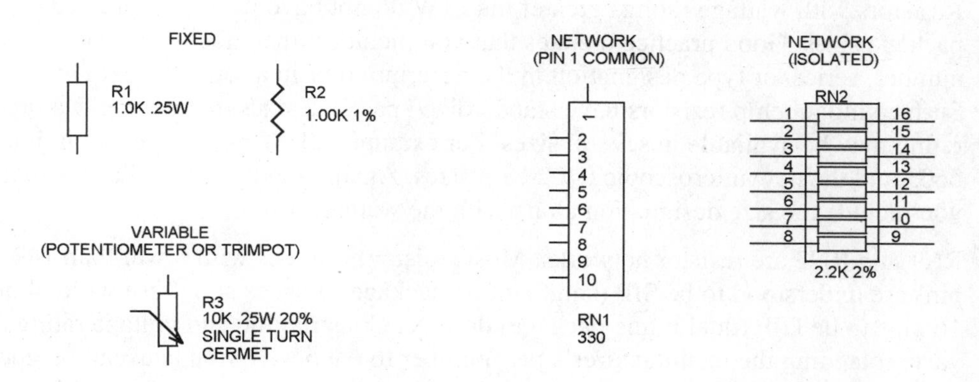

Figure 1-1 Resistor Symbols

The R1 style more closely represents modern film resistors, whereas the R2 style was derived from the wirewound construction of resistors dating back to the turn of the century. Note that Figures 1-1 through 1-12 show the symbols as they would appear in Capture if they were part of an interconnected circuit. Parts with unconnected electrical terminals are normally displayed with connection squares (shown in Figure 1-13 on page 23).

The description of a discrete resistor such as R1 or R2 should include the part value in ohms, wattage, tolerance, temperature coefficient for precision resistors, material, and package size. Materials include wirewound (usually high-wattage power resistors), carbon composition, carbon film, and metal film. Carbon composition resistors have been almost entirely supplanted by low-cost carbon film types that can easily be manufactured in tighter 5 percent tolerance. In fact, they are so widely used that one can safely assume, if no other information is provided, that a fractional watt 5 percent resistor is carbon film. However, relying on such assumptions is not good drafting practice.

Axial lead resistor sizes are well standardized for fractional watt parts, and it is usually not necessary to specify additional package size information when the wattage rating is given. Schematics often have notes such as, ".25W RN55." The RN55 is an old MIL-SPEC package size designation commonly used for 1 percent metal film resistors. Years ago, .25W 1% metal film resistors were the size of .5W carbon resistors (MIL-SPEC RN60 package size) and RN55 parts were rated only .125W. Better processing techniques have raised the wattage rating of RN55 size parts. The ".25W RN55" statement precludes the use of older style parts that may be too large.

Resistors with wattage ratings greater than 1W do not have well-standardized package sizes. Good practice dictates that you include a manufacturer's part number, series, or type designation in the description or in a separate text note. Surface mount chip resistors have standardized package sizes, but a given wattage rating may be available in several sizes. For example, .06W resistors come in both 0603 and the new microscopic 0402 chip sizes. Again, good practice dictates that you include the size designation along with the wattage rating.

RN1 and RN2 are resistor networks. Most resistor networks with fewer than 14 pins are understood to be SIP (single inline package) devices and those with 14 or 16 pins to be DIP (dual inline package) devices. Construction and voltage ratings vary, so adding the manufacturer's part number to the description is usually a good idea. Variable resistors, such as R3, can be either small trimpots used to calibrate analog functions or panel-mounted potentiometers (pots) for user adjustments. There are many different styles, materials, shaft configurations, and pin arrangements; therefore, a specific part number should be included in the description or in a separate text note.

Much of the information about the resistors on a given schematic is redundant and can be summarized in a few text notes. The use of text notes eliminates unnecessary clutter and makes the schematic more readable.

For example, the notes might state:

1. All resistors .25W 5% carbon film unless otherwise specified.
2. All 1% resistors .25W RN55 size metal film with 50 PPM temperature coefficient unless otherwise specified.
3. All 5W resistors Clarostat series VC5E or equivalent wirewound.

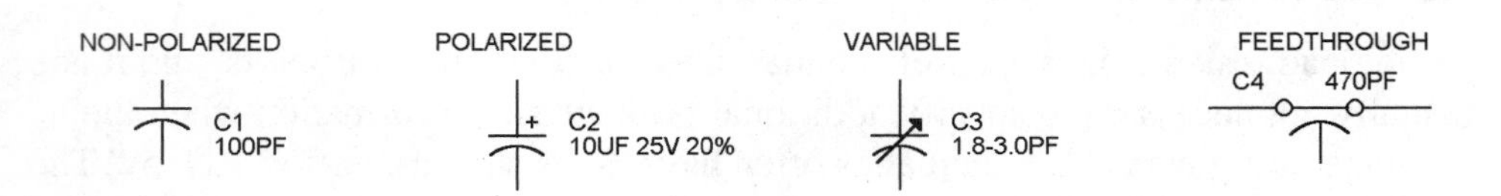

Figure 1-2 Capacitor Symbols

Figure 1-2 shows capacitors. The unit for capacitance is the farad (F), but most capacitors are in the microfarad or picofarad range, so the units UF and PF are used. By convention, the capacitor symbol is always oriented toward ground or the lowest DC voltage. For circuits where the DC voltage is not fixed or known, nonpolarized capacitors are drawn with the convex side facing right or down. Additional information required to describe a capacitor includes the voltage rating, tolerance, and construction. Fixed-value capacitors can be nonpolarized and polarized. Capacitors are classified according to the electrode and dielectric materials used in their construction. Other than surface mount devices, in which a standard size code can easily be added to the part description, there are no real standards for capacitor sizes, and the manufacturer's part number should be included in the description.

Nonpolarized types include ceramic, silver mica, and various metal foil-film and metallized films such as polystyrene, polyester, polypropylene, and polycarbonate dielectrics. Older types with oil or wax impregnated paper dielectrics are obsolete. Polarized capacitors include tantalum and aluminum electrolytic types.

Ceramic capacitors have two- or three-character alphanumeric designations for the dielectric properties that determine initial tolerance and temperature characteristics. Ceramic capacitors used to bypass power at ICs are usually .01UF to .1UF Z5U +80 −20% types. Oscillator, RF, and timing circuits often require more precise and stable ceramic NPO or COG 5% parts with values ranging up to about 1000 PF. Coupling and filtering applications commonly use ceramic Y5P and X7R 10% types ranging up to about .47 UF. It is not uncommon to find circuits in which .1UF Z5U types are used for bypass and identical value .1UF X7R types for timing and filtering. The author can recall one incident in which an improper schematic description resulted in the inadvertent substitution of a Z5U ceramic capacitor into a critical circuit that subsequently malfunctioned.

Film capacitors with metal foil or metallized film are used for specialized applications that require tight tolerance, high-voltage or pulse power handling capacity, or very low leakage. A detailed discussion of the design choices and tradeoffs for the various film capacitors is beyond the scope of this book. However, any incorrect substitutions in high voltage or pulse power applications can have disastrous and life-threatening consequences. In addition, parts that have regulatory agency (UL, CSA, and VDE) approvals are required for many antenna coupling and power supply across-the-line applications. The author suggests that a specific manufacturer's part number be given for such critical parts.

Similar considerations apply to polarized capacitors, especially the aluminum electrolytics. High-frequency switching power supplies with aluminum electrolytic capacitors require that the parts be characterized for load life at high temperature, internal resistance (ESR), and ripple current handling capability. Inclusion of a specific manufacturer's part number is highly recommended. A final note of warning on capacitors has to do with polarity. Reverse polarity on a polarized capacitor can cause fire or explosion. Errors can easily occur when general rules of schematic flow (discussed later in this chapter) are not followed and polarized capacitors are drawn with varying orientations on the same sheet.

Inductive devices are found in RF circuits and power supplies. The unit for inductance is the henry (H), but as with capacitors, practical inductors have much smaller values. Inductors for RF circuits are usually air core or ferrite core (drawn same as iron core) in the microhenry (UH) range; those for power supplies can range up to hundreds of millihenry (MH) and are usually ferrite or iron core. The Capture part library has a very limited number of transformer symbols. Most power transformers have more than two windings or multiple taps on windings, requiring the user to create a custom part with the library editor.

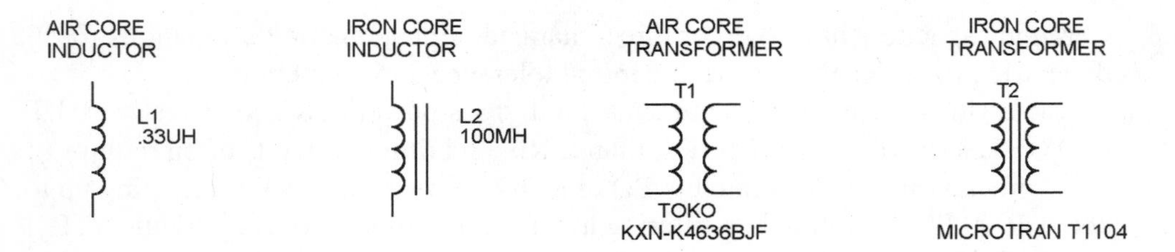

Figure 1-3 Inductor and Transformer Symbols

Almost all applications of inductors and transformers, such as those shown in Figure 1-3, involve consideration of a complex set of electrical parameters (Q factor, leakage inductance, winding resistance and capacitance, and core magnetic properties). A basic description of the device in terms of inductance for inductors or turns ratio, impedance ratio, or voltage levels for transformers is helpful for testing and troubleshooting purposes. A manufacturer's part number is required to specify the device, since there are no standards or generic parts.

Successful postprocessing of the Capture database for generation of a PCB design netlist (list of parts and interconnections), requires careful attention to pin numbers. When a custom transformer or other device is created with the Capture library editor, the Capture pin numbers must correspond to those used in the PCB design.

Capture has only a limited number of switches and relays in the part library. The most common are shown in Figure 1-4. For anything other than the most generic devices, the user will have to create a custom part. Even if it appears that one of the existing parts in the library can be used, careful attention must be paid to pin numbers if a netlist is going to be generated for subsequent PCB design. The manufacturer's part number should be included in the description of any part for which no industry standards exist. Good practice also dictates adding text to label switch functions and positions.

Capture supplies a unique but very graphically descriptive symbol for DIP switches. If text labels are added for on and off positions and the individual switch functions, this symbol makes board setup and troubleshooting more intuitive than individual switch sections scattered throughout a sheet.

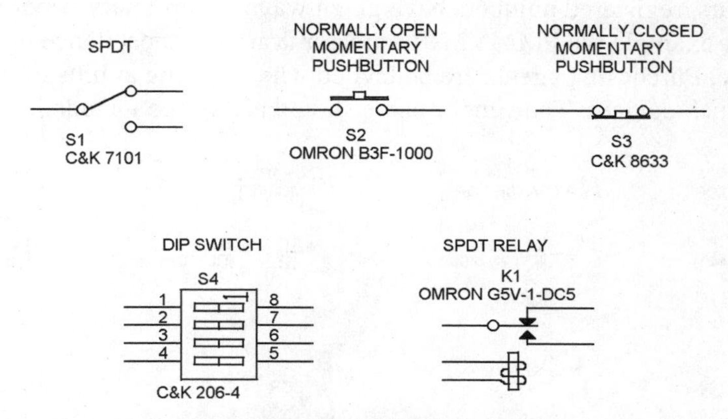

Figure 1-4 Switch and Relay Symbols

The most common semiconductor diodes are shown in Figure 1-5. Optoelectronic devices (LEDs and photodiodes) are covered later in this section. An industry standard numbering and part registration system was started in the 1950s using 1N prefixes for two-terminal diode devices and 2N prefixes for three-terminal transistor devices. In theory, all devices with the same number, for example 1N4148 diodes or 2N4401 transistors, should be fully interchangeable. In general this holds true for low-voltage, low-frequency, or low-power parts manufactured with mature technology. For high-voltage, high-power, or high-speed devices, there are usually significant differences in electrical parameters between vendors' parts. Good practice is to specify at least one qualified vendor.

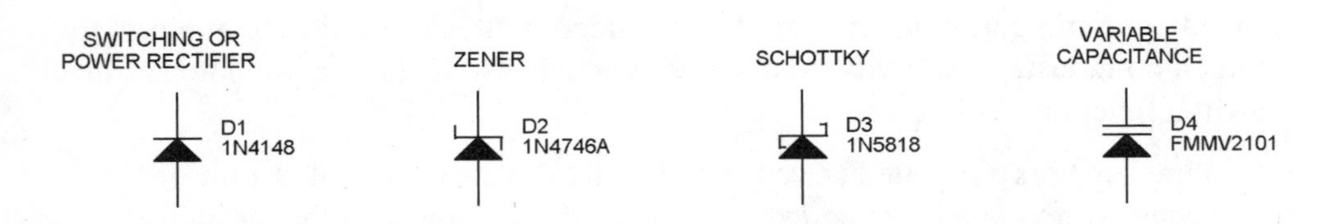

Figure 1-5 Diode Symbols

In recent years, registered numbers have given way to proprietary vendor part numbers, for example the FMMV2101 varactor (variable capacitance diode) shown above. Purchasing agents are bedeviled when looking at bills of materials calling out such devices. The vendor name should always be included.

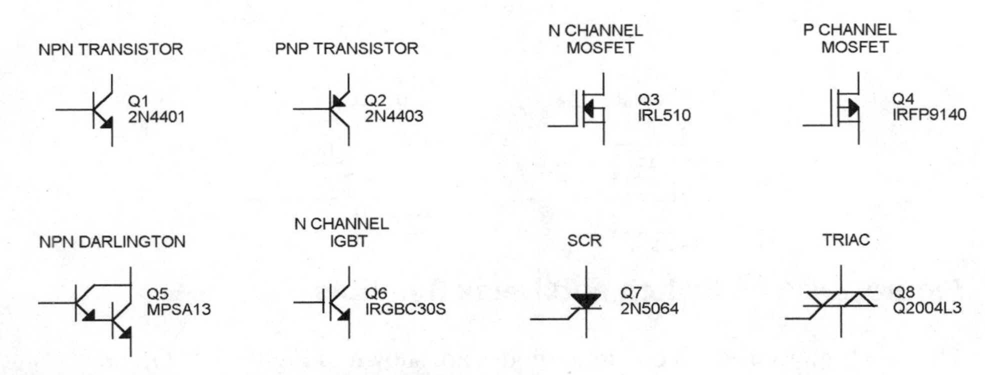

Figure 1-6 Transistor Symbols

Figure 1-6 shows the most frequently used transistors and other three-lead semiconductor devices as they appear in the Capture library. Note that all of the devices in Figures 1-5 and 1-6 appear without an outer circle. Use of an outer circle around semiconductor devices is considered archaic. Registered transistors and SCRs start with 2N for U.S. and 2S for Japanese devices; some older MOSFET devices start with 3N. Any part number starting with a letter is usually a part that originated with a particular vendor. The description guidelines suggested for diodes in the preceding section are also applicable to transistors and other three-lead semiconductors.

A recent development is the introduction of so-called smart power devices. This category includes transistors, MOSFETs, and IGBTs with on-chip features such as current limiting, short circuit protection, gate or output voltage clamping, and over-temperature protection. In a true sense these devices are really three-lead ICs.

Accepted practice has been to use the basic symbol that most closely represents that output device and simply ignore the presence of the other circuit elements. There are pro and con arguments for this approach, but the author suggests that if it is used, an explanatory text note should be included. This note should include the maximum current, clamping voltage, or other applicable parameters.

The devices in Figure 1-6 are shown in their preferred orientation, with current flowing from top to bottom. If at all possible, the schematic should be drawn with the devices in this orientation. The resulting schematic will be easier to understand.

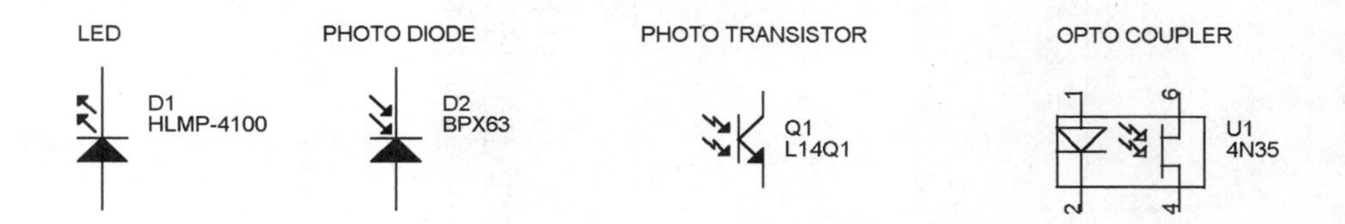

Figure 1-7 Optoelectronic Symbols

Considerations discussed for diodes and transistors apply to the optoelectronic devices shown in Figure 1-7. LEDs, photo diodes, and photo transistors have an optical function, such as a power-on indicator or an IR (infrared) communications link. Good practice is to include a text label explaining the intended function.

The term *analog* or *linear* refers to circuitry with signals that can vary over a range of voltages. Analog ICs, such as the examples shown in Figure 1-8, require certain unique considerations when drawn on a schematic. Capture comes with several analog part libraries, so a very wide range of parts is available. A triangular symbol has been in use for many years to indicate operational amplifiers (op amps) and comparators. All op amps and comparators use the same symbol, but the pinouts vary. Many op amp and comparator ICs contain multiple devices. Capture automatically handles multiple device parts and allows the user to determine which device is being placed (drawn) onto the schematic. Capture uses the convention of numbering the individual devices with an alphabetic reference designator suffix starting with the letter A. For example, if an LM324 quad op amp is assigned reference designator U1, Capture would label the individual devices U1A, U1B, U1C, and U1D. Power and ground pins appear only on the first device, U1A in this case.

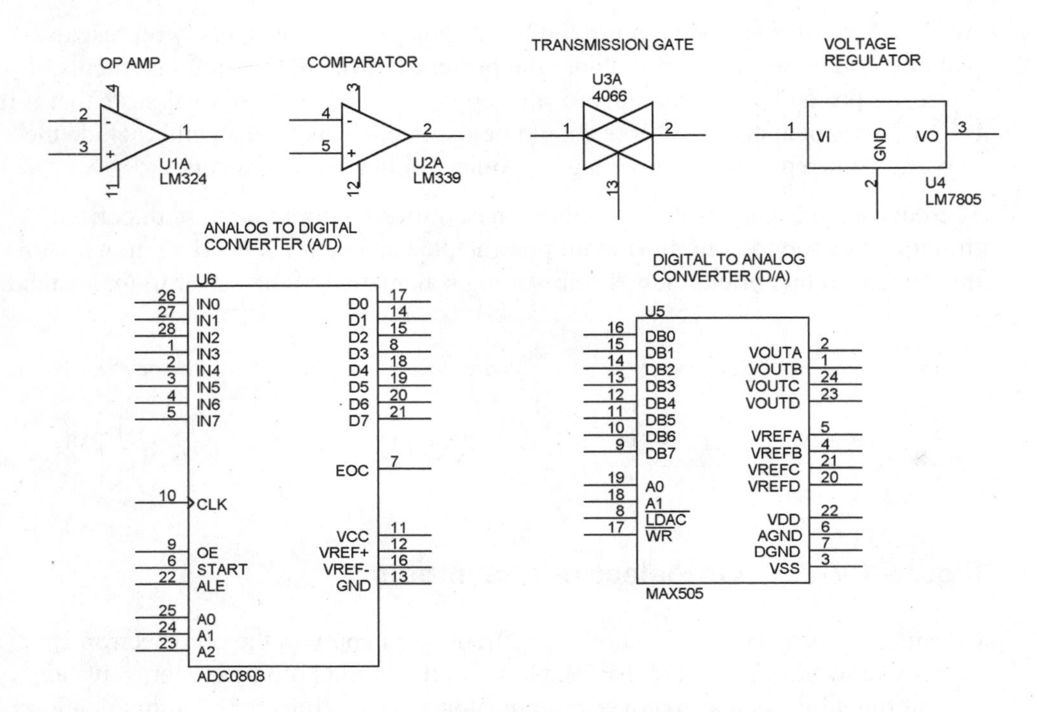

Figure 1-8 Analog Integrated Circuit Symbols

When placing parts on the schematic, Capture allows the parts to be rotated in 90-degree increments and mirrored horizontally and/or vertically. This high level of flexibility can be a disadvantage in that it allows the user to draw schematics that are difficult to interpret. Proper schematic "flow" will be discussed in detail later in this chapter. Two simple rules that should be followed for schematic flow when drawing op amps and comparators include:

- Draw op amps and comparators with the triangle symbol pointing from left to right as shown in Figure 1-8. This assures compatibility with the requirement that signals should flow from left to right.

- Draw the "–" inverting input on the top. Most electronics engineering textbooks and references show classic op amp feedback networks and comparator circuits with the inverting input terminal on top. Failure to follow this rule can lead to confusion and difficulty understanding circuit operation.

Transmission gates, once a novel part, are now widely used. The author has seen the symbol for the CMOS transmission gate, U3A, in schematics for industrial electronics. The symbol may originate from the similarity of the transmission gate to bidirectional buffers, or it may have evolved from a piping symbol for a valve.

No standard symbols exist for more complex analog ICs. A simple rectangular block with labeled and numbered pins is often used. Good practice is to include every pin on the device, even if it is not used or not internally connected. Pins should be arranged with inputs on the left, outputs on the right, power on top, and ground at the bottom. Multiple signals such as data or address should be arranged in order from top to bottom. An older practice is to arrange all pins in the same physical order as on the actual IC. This archaic practice usually results in a messy arrangement of signal interconnections.

Most analog ICs have part numbers that provide an adequate description. Many part numbers have alphabetic prefixes that identify the original vendor, for example CA for RCA (now Harris), LM for National, MAX for Maxim, and TL for Texas Instruments. Often second source vendors use the same prefix. The LM series is widely sourced. Including a vendor name with the part description would only be a requirement for relatively new or unique parts. Part numbers usually have an alphabetic suffix that identifies the package. The suffix may also identify the temperature range or other electrical parameter such as offset voltage, speed, or power consumption. An LM324N comes in a 14 pin DIP, an LM324M comes in a SOIC (small outline IC) surface mount package, and an LM324AN is a low-power DIP version.

Schematics are sometimes drawn as part of a reverse engineering project, and an entire book could easily be written on this subject. Determining the correct part number for a device can be difficult. Date codes can be mistaken for part numbers. The electronics industry uses four-digit date codes for most ICs. The first two digits are the year and the last two digits are the week. A part manufactured the 12th week of 1994 would be stamped 9412. Products manufactured in very high volume often have in-house part numbers, even on generic devices. Identifying these devices and finding an equivalent commercial part number can be difficult.

The term *digital* logic refers to circuitry with signals that are restricted to a limited number of logic states. A detailed description of digital logic functions and theory is beyond the scope of this book. Logic families in widespread use today include TTL and CMOS. Both use standard Boolean logic with 0 and 1 states. Zero is represented by a low voltage (near ground) and 1 by a high voltage (2.5 volts or higher depending on the device). Some devices have additional high-impedance or "off" state. This allows the outputs of multiple devices to be connected together, such as on a computer data bus, with only one device enabled at a time and the others in the off state.

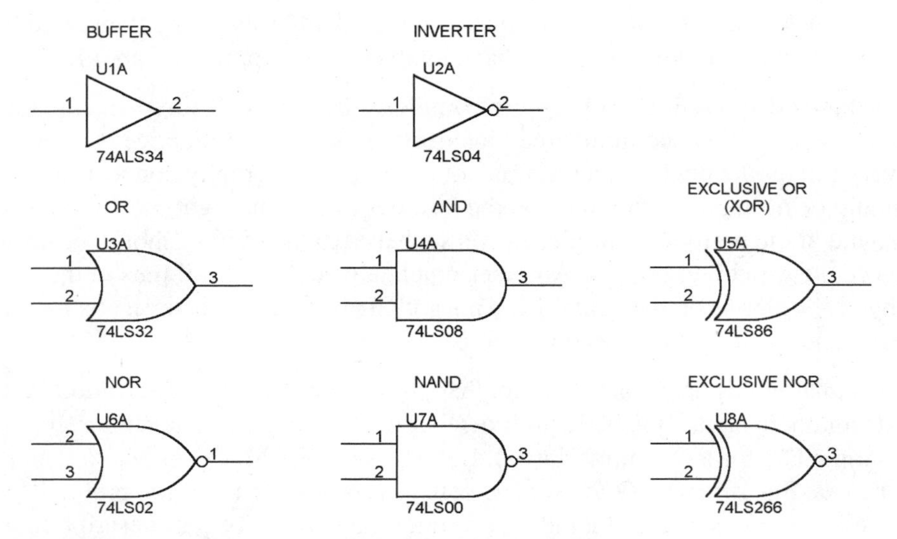

Figure 1-9 Digital Logic Gate Symbols

The four basic digital logic functions include the buffer, OR, AND, and XOR (exclusive OR). Symbols for these logic functions are shown in Figure 1-9. The four basic functions can also have inverted outputs. In this case they are referred to as the inverter (NOT), NOR, NAND, and XNOR (exclusive NOR), and the symbol is drawn with a small circle on the output. The small circle indicates an inverted signal, either input or output. Logic functions, when implemented in electronic circuitry, are referred to as *gates*. Buffer and inverter gates can only have a single input and output. XOR and XNOR gates, by nature of their logic function, can only have two inputs. All other gates can have two or more inputs.

Digital logic ICs may contain from one to six gates. Capture handles digital ICs with multiple gates using the same principles as discussed in the previous section for analog ICs with multiple devices. Likewise, a part number description is generally sufficient for digital logic, since multiple vendors source most parts. TTL is the most widely used logic family. Some TTL part numbers are shown in Figure 1-9, such as the 74LS02 NOR gate. The 74 prefix identifies the part as commercial temperature range (0 to 70 degrees C); a 54 prefix is used for military temperature range parts (–55 to 125 degrees C). The LS designates the part as belonging to the low-power Schottky logic family. The original TTL logic had no special designator. The part number for the quad NOR gate would have been 7402. Many different families have been introduced, each with its own two- or three-letter designator. In fact many of these new families are not even based on transistor technology. CMOS variants, such as a 74HC02, entail MOSFET technology with somewhat different logic levels but retain the popular device functions and pinouts originated with TTL.

Capture makes use of the fact that pinouts are the same for a given device function in all TTL logic families. Only a single part definition is used in the TTL library, and the various logic families are listed as optional part names. This is an advantage for the user, because part names can be edited. If the requirement is for a new 3.3V logic device such as a Phillips 74LVT244, which does not yet appear in a Capture library, one can use a 74LS244 and simply edit the name. This is much more convenient than creating a new part.

Figure 1-10 shows advanced digital ICs. As with advanced analog integrated circuits, these devices are drawn as rectangular blocks with named and numbered signal pins. Note that U2 and U3 both have inverted signal inputs and outputs denoted by a small circle. U1A also has noninverted and inverted outputs. Here a bar above the signal name denotes the inverted output on pin 2. Both the small circle and bar above the signal name conventions are used with Capture library parts on a rather arbitrary basis. In some cases, the choice of convention appears to have been based on how the original vendor represented the part in the databook. Another signal convention is the use of a small triangle to represent edge-triggered clock inputs, as shown on U1A pin 3 and U2 pin 15.

U5 and U6 introduce another special category — programmable memory and microcontroller devices. These devices, and others such as the broad class of programmable logic devices, contain firmware. Firmware is the same as software, except it is more or less permanently programmed into the device. With the advent of electrically erasable memory devices, the distinction between software and firmware has become more blurred, but convention is still to refer to any code loaded into nonvolatile memory as firmware.

Whenever devices requiring or utilizing firmware appear on a schematic, it is good practice to add a note, either to the device description or as a separate text note, that gives the firmware name and the earliest compatible version. Why note the earliest compatible version? Today schematics are usually generated at the start of a design project — not as part of the documentation created after the fact. Often earlier revisions of a product will exist. Firmware used with earlier board revisions may not be compatible. Firmware seems to go through rapid changes and there are usually multiple firmware versions released before the board hardware is updated. Noting the earliest compatible version can help prevent mistakes during manufacturing, final testing, and field servicing.

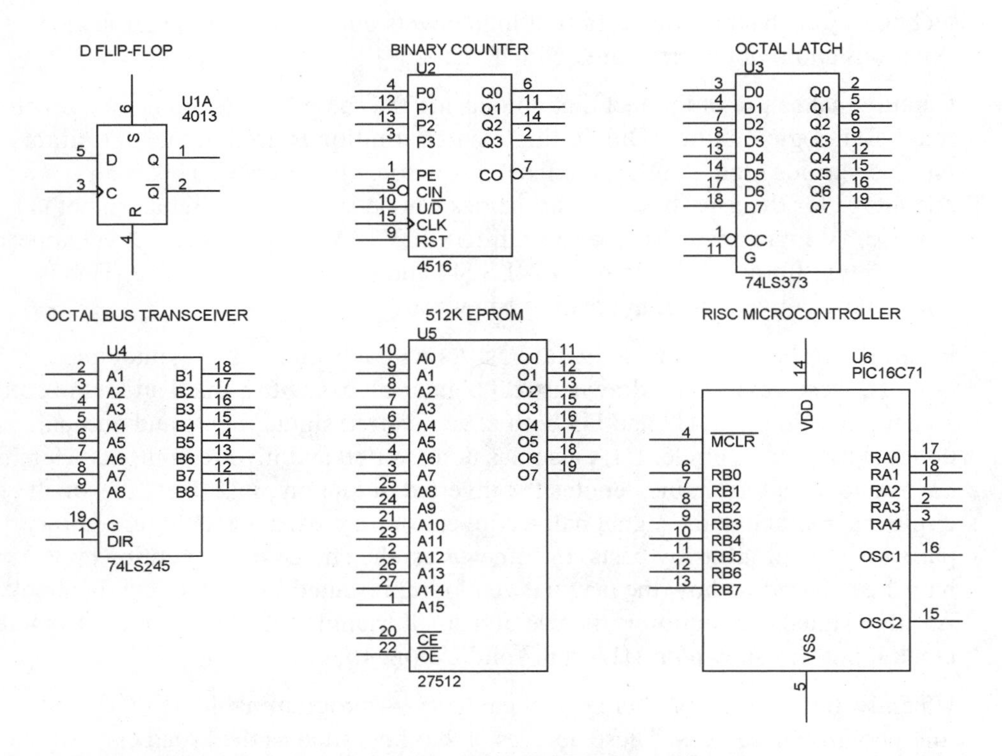

Figure 1-10 Advanced Digital Logic Symbols

Memory devices usually have a read or write access time, which is a very critical parameter in high-speed computer systems. Often the part number has a dash suffix, such as –150, where the number identifies the access time in nanoseconds. If this parameter is critical and the device is available in multiple speed grades, this information must be included on the schematic. A similar consideration applies to programmable devices, in which the programming voltage often varies between different vendors. If relevant, programming voltage information should also be included.

Figures 1-11 and 1-12 complete the overview of circuit symbols supplied with Capture. There are some inconsistencies between the Capture library parts and ANSI recommended reference designators. The ANSI designator for a microphone is MK, yet Capture uses X. Reference designators can be edited once the part has been placed onto the schematic, so this is not a serious problem. The reasoning for the letter A on the motor symbol is not clear.

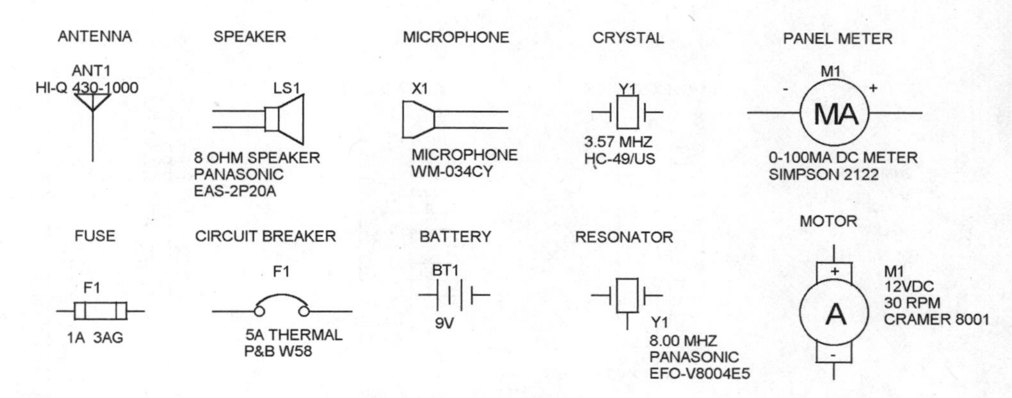

Figure 1-11 Miscellaneous Symbols

Batteries are represented by the symbol shown for BT1. Again there is a minor inconsistency. BT1, as drawn, shows a two-cell battery. To be technically correct, a 9V alkaline or carbon-zinc battery should be drawn showing six 1.5V cells. In today's environment of complex computer systems, such details often are glossed over. As long as the battery is correctly described, using the Capture-supplied symbol shown above for any multicell battery should not cause any confusion.

Some of the most commonly used connector symbols are shown in Figure 1-12. By convention, any fixed connector attached to a panel or motherboard is referred to as a *jack* and uses reference designator J. A removable connector, including the card edge connector on a daughterboard or PC bus expansion board, is referred to as a *plug* and uses reference designator P or PL. With the advent of user-configured computer expansion boards, the jumper or jumper plug has become ubiquitous. Jumpers are typically incorporated onto a circuit board as paired pads on .1 inch centers. For user-configurable board options, a header can be installed and a removable jumper provided to make the connection.

Jumpers used for manufacturing options or calibration functions often are normally closed with a small trace between the pads. The trace can be cut with an X-Acto knife to open the jumper. If required, a wire can be soldered between the pads to re-establish the jumper connection. It is very important to clearly describe the function of jumpers and whether the function is asserted with the jumper open or closed. For complex functions set with multiple jumpers, a table listing is good practice. Normally closed jumpers should be shown with a line drawn between the circles, similar to a normally closed switch.

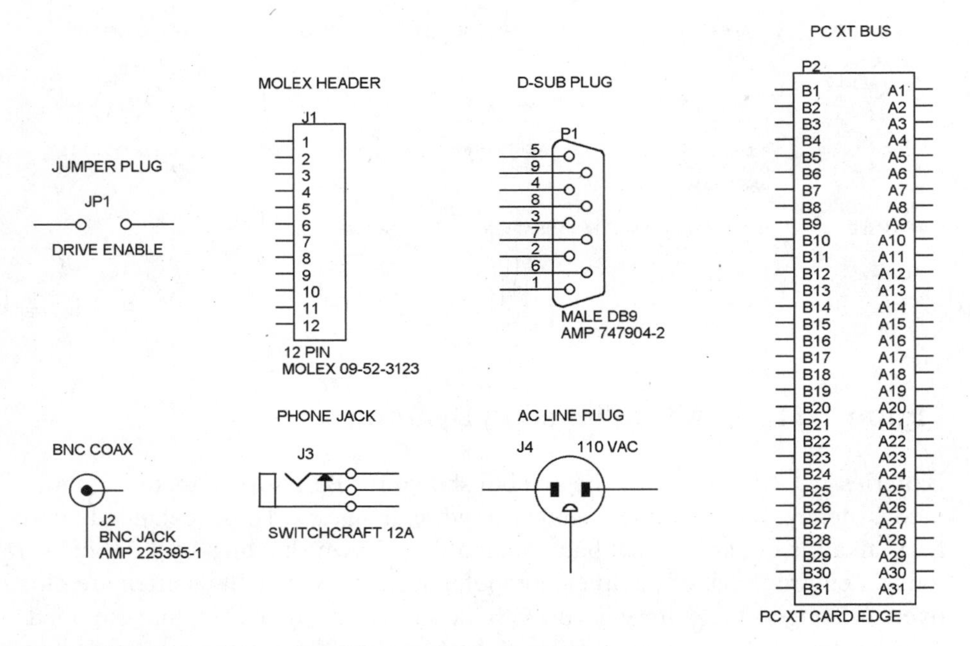

Figure 1-12 Connector Symbols

Good practice is to describe the physical configuration of all connectors, give a vendor name and part number if applicable, and add a text note that describes the function served or the connecting device.

Special Schematic Symbols

Capture and other schematic capture programs use certain special symbols to handle signal and power flow. These symbols are shown in Figures 1-13 and 1-14. The most basic symbol is a wire that interconnects part pins. Placing a wire between two pins causes a line to be drawn. At the same time, the connection is stored in the schematic database. A connection between two or more pins is referred to as a *net*. After the schematic design is completed, the connectivity information in the schematic database can be postprocessed into a netlist. The netlist is used to transfer the connectivity information into other programs and systems, as might be used for PCB design.

R1 in Figure 1-13 shows connection squares that normally appear on part pins before they are connected. Once a wire or another part pin is connected, the connection square disappears. Pins that are not intended to be part of any net must have a no connect symbol drawn at the pin (as shown at PL1 pin 1 on Figure 1-13). Otherwise Capture will flag the pin as an error.

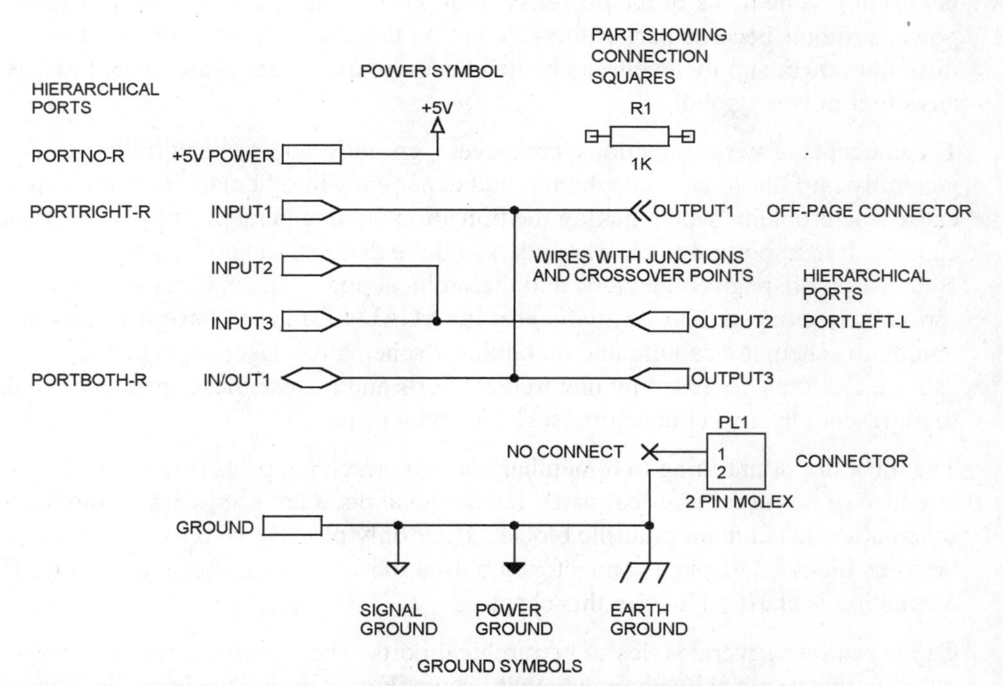

Figure 1-13 Capture Signal and Power Symbols

Wires that cross one another are not connected or considered part of the same net unless a junction symbol is placed at the crossover point. Capture junctions appear as small, filled-in squares. Manual drafting techniques often use a small dot to represent a junction and a break in the line to represent a crossover point where no connection exists. In Capture, wires must run continuously from pin to pin. The only exceptions are when a signal label or hierarchical port is associated with the wire.

Power symbols are used to represent connections to power and ground. Capture offers multiple ground symbol styles, including signal ground, power ground, and earth ground. These three ground symbols are electrically identical. They are *not* isolated from one another, because they all share the same signal name, GND.

Capture provides an undefined power symbol. This can be drawn in several styles and rotated in 90-degree increments. The arrowhead pointing up style is most widely used and recommended. The undefined power symbol can be given any name when it is placed on the schematic. Pins and wires connected to power symbols with the same signal name become part of the same net. The name of the net is simply the name of the power symbol. For example, all wires tied to +5V power symbols become part of the +5V net. In this manner, any number of different power supply levels can be handled. Ground symbols are treated just as any other power symbol.

The concepts of wires, junctions, crossovers, grounds, and power symbols are generally familiar to anyone who has had experience in schematic drafting and cause few problems when making the transition from manual drafting to schematic capture. It is important to clearly understand the distinction between physical connectors, off-page connectors, and hierarchical ports. The hierarchical port (previously referred to as a *module port* in OrCAD SDT) is a concept somewhat unique to schematic capture and hierarchical schematics. Users new to schematic capture are often confused by hierarchical ports and misuse these special symbols to represent physical connectors (such as a plug or jack).

The most important thing to remember about hierarchical ports is that they are not intended to represent physical parts. Hierarchical ports are used with hierarchical schematics that contain multiple blocks. Their only purpose is to route signals between blocks. Off-page connectors are used to route signals between sheets. This distinction is clarified later in this chapter.

Capture allows several styles of hierarchical ports. The *sense* or *direction* of a port refers to the circuitry block on which it occurs. For example, the input ports on Figure 1-13 represent signals coming from other circuitry blocks. Bidirectional ports are used for data buses in which signals can flow in both directions. Unspecified ports are used for power and reference voltage connections between blocks. Special rules apply to unspecified ports.

Recommended practice is to place input ports on the left and output ports on the right side of the sheet. Placement of bidirectional and unspecified ports is arbitrary. Suggested styles for the various ports are shown in Figure 1-13. However, Capture allows any style to be used for any type of port.

As with power symbols, any wire tied to a port with the same name will become part of a net with that same name. The style or type of port is irrelevant, only the name matters.

As mentioned earlier, special rules apply to unspecified ports. The term *unspecified* refers to signal flow. Unspecified ports are generally used only for analog signals and power supply connections. Signal flow direction is assumed to be unknown or inapplicable. The most common use of an unspecified port is to isolate power and ground for a particular circuit block. The need for isolated power often arises in systems with logic running at multiple voltage levels. Examples include core logic for a Pentium processor at 3.3V interfaced to an ISA bus at 5V or a RAM subsystem with battery backup.

For now, just remember that power symbols on a given circuit block tied to an unspecified port become part of a net with the name of the port, not with the name of the power symbol. This may sound very confusing. The rationale will become apparent as you learn more about concepts such as invisible power pins and work on designs with isolated power supplies in the tutorial exercises.

A no-connect symbol should be placed at pins or wires that are not connected to any other circuit element. While this may seem redundant, it is very important and required by Capture. One of the most useful features of Capture is the electrical rules checking routine. This routine has a list of rules, most of which relate to types of connections that are allowed. Violations are flagged to alert the user to possible problem areas and errors. Unconnected pins are flagged as an error unless a no-connect symbol is placed at the end. Unconnected pins will also display a connection square.

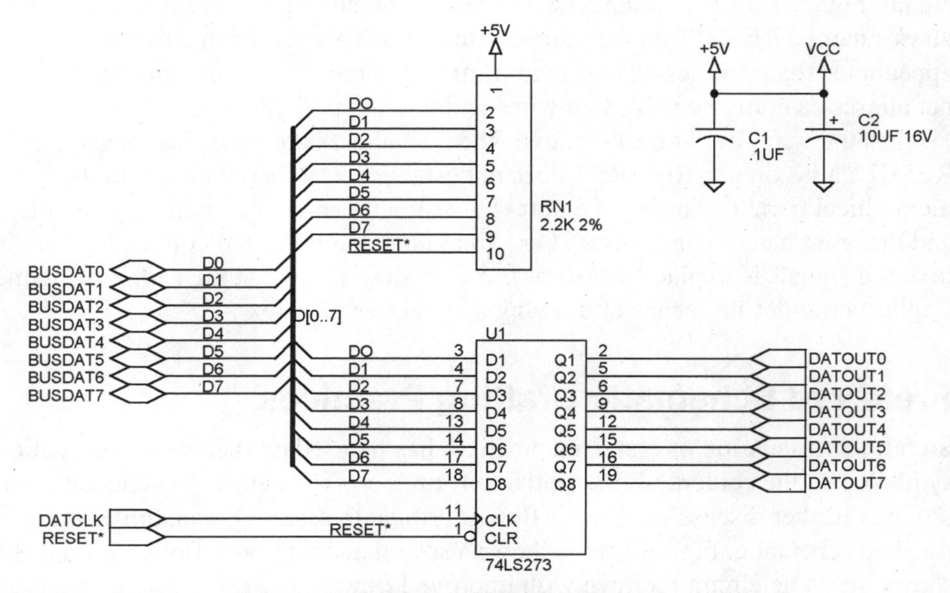

Figure 1-14 Advanced Capture Objects

Figure 1-14 introduces two more Capture schematic objects: bus structures and net aliases. Capture uses the term *aliases* in place of *name* (OrCAD SDT used the term *label*). The use of bus structures to represent a group of related signals, such as a data or address bus, became widespread practice after microprocessor technology was introduced in the early 1970s. Representing an eight-bit data bus or 16-bit address bus with a single broad line is more convenient than drawing out each signal. Besides the obvious time savings in drafting, the signal flow becomes much clearer and easier to understand. With the advent of 32- and 64-bit bus structures, drawing out each signal is no longer an option.

Capture allows bus structures with any number of signals. Certain conventions must be followed. Signals enter (or exit) a bus via a 45-degree diagonal bus entry symbol. Merely drawing a wire to a bus will not result in a connection. A uniform name must be used for all signals on a bus, for example D0, D1,...D32 or ADDR16, ADDR17,...ADDR24. The name always consists of an alphabetic prefix and a numerical suffix. Corresponding bus aliases of the form D[0..32] or ADDR[16..24] must be placed on the bus. Individual net aliases such as D2 or ADDR17 must be placed on wires entering the bus.

Capture net aliases are symbols. Net aliases can be used not only with buses but also wherever it is convenient to identify and connect wires representing the same signal. Figure 1-14 shows the RESET* signal routed to pin 10 of RN1 by placing aliases named RESET* on the appropriate wires. This results in a neater appearance than if wires were routed from end to end. It is important to note that net aliases can only be placed on wires and that the first character of the alias touches the wire. In Figure 1-14, two RESET* aliases are used. Just placing a RESET* alias on pin 10 of RN1 does not establish a connection back to the hierarchical port. The name of a port is not directly associated with any net alias, and different names can be used. On a final note, the preferred convention for inverted signals is to place an asterisk or backslash (* or \) at the end of the name. Capture provides no means of drawing a bar above alias text.

Preferred Schematic Drafting Practices

So far, the discussion of preferred practices has focused on the use of schematic symbols and the content of descriptive information. The subject of schematic flow requires further discussion. Proper flow is critical to assuring readability of the finished schematic. Figure 1-15A shows a schematic with poor flow. Figure 1-15B shows the same circuit redrawn with improved flow.

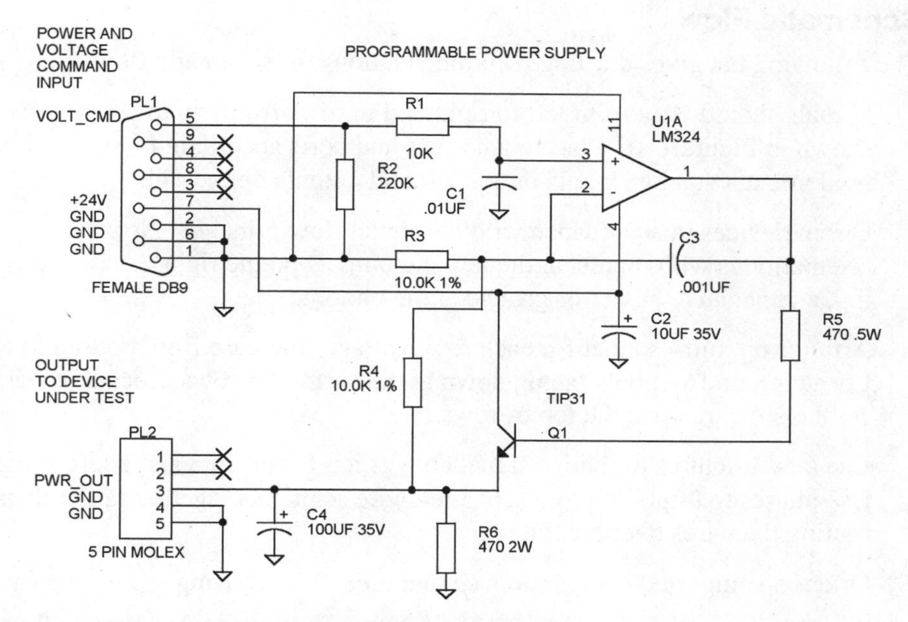

Figure 1-15A Schematic with Poor Flow

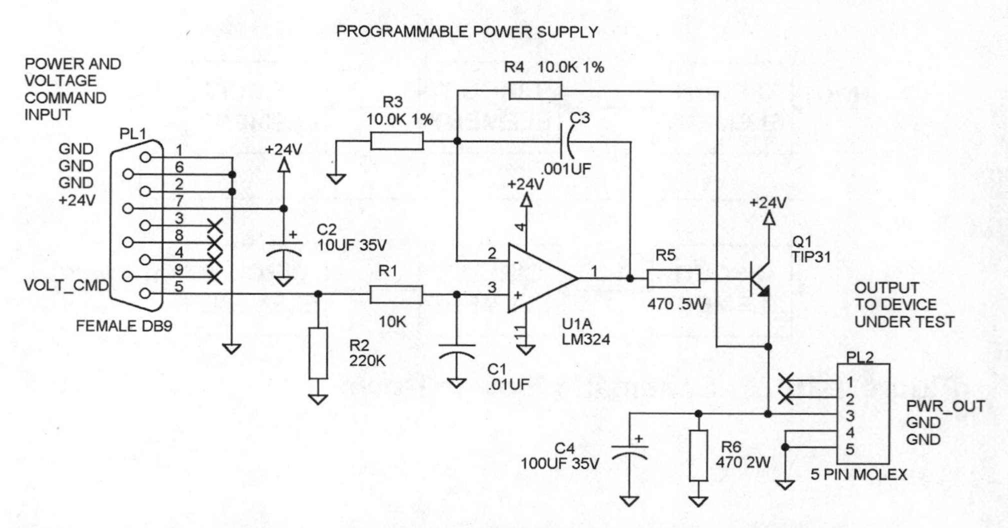

Figure 1-15B Schematic Redrawn with Improved Flow

Schematic Flow

The following list gives drafting recommendations for schematic flow:

- Signals should flow from left to right and in rows from top to bottom, as shown in Figure 1-16. Place connectors and ports according to the predominant signals: inputs on the left and outputs on the right.
- Orient devices in accordance with the signal flow principle. Normal orientation is with inputs on the left and outputs on the right, positive power on top, and negative power or ground at the bottom.
- Arrange circuitry so that for each row, voltages increase from bottom to top. Locate ground symbols facing down at the bottom of rows. Locate power symbols facing up at the top of rows.
- Use bus structures to relieve signal congestion for all data and address signals. Use aliases to implicitly join separated wire segments together rather than routing the wires over great lengths.
- Orient op amps and comparators so that their "–" inverting inputs appear at the top.
- Connect device power and ground pins to individual power and ground symbols. Do not tie many leads together at a single power or ground symbol.

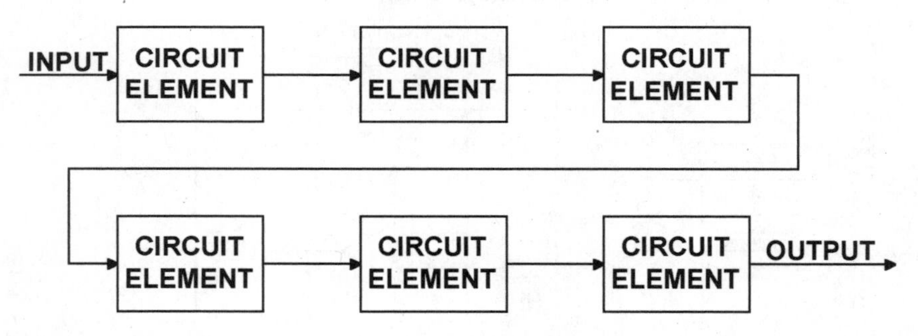

Figure 1-16 Schematic Flow in Rows

Signal Naming Conventions

Certain conventions should be observed for naming signals and power. As previously discussed, an asterisk or backslash (* or \) following the signal name represents inverted signals or signals that are active low (logic function enabled or asserted at a zero logic level). Bus signals must be numbered in sequence, such as D0, D1, ...D7. While Capture does not impose any limitations, data and address signals are always numbered starting from zero. Signal names containing D, DAT, or DATA are common for data lines and A or ADDR for address lines. Edge-triggered signals used to clock counters or flip-flops should contain CLK in the name, such as BUSCLK1. In general, signal names should be descriptive. The underline character (_) can be used to join segments of a signal name for clarity. For example, the signal name A2D_D0 (analog to digital converter data line 0) reads better than A2DD0. Avoid picking arbitrary signal names. If in doubt, construct an abbreviation from the plain language description of the signal.

Designs frequently utilize multiple power supplies and reference voltages. Choose appropriate names for power symbols; +5V is more descriptive than VCC. Capture library parts have invisible power pins that use vague names such as VSS, VDD, and VCC since the actual voltage levels may vary from design to design. Do not use these types of names for known and well-defined power levels!

Unless you are designing battery-powered vacuum tube equipment, do not use B+ as a power name. B+ is an archaic term for the "B" or plate supply battery. For battery-powered portable equipment, varying battery voltages can be represented by adding BAT to the power name, for example +9VBAT. Reference voltages appear in analog circuits, especially digital to analog and analog to digital converters. Precision reference circuits used to generate a reference voltage are limited to a low current. The naming convention should preclude mistaking a reference for a power supply. A name such as +5.00VREF identifies a precise reference voltage that can be expected to vary less than .01 volt from the nominal +5.00 volt value.

Title Block and Notes

All schematics should include a title block. Capture provides various predefined title block formats that are suitable for most organizations. User-defined formats are also possible. Additional text notes are almost unavoidable in modern schematics. Capture offers a flexible text-editing tool for this purpose. In the case of text notes, too much detail is better than not enough. A limitation of Capture is the inability to adequately draw nonschematic elements, such as signal waveforms.

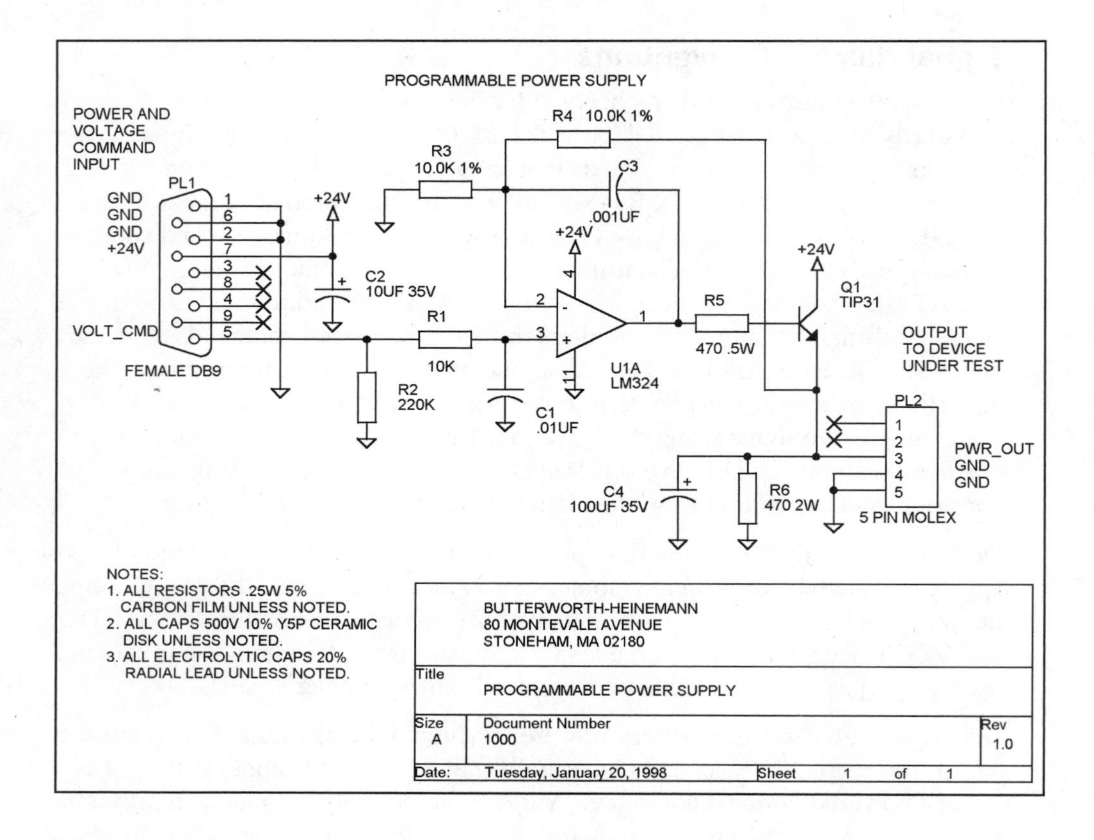

Figure 1-17 Finished Single Sheet Schematic

Figure 1-17 shows the complete version of the earlier schematic redrawn for improved flow. A title block and notes have been added. The title block for a schematic typically contains the following elements:

- **Company name and address**. Self-explanatory.
- **Title**. The plain language description of the device or circuit.
- **Size**. Letter designators A-E are used for standard drawing sizes. Most schematics are now laser printed on A size paper (8.5 by 11 inches).
- **Document number**. Most companies track documentation by a document or drawing number.
- **Revision code**. Either a letter or number can be used. Decimal numbers are common in the electronics industry because of the rapid pace of changes. Development prototypes are given revision codes less than one, such as 0.9 for

the final prototype. First production is usually revision 1.0. A major change such as a new circuit board causes the revision to jump to the next integer; for example from 1.5 to 2.0. Minor changes, whereby only a few component values are modified or jumpers and cuts are made on the board, cause the revision code to increment by .1; for example from 2.0 to 2.1.

- **Date**. Automatically assigned and updated by Capture whenever any editing is done.
- **Sheet number and number of sheets**. Self-explanatory.

Text notes complete the schematic. Back in the days of manual drafting on paper, editing text notes meant using an eraser. To avoid having to repeatedly erase and rewrite everything whenever a change or addition occurred, notes started at the bottom and ran in sequence going up. With the ease of CAD editing, this practice has become archaic. Notes are now written and numbered in normal sequence from top to bottom. Some of the items that should appear in the text notes include:

- Default values for components, such as resistor wattage rating and tolerance.
- Preferred or approved vendors for critical or unique parts and any other important information not contained in the part descriptions.
- Engineering change descriptions and history, including compatibility of previous revisions.
- Brief calibration and test instructions or a reference to a separate document.
- Data on firmware (such as the name of the code file and checksum).

Hierarchical Schematics

In the last decade, the trend for schematics has been away from large C, D, and even E size pen-plotted or hand-drawn sheets. The large drawing sizes are cumbersome to handle and expensive to reproduce. Products have become very complex — to the point that even several E size sheets may not suffice for the schematic of a new VME bus computer board. Multiple-sheet A size schematics are the answer. They can be quickly run off on a laser printer and inexpensively reproduced on any copying machine. The only problem is providing some means of oversight and continuation between sheets. The simplest approach is to use off-page connectors (see Figure 1-13). This is feasible with schematics containing a few sheets. Larger schematics require a better means of oversight for signal routing between circuit blocks. Capture solves this problem by using the concept of a hierarchical schematic, illustrated by the example in Figure 1-18A through D.

The first sheet provides an overview of the organization or hierarchy of the design. Blocks, each representing additional circuitry (one or more sheets), are shown along with the interconnections between them. Capture uses the somewhat confusing terms "parent" and "child." The parent schematic (Figure 1-18A) shows interconnections between the child schematics (Figures 1-18B through D). The filled square symbols on the circuit blocks shown in Figure 1-18A are called *hierarchical pins*. These hierarchical pins correspond to the hierarchical ports on Figures 1-18B through D. An alias (name) identifies each hierarchical pin. The most important point to remember is that a hierarchical pin on the parent schematic must be created for every hierarchical port that appears on the child schematic. Wires between the hierarchical pins interconnect the blocks.

At first all this terminology may appear complex and confusing. Just think of it as dividing circuitry into functional blocks and showing the interconnections between the blocks.

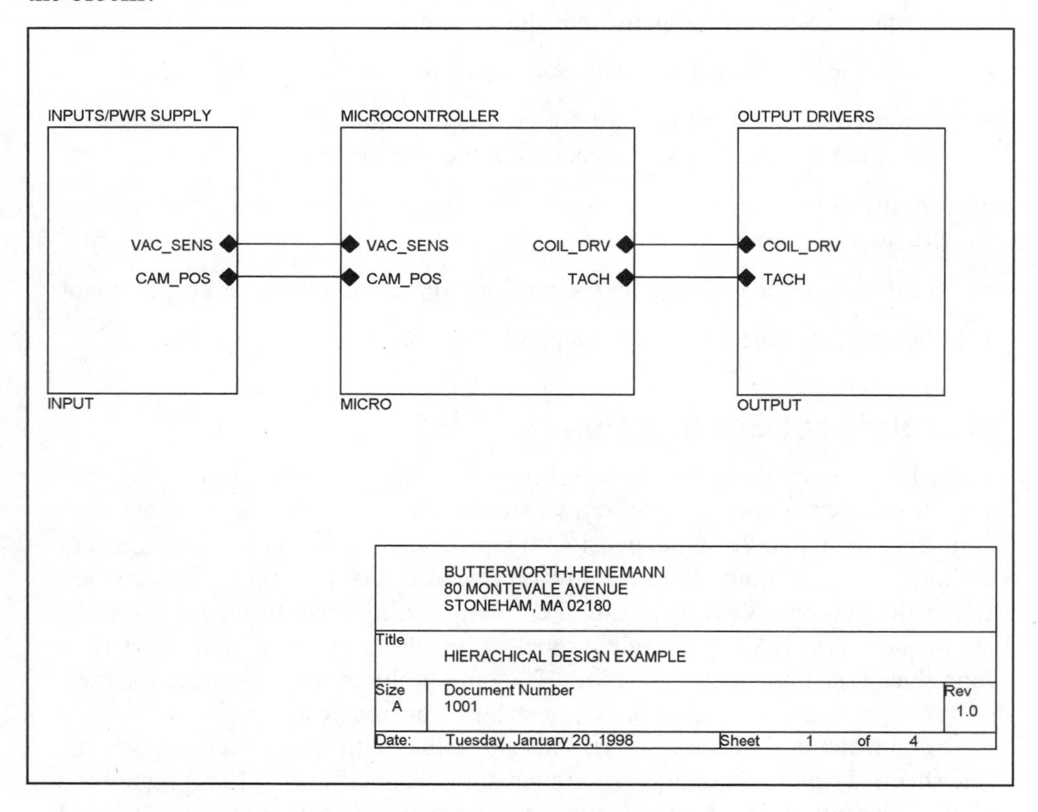

Figure 1-18A Hierarchical Design Example Sheet One

Sheet 1 of the hierarchical design gives an overview of the major circuit blocks and interconnections. In this context, the terms *block* and *sheet* are somewhat interchangeable. Most designers think in terms of circuit blocks. It is useful to think of the first sheet as a block diagram of the circuitry. In this example, every hierarchical pin is connected to another pin with the same name, such as VAC_SENS from the input block to VAC_SENS on the microcontroller block. If required, VAC_SENS could be connected to a hierarchical pin with a different name. Hierarchical pins can also be connected directly to device pins.

There is no limitation on the interconnection of blocks. The example in Figure 1-18A shows a single-level hierarchy. Sheet 1 (parent) shows three lower-level blocks (children). Capture supports multiple-level hierarchies. For example, sheet 2 could contain several circuit blocks, each represented by additional sheets. Designs with more than a few levels can become confusing. Two or three levels seem to work out for most practical designs.

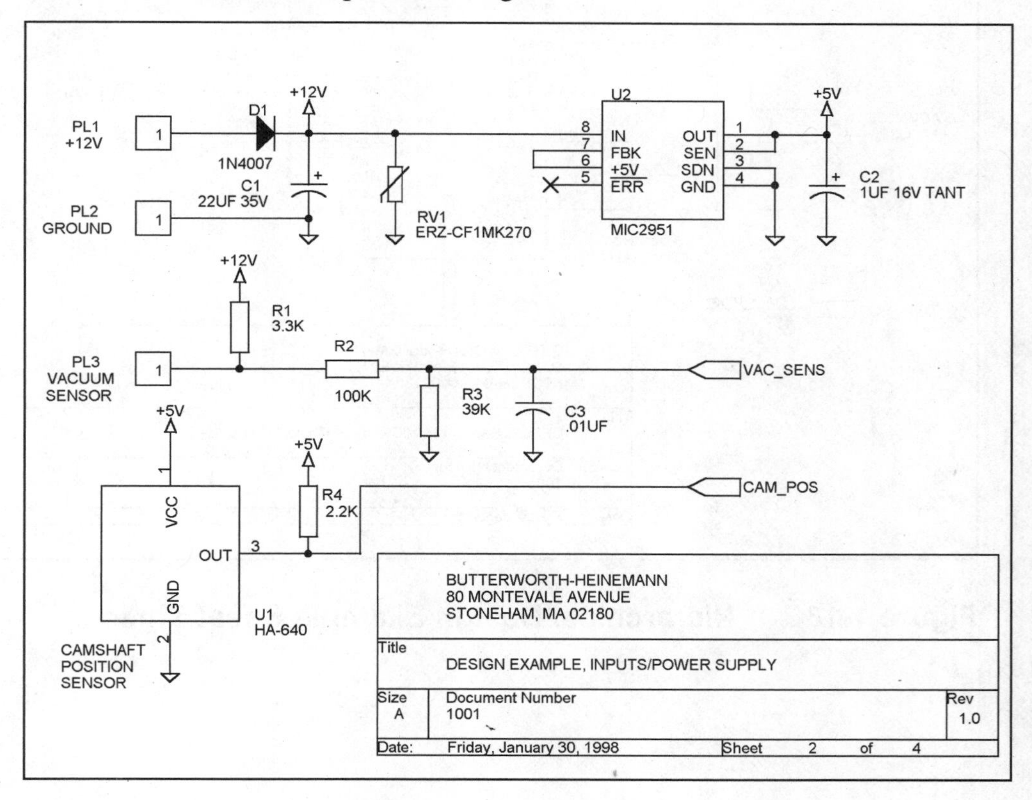

Figure 1-18B Hierarchical Design Example Sheet Two

Power is automatically carried across blocks unless special steps are taken to isolate it. In the example, +12V power comes in at PL1. The power supply is connected to +12V and +5V power symbols. Power is automatically routed to any other block that contains +12V or +5V power symbols. Ground connections are treated the same. As mentioned before, Capture provides multiple ground symbols. All of these ground symbols are tied together in a single ground plane that runs throughout the design unless special steps are taken to isolate one section of the circuit.

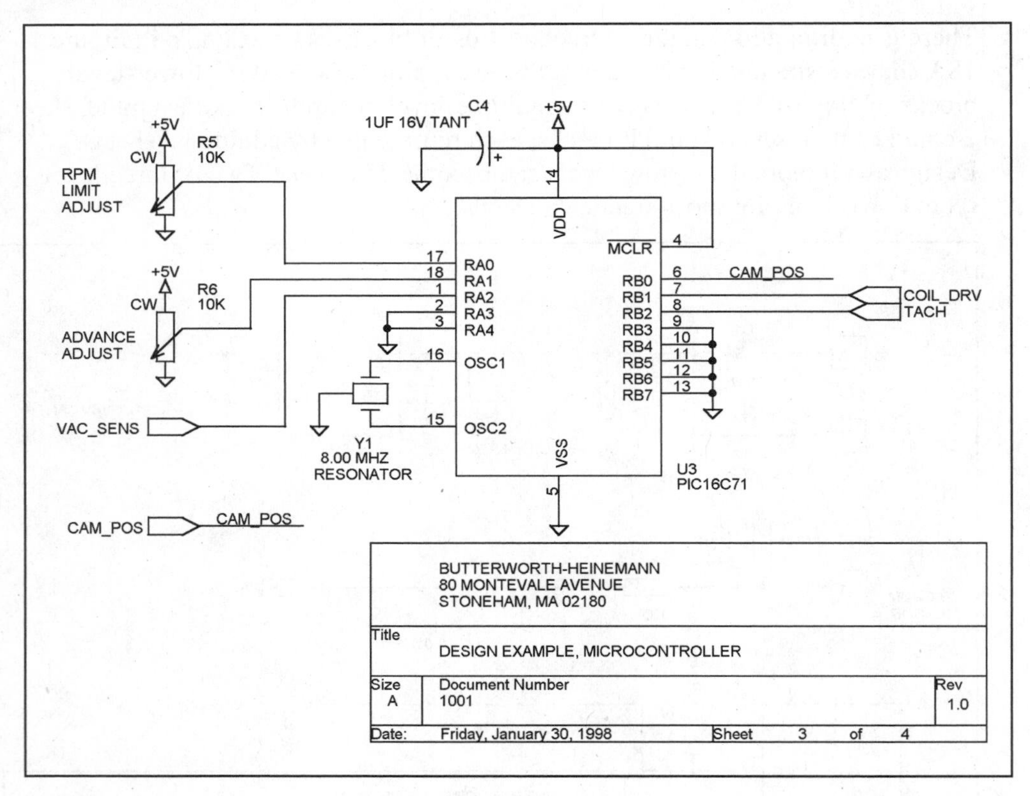

Figure 1-18C Hierarchical Design Example Sheet Three

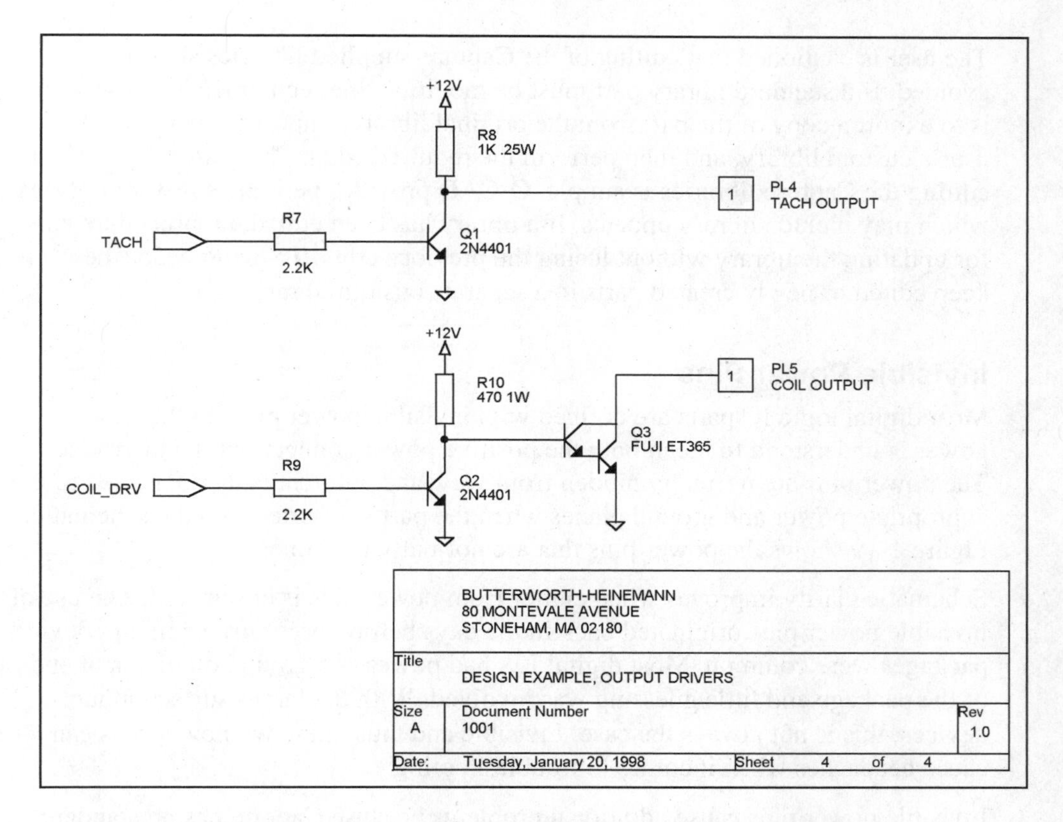

Figure 1-18D Hierarchical Design Example Sheet Four

Capture Part Libraries

One of the great strengths of Capture is the comprehensive set of part libraries supplied with the program. Basic electronic components such as those shown in Figures 1-1 through 1-7, 1-11, and 1-12 appear in a general purpose "DEVICE" library. Generic "CMOS" and "TTL" libraries include the parts shown in Figures 1-9 and 1-10. Capture includes many additional vendor-specific libraries for parts from Intel, Maxim, Motorola, National, TI (Texas Instruments), and other large companies. All together, thousands of parts are available.

In addition to the libraries supplied with Capture, the user can easily create and edit custom part libraries. Since complex ICs are represented by a rectangular block with numbered and named pins, creating a new part definition and adding it to a custom library is not a difficult or time-consuming task.

The user is cautioned that editing of the Capture-supplied libraries should be avoided. If a standard library part must be modified, the recommended procedure is to export a copy of the part from the original library, import the part into the user's custom library, and then perform the required edits. The reason for not editing the Capture libraries is simple. OrCAD provides periodic software updates, which may include library updates. If a library has been edited, no procedure exists for updating the library without losing the previous edits. To avoid headaches, keep edited or newly created parts in a separate custom library.

Invisible Power Pins

Most digital logic IC parts are defined with invisible power pins. In this context power is understood to mean both the positive power connection(s) and ground. The power pins are normally hidden from view and automatically connected to the appropriate power and ground planes when the part is inserted into the schematic. Figure 1-19 shows the power pins that are normally invisible.

Schematic clarity improves when clutter from power pins is eliminated. The use of invisible power pins originated back in the days before large surface mount device packages were common. Most digital ICs had power and ground on diagonal ends of the package and little guessing was involved. With the larger surface mount devices, this is not always the case. Invisible and thus unknown power pins can cause headaches when it comes to troubleshooting.

Invisible power pins cause additional problems because Capture has no standard naming convention for these pins. TTL circuits use the names VCC and GND. CMOS circuits use VSS and VDD. There is no way to tell which names are used other than examining the part definition with the library editor. A further and even more serious complication arises with designs that require multiple voltage levels. Multiple voltage designs, such as PC motherboards with 3.3V and 5V logic are now commonplace. Only two options exist for such designs: create custom parts with visible power pins, or separate all 3.3V and 5V devices onto different blocks and isolate power going to these blocks.

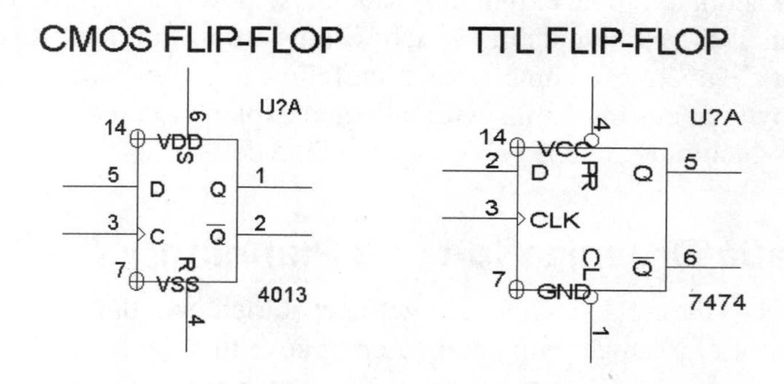

Figure 1-19 Invisible Power Pins

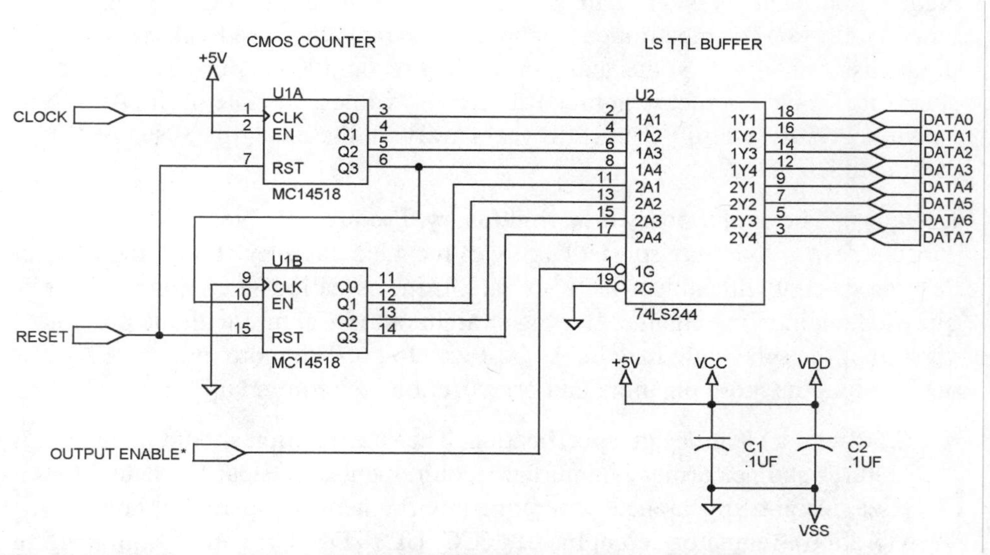

Figure 1-20 Design with CMOS and TTL Logic

The invisible power pins are automatically tied to nets with the same name. In Figure 1-20, the TTL part is already connected to ground since the invisible ground pin is named GND. The CMOS part has an invisible ground pin named VSS. In order to connect this pin to ground, you must place a VSS power symbol on the sheet and then tie that symbol to ground. The TTL part uses VCC for positive power, the CMOS part uses VDD. To connect these pins to +5V, you must add power symbols with the names VCC and VDD and then tie them to +5V.

You should add a text note explaining what these power and ground ties represent. Then add another text note listing which IC pins are power and ground. Do not assume that the average technician or even fellow engineer will understand invisible power and ground pins without some explanation or that they can remember which pins are used on every possible device.

Schematic Organization and Planning

Two schools of though exist about how to get started on a design using schematic capture software. Younger engineers often believe that the computer screen is the new demigod of design. Start with an empty screen and try to organize your thoughts as you go along. Use the computer as an electronic sketchpad. Older engineers, on the other hand, tend to view schematic capture as a drafting tool useful for preparing neat documentation at the completion of a design project.

Neither approach gives optimum results. Traditional desktop PCs and engineering workstations are not electronic sketchpads. Attempts to make PDAs (personal digital assistants) with stylus and handwriting recognition for use as electronic sketchpads have not met with much success — witness the fate of the Apple Newton. A truly useful PDA is still years away. In the meantime, use an old-fashioned lab notebook.

Designs can be evolutionary or revolutionary. Products evolve, and even an entirely new design may consist largely of recycled circuitry. Truly revolutionary designs starting with a clean slate are rare. If the so-called new design is merely a glorified engineering change, starting straight out in Capture with the previous version of the schematic may be the best way to go. The more revolutionary the new design, the more planning and organization are required up front:

- Start with a clear design specification. The specification should include features and performance, interface requirements, physical characteristics (size, weight, appearance), operating environment (temperature, humidity, vibration), regulatory compliance (FCC, UL, VDE, CE), and manufacturing cost target.
- List all ICs and other specialized parts required for the design. Run copies of databook pages showing pinouts for all parts.
- List signal names and buses. Interface signals are often established ahead of time from the design specification. The choice of processor will determine data, address, and bus control signals.
- Plan power usage and representation.

- Create any required new parts in your custom library.
- Sketch out a rough block diagram and use it to plan out the design hierarchy.
- Draw the schematic using Capture. Add copious text notes. Run the electrical rules checking routine to catch errors that might slip by, then print out the schematic.
- Run postprocessing routines to generate a bill of materials for manufacturing and a netlist for PCB design. Carefully examine the output. It will often reveal subtle errors.
- After the PCB design and board debugging are completed, go back and incorporate any engineering changes and other changes necessitated by PCB layout considerations.

Conclusion

In this first chapter you have learned the background information required for generating schematics with OrCAD Capture. A clear understanding of this information is a prerequisite for successfully using Capture. In subsequent chapters, the focus will shift to the internals of Capture.

Review Exercises

1. Explain the difference between a reference designator and a part value.
2. What are the standard reference designator prefixes for capacitors, diodes, inductors, transistors, resistors, switches, transformers, and integrated circuits?
3. Explain the correct usage of the reference designators J, JP, P, and PL for various types of connectors.
4. Write down the multiplier prefixes for 10^{-12}, 10^{-9}, 10^{-6}, 10^{-3}, 10^{3}, and 10^{6}.
5. Write down the engineering units for capacitance, inductance, and resistance.
6. As explained on page 9, the resistor symbol closely represents the physical construction of actual resistors. Describe how several other basic electronic symbols relate to the physical construction of the devices they represent.
7. Why does good practice dictate that schematic notes describe the package type and size of resistors?
8. List some of the parameters required to completely specify a capacitor.

9. Describe the recommended rules for drawing op amps and comparators.
10. Explain the correct usage for off-page connectors and hierarchical ports in multiple-page schematics.
11. What special Capture symbol must be placed on unconnected part pins?
12. List the basic elements that should appear in a schematic title block.
13. Hierarchical schematic structure reduces complex schematics into easily understood circuit blocks. For schematics not exceeding several pages, a multi-sheet structure using off-page connectors also is effective. Using pencil and paper, redraw the schematic in Figures 1-18A to 1-18D as a three-sheet schematic with off-page connectors.
14. What are the advantages and disadvantages of invisible power pins?

2

Installation and Configuration

This chapter covers installation and configuration of OrCAD Capture for Windows schematic drafting software. You will learn how to install and configure the software in preparation for the tutorial exercises in the following chapters. Please read this chapter very carefully.

Capture runs under Windows. Capture version 7.11 (release date October 15, 1997) is the last version to support the 16-bit Windows environment. Capture version 7.20 (current at the publication date of this book) will run only under Windows 95 or Windows NT. Programs that run under Windows 95 should also run under the new Windows 98. The author presumes that most readers will be using Capture version 7.20 or later and Windows 95 or Windows 98.

The reader is assumed to be familiar with Windows and basic PC operations. Sometimes questions do arise. An excellent source of information is:

> *Microsoft Windows 95 Resource Kit*, Redmond, Washington: Microsoft Press, 1995.

Capture has a much-improved user interface compared with the older DOS based OrCAD SDT 386 versions. Command selection via the mouse is now clean and efficient. While Capture supports macros, users find that the improved menu structure and available hot keys negate the necessity for user-defined macros in most situations.

System Requirements

OrCAD Capture runs in an IBM-compatible PC environment under Windows. Recommended minimum system requirements for reasonable performance include:

- Pentium processor. For educational purposes, Capture will run on a 486, but with degraded performance.
- Windows 95 (or 98) recommended.
- 16 MB RAM.

- 100 MB free hard disk space (Capture and support files occupy about 50 MB).
- SVGA graphics with 1024 × 768 resolution. A 17-inch monitor is suggested for all professional use.
- Microsoft compatible, two-button mouse.
- CD-ROM drive.
- Laser or inkjet printer for hardcopy.

Capture (version 7.11 and later) appears to be stable if multiple programs are running under Windows and in a networked environment.

The trend in modern schematic drafting is away from large-format, single-sheet drawings. Most companies have standardized on the use of multiple-sheet, A-size (8.5 × 11 inch) schematics. This solves many problems related to printing, copying, and distributing technical information. The use of color is not generally required for schematics; thus a laser printer is the ideal hardcopy device.

Installation

Installation consists of loading and configuring the software, which is usually supplied on CD-ROM. OrCAD can supply Capture on floppy disk, but most PCs running Windows 95 are now equipped with a CD-ROM drive.

Insert the Capture CD-ROM into your CD-ROM drive, typically the D: drive. If your system is not configured for autorun, use the Windows Explorer, change to the CD-ROM drive and then double click on the SETUP.EXE icon. The InstallShield Wizard loads, starts the installation process, and displays a welcome screen. Click on Next to continue the installation process. The OrCAD License Agreement screen appears. You can scroll through the license information. In general, the license restricts installation and use of OrCAD Capture to a single PC. Special considerations apply for networked environments; contact OrCAD for details. Click on Yes to accept the license agreement. The next screen that appears allows you to select the operating system, as shown in Figure 2-1. InstallShield usually picks the correct operating system, in this case Windows 95. Verify that the correct operating system is selected for your PC. Then click on Next to continue.

The next screen, shown in Figure 2-2, allows you to select the elements to install. In most cases you will install both Capture and the Supplemental Libraries. The libraries contain electronic part symbols.

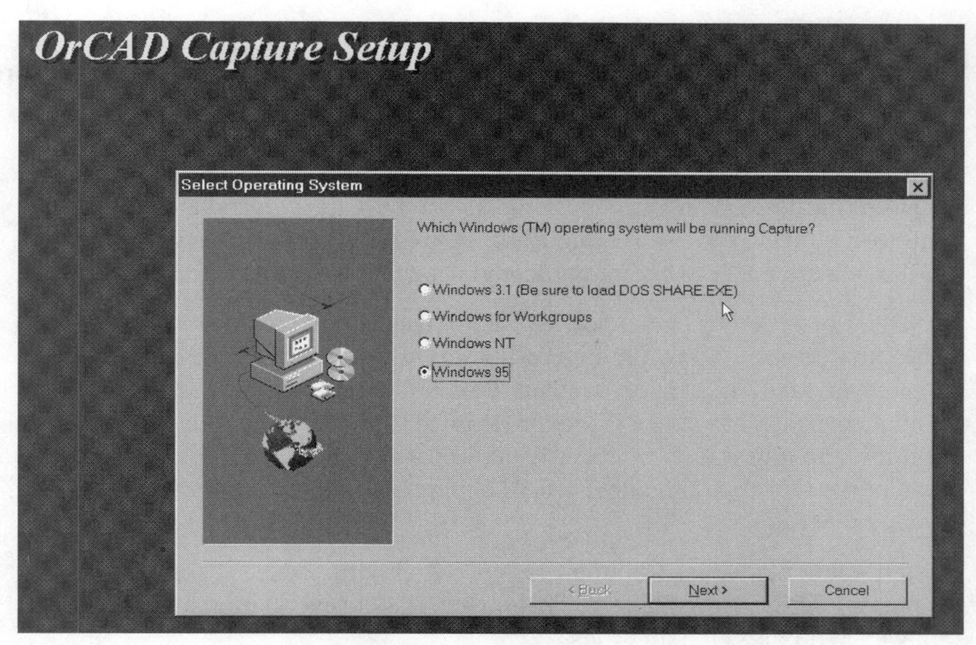

Figure 2-1 Select Operating System Screen

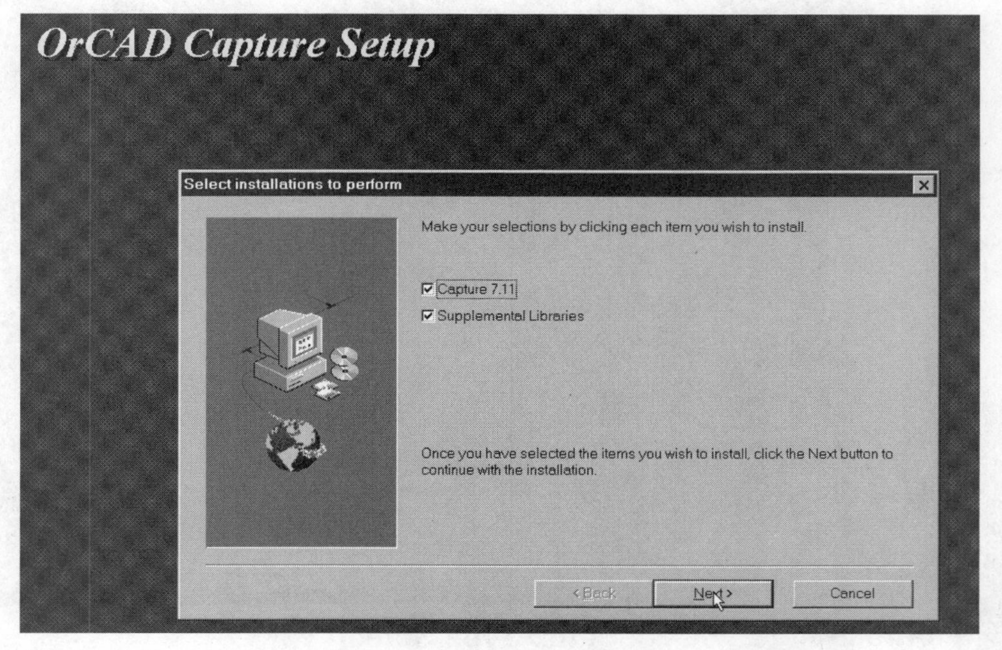

Figure 2-2 Select Installations to Perform Screen

The screen shown in Figure 2-3 appears. This screen allows you to select a typical or custom installation and determine the destination directory into which Capture files will be installed. Unless you have a specific reason, such as installing a second version of Capture, you should accept the default C:\ORCADWIN\CAPTURE destination directory. Otherwise problems may occur later on. In most instances, you should select the Typical installation. This option will load all the Capture program files required for the exercises in this book. When you are ready to accept the selections, click Next to continue.

The next two screens that appear allow entry and confirmation of user information. The entry screen is shown in Figure 2-4. Enter your name, company name (if applicable), and registration number. The registration number can be found on the product package. Be sure to write this number down in a location where you can find it in the future such as the inside cover of the OrCAD User's Guide. You will require the registration number when logging onto the OrCAD web site or calling for technical support.

InstallShield now loads Capture files onto your hard drive and sets up the Windows Registry for Capture. This process takes several minutes. The screen shows the progress loading files.

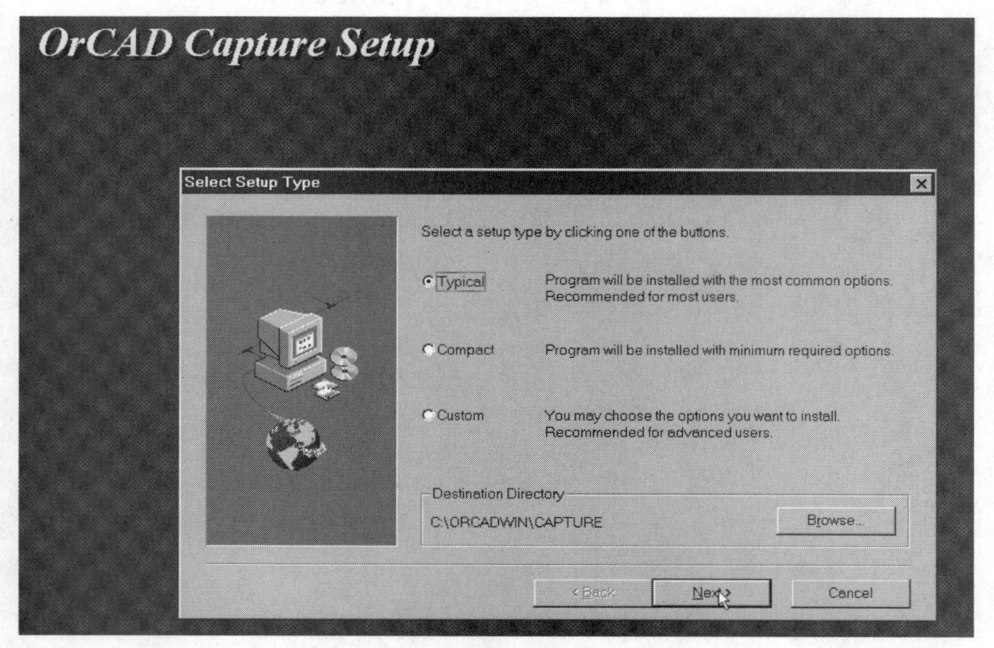

Figure 2-3 Select Setup Type Screen

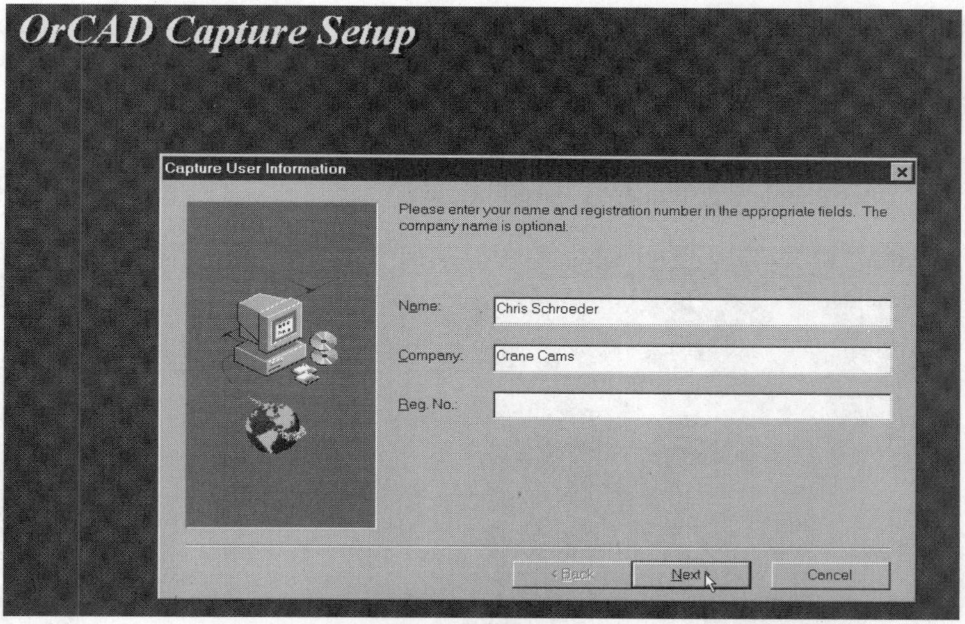

OrCAD Capture Setup

Capture User Information

Please enter your name and registration number in the appropriate fields. The company name is optional.

Name: Chris Schroeder

Company: Crane Cams

Reg. No.:

< Back | Next > | Cancel

Figure 2-4 User Information Entry Screen

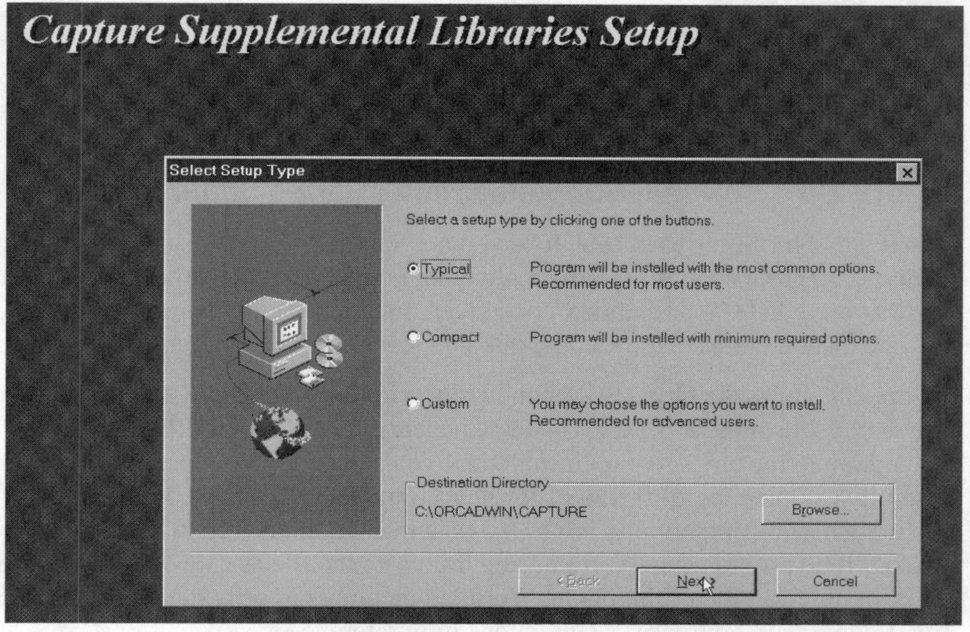

Figure 2-5 Supplemental Libraries Setup Screen

The Supplemental Libraries Setup screen shown in Figure 2-5 appears next. Select Typical setup. This loads the all the available libraries. Custom allows the user to select libraries. Unfortunately the library names are often somewhat unclear, for example Moto1 contains Motorola 1N series diodes. When you are starting out, you should load all the available libraries. In time, you will find which are useful and you can delete infrequently used libraries.

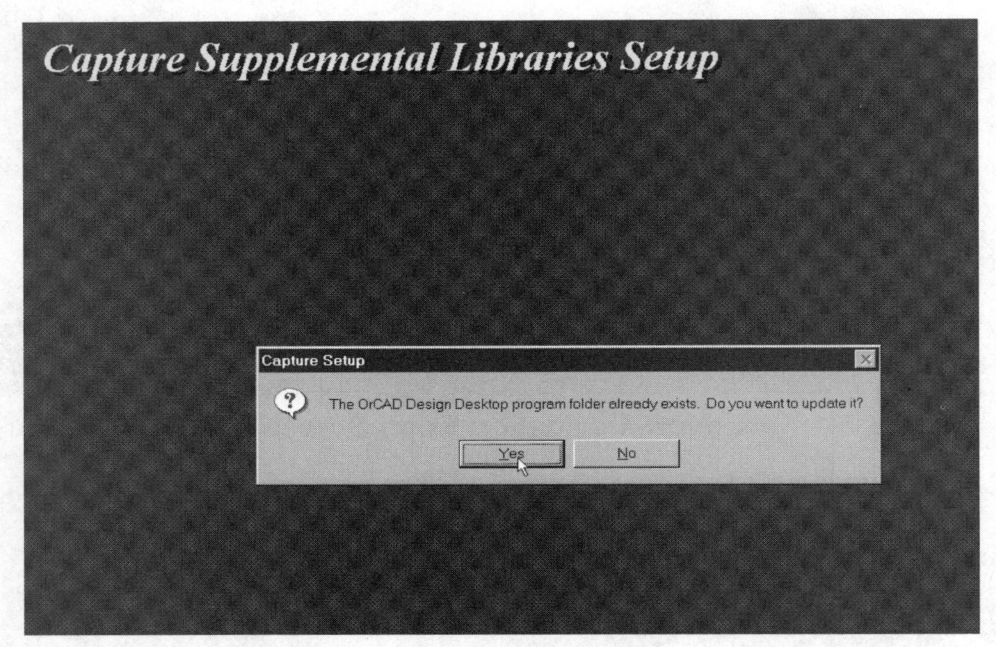

Figure 2-6 Final Supplemental Libraries Setup Screen

When the library files have been loaded onto your hard disk, the screen shown in Figure 2-6 appears. Click on Yes to add an option to your Windows desktop that allows you to list Capture library contents. A final screen will appear indicating that the installation is complete and that you may remove the Capture CD-ROM.

Figure 2-7 shows the OrCAD Design Desktop for Windows 95. Late information that is not contained in the User's Guide appears in the Capture README file. After completing any software installation, you should always take time to carefully examine the README file. New commands, bug fixes, known limitations, and compatibility issues are documented in this file. Click on Capture README. This launches WordPad and brings up the file as shown in Figure 2-8. Click on File and then Print on the WordPad menu bar to print out the README file. Keep it with your User's Guide for future reference.

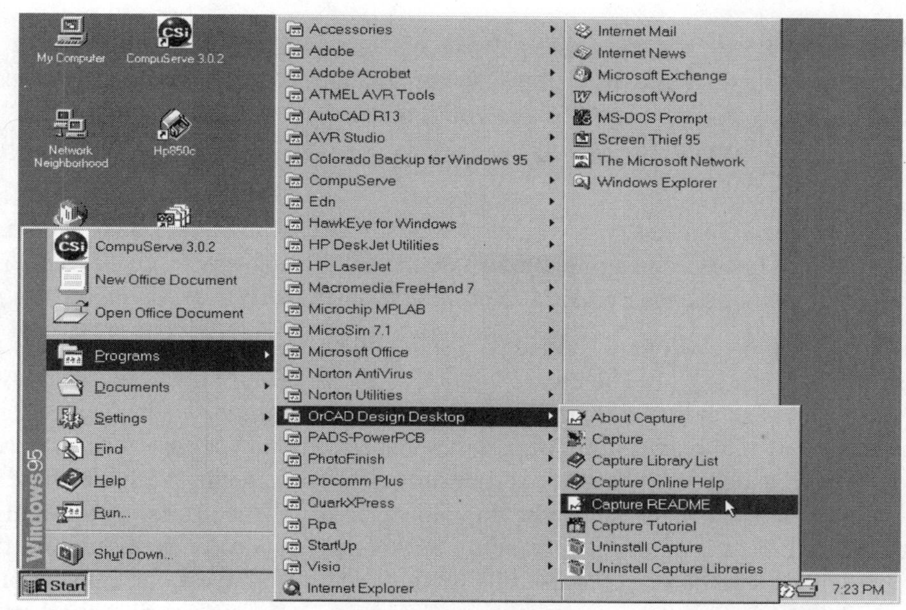

Figure 2-7 OrCAD Design Desktop in Windows 95

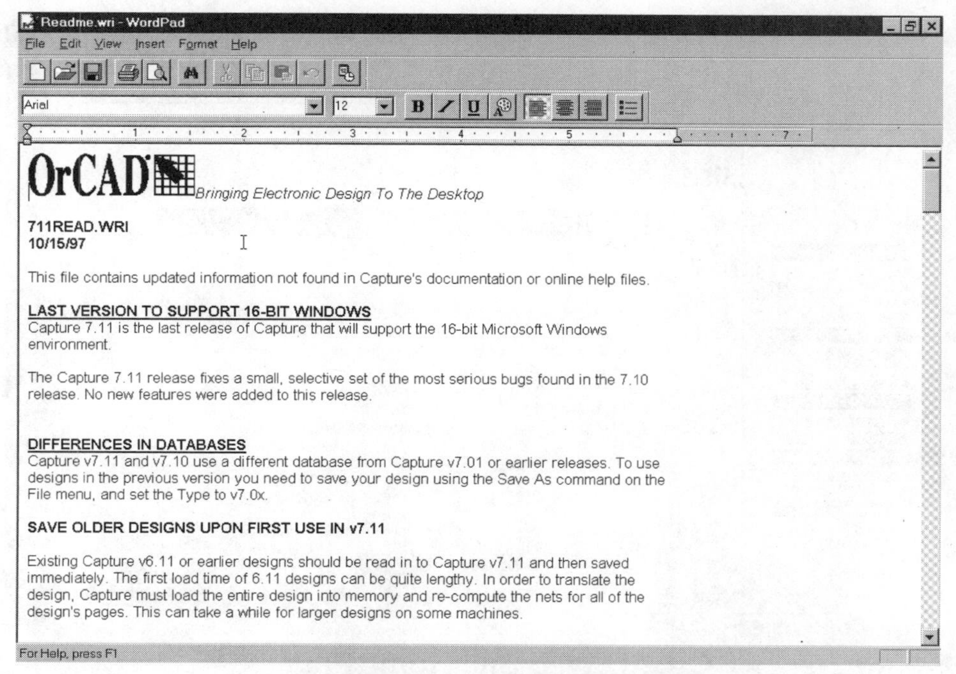
Readme.wri - WordPad

File Edit View Insert Format Help

Arial 12

OrCAD *Bringing Electronic Design To The Desktop*

711READ.WRI
10/15/97

This file contains updated information not found in Capture's documentation or online help files.

LAST VERSION TO SUPPORT 16-BIT WINDOWS
Capture 7.11 is the last release of Capture that will support the 16-bit Microsoft Windows environment.

The Capture 7.11 release fixes a small, selective set of the most serious bugs found in the 7.10 release. No new features were added to this release.

DIFFERENCES IN DATABASES
Capture v7.11 and v7.10 use a different database from Capture v7.01 or earlier releases. To use designs in the previous version you need to save your design using the Save As command on the File menu, and set the Type to v7.0x.

SAVE OLDER DESIGNS UPON FIRST USE IN v7.11

Existing Capture v6.11 or earlier designs should be read in to Capture v7.11 and then saved immediately. The first load time of 6.11 designs can be quite lengthy. In order to translate the design, Capture must load the entire design into memory and re-compute the nets for all of the design's pages. This can take a while for larger designs on some machines.

For Help, press F1

Figure 2-8 OrCAD Capture README File

Web-based Support

OrCAD has an excellent web site at www.orcad.com. If you have Internet access, you should visit this web site. You can obtain free demo software for new products, obtain product information, and access technical services. Figure 2-9 shows the OrCAD home page.

Under Technical Services, you can access the OrCAD Knowledge Base shown in Figure 2-10. This search engine allows you to look up answers to many problems and questions that may arise using Capture and other OrCAD products.

You can download software service packs and other useful utilities and programs, as shown in Figure 2-11. The complexity of software development in a Windows environment is such that all commercially available programs will have a certain number of bugs. In between major releases that are shipped to customers covered under a technical support agreement, interim releases are made available in the form of service packs. These can be downloaded from the web site. You can also submit a technical support request (scroll to this menu option at the bottom of the Technical Services screen). To use this service, you must have a current support agreement, registration number, and E-mail address. Your registration number serves as a password. You can expect an answer via E-mail within 72 hours.

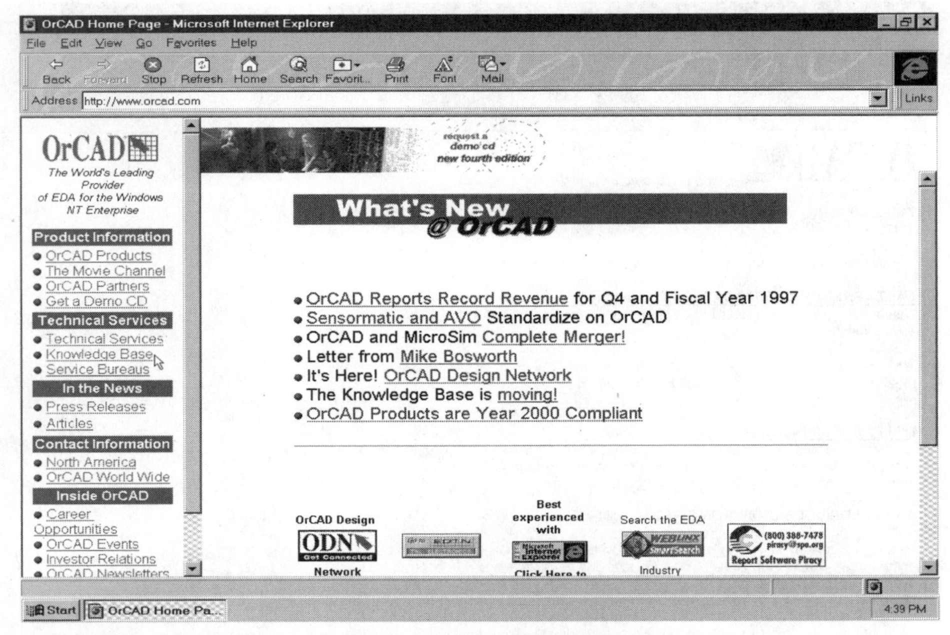

Figure 2-9 OrCAD Web Site Home Page

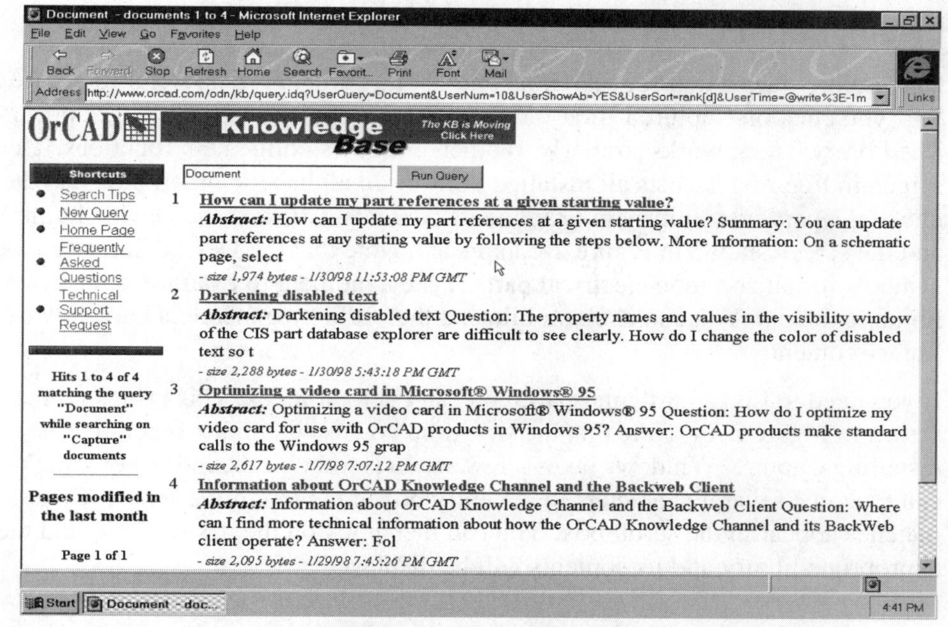

Figure 2-10 OrCAD Web Site Knowledge Base

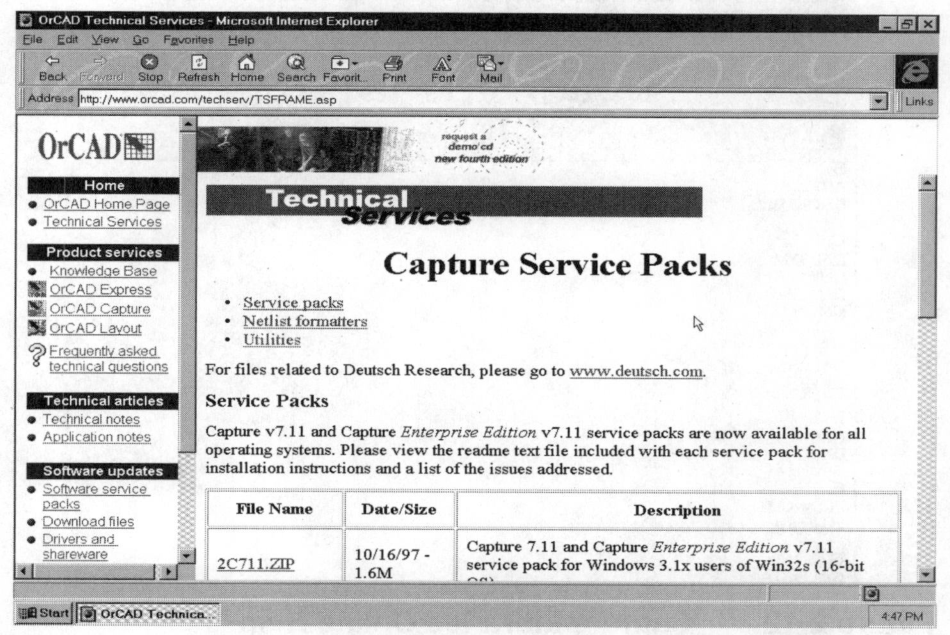

Figure 2-11 OrCAD Web Site Technical Services Area

Listing Capture Libraries

Capture provides a tool for searching and listing parts libraries. Refer to Figure 2-7. If you click on Capture Library List, the screen shown in Figure 2-12 appears. The library listing works similarly to other Windows online help functions. The screen in Figure 2-12 lists all installed libraries in alphabetic order. You can then click on a particular library to examine its contents. Click on the Device library and the screen shown in Figure 2-13 appears. Note that the Device library contains symbols for all common electrical parts. Take a moment to examine the Device library contents. You can click on File and then Print to generate a hardcopy of the library contents.

If you need to find a particular part, you can click on Index. This brings up the screen in Figure 2-14. Note that the first time you use the index function after installing Capture, Windows takes a few seconds to build the index key file. You can type in a several characters in the entry box at the top of the screen. Close matches appear in the scroll box. You can then click on the desired part, and the appropriate library and its contents will be displayed.

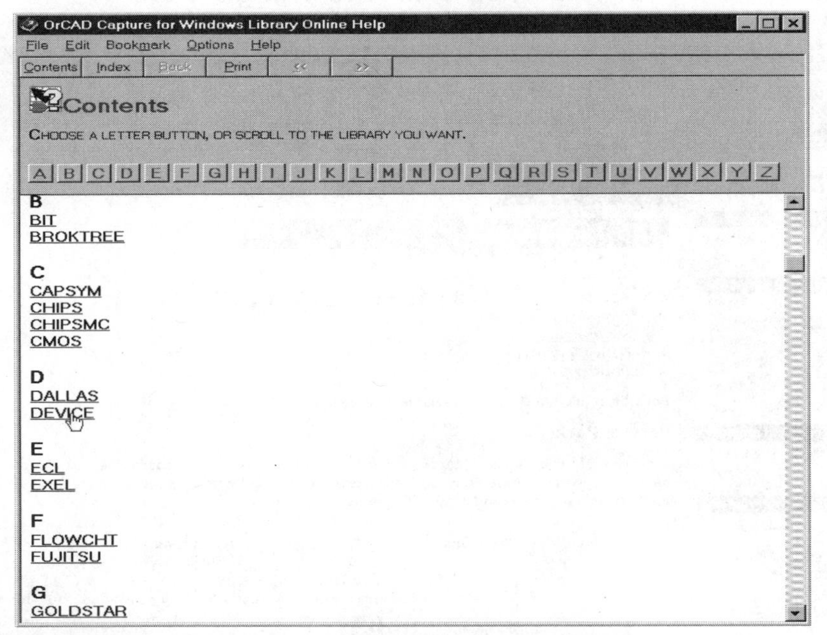

Figure 2-12 Capture Libraries Online Help

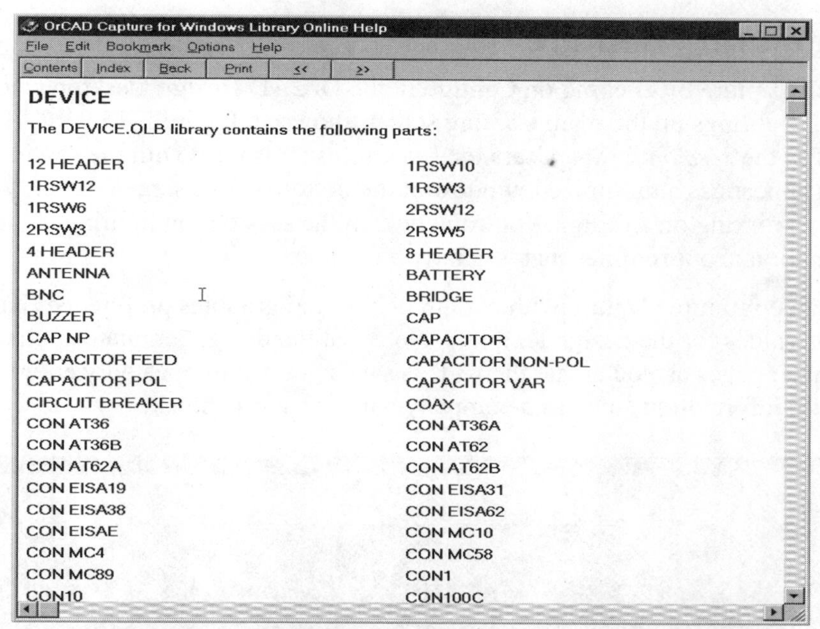

Figure 2-13 Examining the Device Library Contents

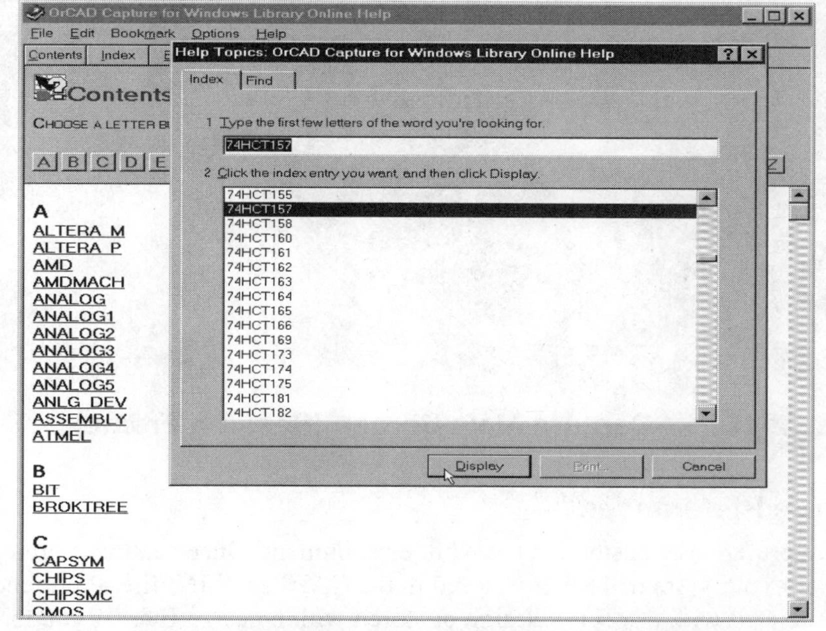

Figure 2-14 Searching Capture Library Contents

Configuring Capture

Launch Capture by clicking on Capture in the OrCAD Design Desktop (see Figure 2-7). This brings up the main Capture screen shown in Figure 2-15. OrCAD refers to this as the *session frame*. Users tend to confuse this term with the session log, which appears as a minimized window at the bottom of the screen. Later, when you are working on a design, you can click on the session log to bring up status information about routines that you have executed.

You can configure Capture with certain design and graphics preferences and with default values for the design template. Note that the design template applies to new schematic pages as you create them. This saves you from repeatedly entering standard information, such as a company name on a title block.

Figure 2-15 Capture Main Screen (Session Frame)

Capture Preferences

Capture preferences customize the work environment. Once set, these preferences persist on your system. They are stored in the CAPTURE.INI file. Preferences are not associated with a specific design or stored within design files. If you load a design created on another system running Capture, your preferences are not affected.

Color and Print Preferences

To set preferences, click on Options and then Preferences on the main Capture screen. As with other Windows programs, you can also use the keyboard shortcut ALT, O, P. Note that keyboard shortcuts can be used for most Capture commands. This brings up the preferences window shown in Figure 2-16. The color and print preferences tab is always displayed first. You can click on the other tabs at the top of the window to set additional preferences.

The color and print preferences affect screen display and printing. You can select colors for the various entities that Capture displays. In most cases, the default colors are acceptable. To change the color of a particular entity, click on the color box. This brings up a color palette window. You can then click on a new color. Note that borders and grid references are displayed in the same color as the titleblock. When you print a schematic page, only those objects with a check mark appear on the hardcopy. Background and selection (a highlight color for selected objects) cannot be printed. Grid dots are normally not printed.

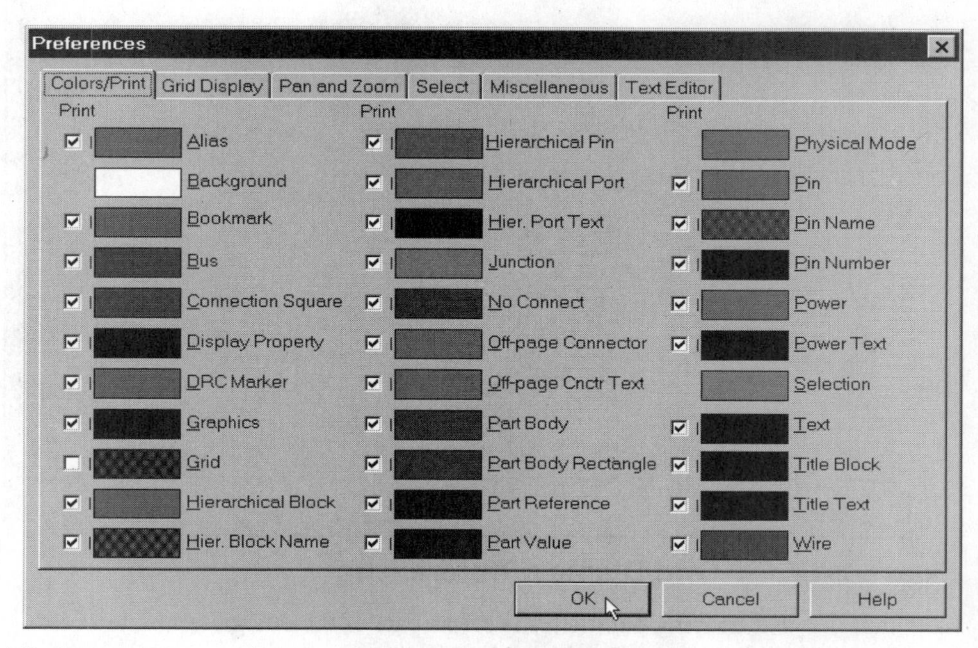

Figure 2-16 Colors and Print Preferences

Grid Display Preferences

Click on the Grid Display tab. This brings up the grid display preferences window shown in Figure 2-17. You can select the type of grid (dot or line), turn grid display on or off, and determine whether the pointer snaps to a grid location when placing or moving objects. Independent selections apply to schematic page and part and symbol grids.

The default dot grid is recommended. The author cannot recall any situation in which it was advantageous to turn off the grid display or to select a line type grid. Precisely aligning text, such as reference designators, part values, and pin and port names may require turning off the snap to grid feature. However, snap to grid **must be enabled** when placing or moving any electrical objects such as parts, pins, ports, wires, buses, and net aliases.

The hotkey CTRL + T toggles snap to grid on and off. When drafting a schematic, use of this hotkey is more efficient than clicking on the grid display preferences window.

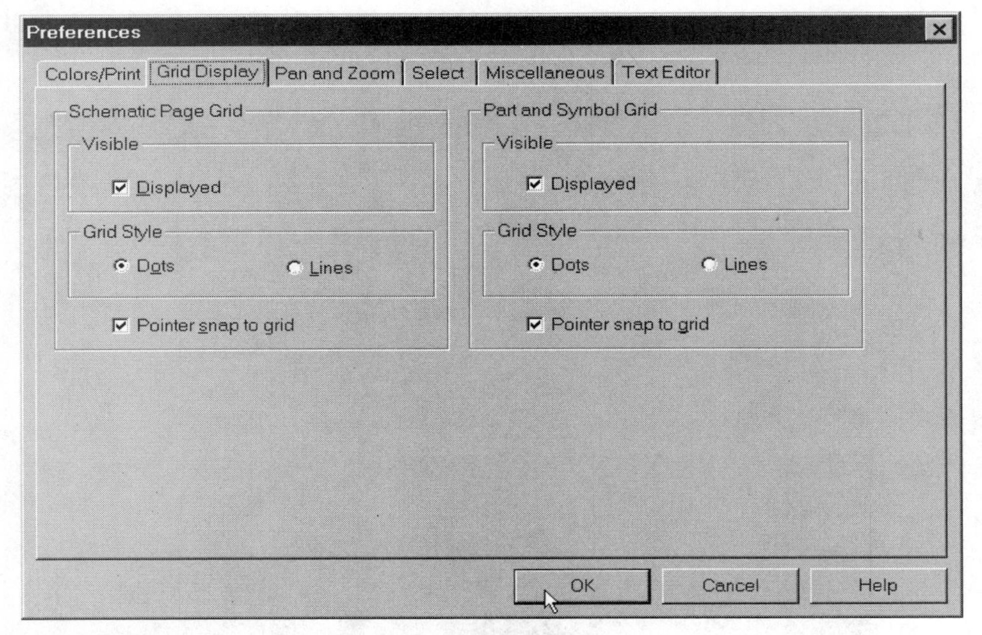

Figure 2-17 Grid Display Preferences

Note that the grid-spacing parameter is set as part of the design template configuration, which is explained later in this chapter.

Pan and Zoom Preferences

Click on the Pan and Zoom tab. This brings up the pan and zoom preferences window shown in Figure 2-18. You can select the zoom factor and auto scroll percentage. The zoom factor sets the magnification or reduction when you use zoom commands. The term *auto scroll* actually refers to automatic panning across the display. Automatic panning occurs whenever you hold the left mouse button down. The percentage value determines what percentage of the total drawing area the display jumps when the mouse pointer reaches the display edge. In most cases the default values for pan and zoom operations give good results and do not require modification.

Capture allows the use of certain hotkeys as keyboard shortcuts for commands. The I key commands zoom in and the O key commands zoom out. Note that hotkeys are not case sensitive.

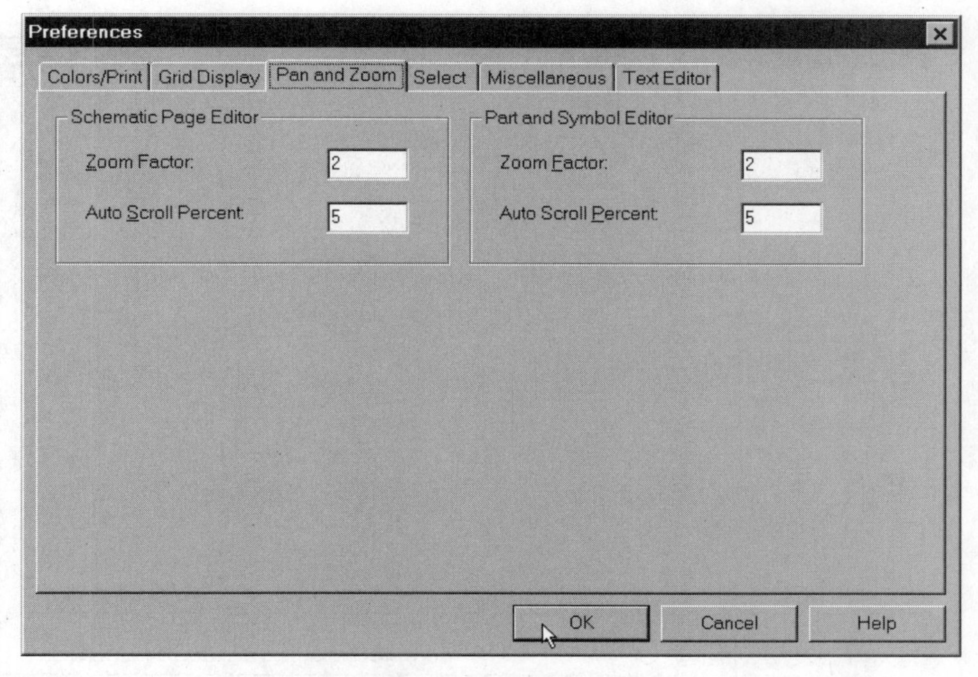

Figure 2-18 Pan and Zoom Preferences

Select Preferences

Click on the Select tab. This brings up the select preferences window shown in Figure 2-19. You can chose area select options and set visibility of the tool palette for both schematic and part-editing operations. Area Select refers to selecting objects on the screen while dragging the mouse pointer with the left button held down. Intersecting mode selects all objects that touch the selection box created by dragging the pointer. Fully enclosed mode selects only those objects within the selection box. Intersecting selection mode works best for most editing operations. When objects overlap, fully enclosed selection mode is often advantageous. You can also set the maximum number of objects to display at high resolution while dragging. The default value of 10 is adequate for most users. A higher value may be required when editing very complex and dense schematics, but noticeably reduces response on slower systems. If you use a fast Pentium processor, you can experiment with higher values.

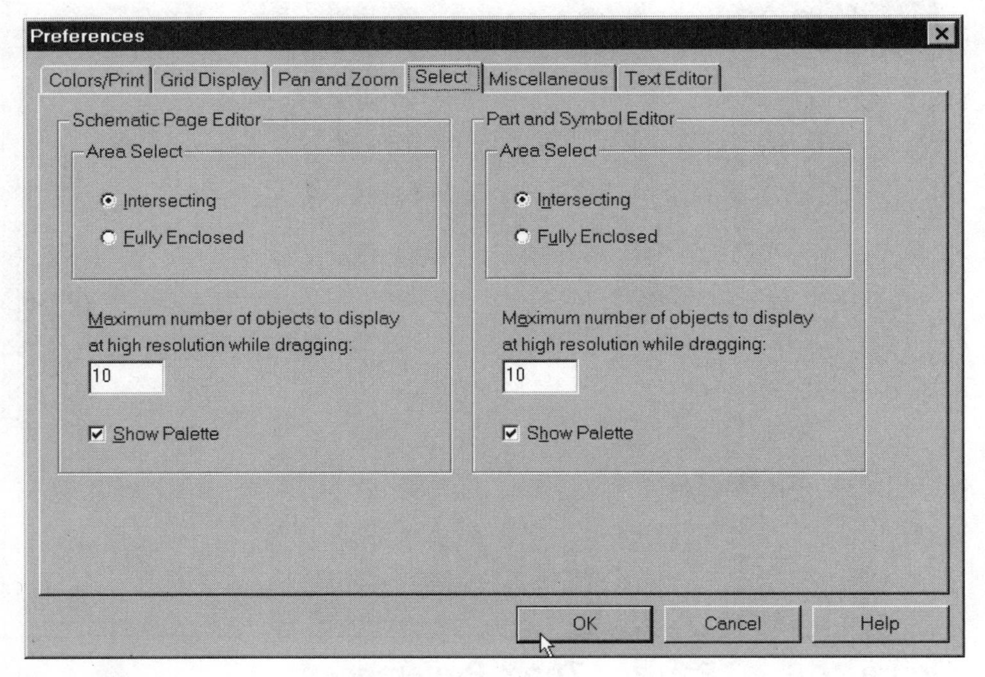

Figure 2-19 Select Preferences

The tool palette appears on the right side of the screen when editing a schematic or part. Clicking on tool icons to select commands works the same as with other Windows programs. The tool palette includes the most commonly used commands. All the commands on the tool palette and additional, less frequently used commands can be accessed by clicking on the pull-down menu bar at the top of the display or via keyboard shortcuts. Disabling the tool palette slightly increases the display area available for editing. However, most users find that clicking on the tool palette icons is the fastest means of command entry. For this reason, the author suggests the tool palette remain enabled.

Miscellaneous Preferences

Click on the Miscellaneous tab. This brings up the miscellaneous preferences window shown in Figure 2-20.

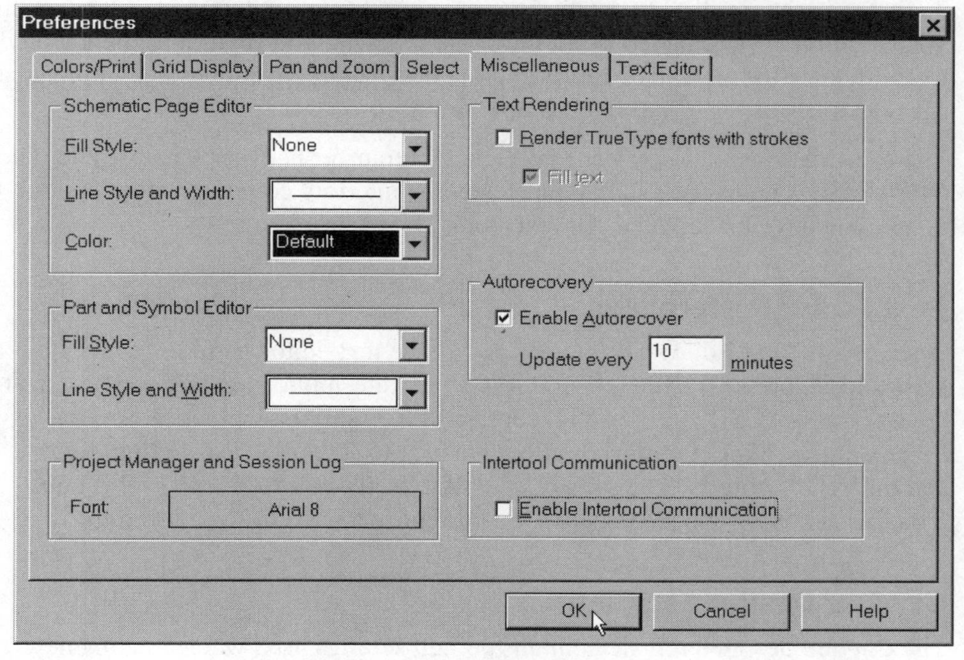

Figure 2-20 Miscellaneous Preferences

Fill, line, and color settings on this tab apply only to certain nonelectrical objects:

- Fill styles apply to rectangles, ellipses, and polygons.
- Line styles and widths apply to lines (not to be confused with electrical wires), rectangles, polygons, ellipses, and arcs.
- Color settings apply to rectangles, ellipses, and polygons. Note that the default color selection is that which you have defined for graphics objects on the color/print preferences tab. If you select a particular color, it applies only to

new objects that you draw after you make the color selection. Objects drawn in a particular color maintain that color. Objects drawn in the default color track the color selection for graphics objects.

The project manager and session log font defaults to Arial 8 (the 8 refers to a text height of 8 points).

Text rendering (Render True Type fonts with strokes) improves text quality when creating hardcopy on older-generation pen plotters. Do not select this option if you are using a laser or ink jet printer. If you select text rendering, you should also select the fill text option.

Autorecover allows design data to be recovered if Capture terminates abnormally. While Autorecover enabled is not the default setting, you should enable it. Use 10 to 15 minutes for the update interval. Note that Autorecover does not automatically save data. If you exit Capture normally, the Autorecover file is erased.

Intertool communication refers to design data transfer between Capture and other OrCAD EDA tools. You should not select the intertool communication option unless you have loaded other OrCAD tools.

Text Editor Preferences

Click on the Text Editor tab. This brings up the text editor preferences window shown in Figure 2-21. The text editor is a separate application that can be used to create or edit ACSII text files that contain SPICE simulation or VHDL data. Syntax highlighting preferences allow the user to set colors used to highlight certain VHDL objects. The subject of VHDL is beyond the scope of this book. You can also enter tab settings and the current font used by the text editor.

Capture Design Template

The Capture design template contains default settings used when creating new designs. These settings include font usage, title block, page size, grid reference, hierarchy, and SDT compatibility. Some design template settings can be modified once a design or schematic page has been created. The design template settings are stored in the design file. If you load a design created on another system running Capture, the design comes up with its own template settings, not those currently defined in your system.

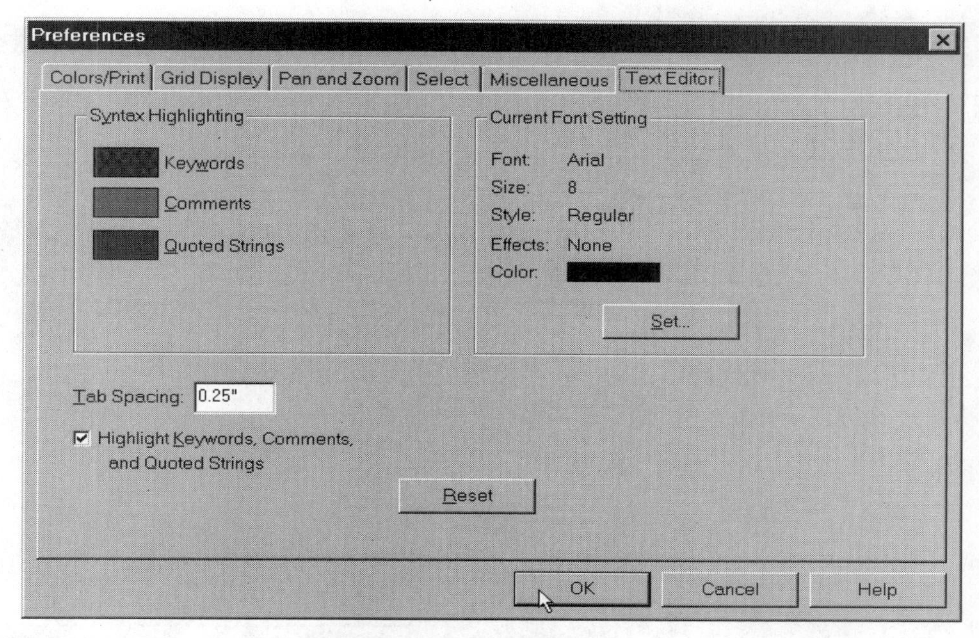

Figure 2-21 Text Editor Preferences

Design Template Font Settings

To change design template settings, click on Options and then Design Template on the main Capture screen. You can also use the keyboard shortcut ALT, O, D. This brings up the design template window shown in Figure 2-22. The font settings tab is always displayed first. You can click on the other tabs at the top of the window to set additional design template parameters.

The design template font settings affect screen display and printing. You can select fonts for the various text objects used in Capture designs. In most cases, the default Arial 5-point font gives good results. To change the font setting for a particular text entity, click on the font box. This brings up a standard Windows font dialog box that allows you to set the font name, style (regular, italic, bold, and bold italic for most fonts), and font size in points. Note the point size is a fundamental unit of measurement in typography that gives the height of text characters (72 points equal 1 inch). Thus the default 5-point font results in a text character height of approximately .07 inch. This size works well for schematics drawn on the standard .1-inch grid, as it allows adequate space between adjacent text objects.

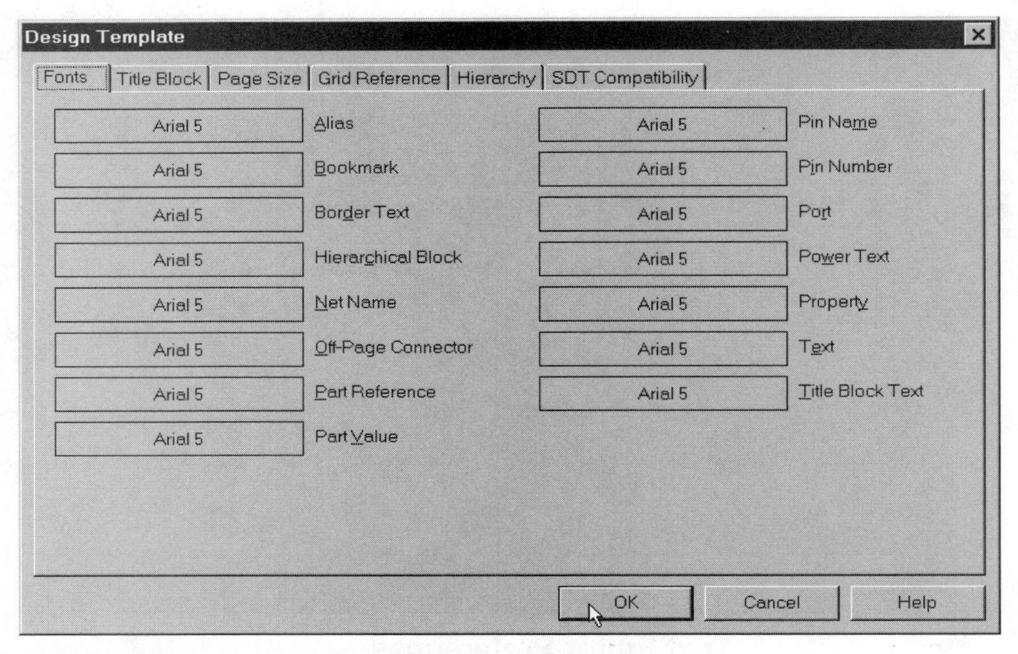

Figure 2-22 Design Template Font Settings

Sans serif fonts such as Arial are recommended for technical drawings. Avoid using serif fonts (such as Times New Roman or Courier). Maintain a consistent point size for all text objects; the only exceptions might be caution or warning notes or in the title block area.

Note that font settings are used for both display and printing. The default font settings are used whenever you draw new objects. You can individually edit the font settings for existing objects. If you change the default settings, the change does not affect existing objects; their original font settings remain unchanged. You cannot globally change settings in an existing design by changing the design template. This principle applies to **all** design template settings, not just the font settings.

Design Template Title Block Settings

Click on the Title Block tab. This brings up the design template title block settings window shown in Figure 2-23. You can enter information used in the default title block that Capture places onto all new schematic pages. The same information also appears in as a header in reports, such as a bill of materials listing, automatically generated by Capture.

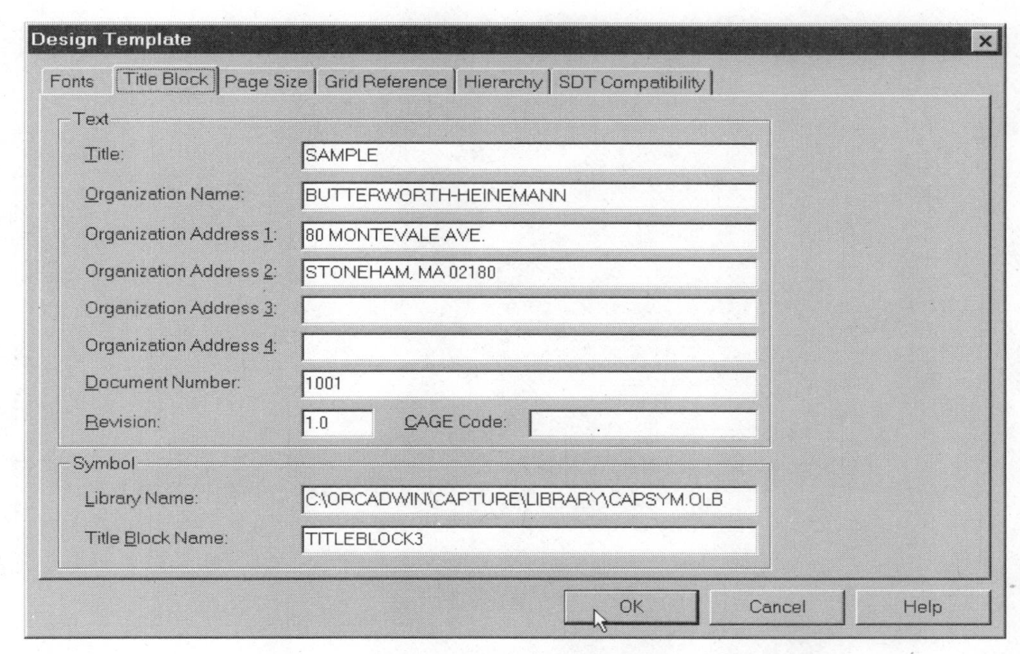

Figure 2-23 Design Template Title Block Settings

Refer to Chapter 1, page 30 for more information about the different fields used in a title block. Note that the CAGE (commercial and government entity) code refers to a federal identification number primarily used by aerospace and military contractors. The CAGE code appears in ANSI (American National Standards Institute) title blocks.

The symbol fields allow you to specify the library path and name of the title block symbol that Capture automatically places on all new schematic pages. The CAPSYM (Capture symbols) library includes a selection of standard title blocks or you can define your own. Titleblock3 is a predefined title block with three lines reserved for the organization name and of address (see Figure 1-17 on page 30 for an example). If you need to enter additional organization (company) address lines, then you can use the Titleblock4 or Titleblock5 symbols included in the CAPSYM library.

Design Template Page Size Settings

Click on the Page Size tab. This brings up the design template page size settings window shown in Figure 2-24. You can enter page size settings that Capture uses for all new schematic pages.

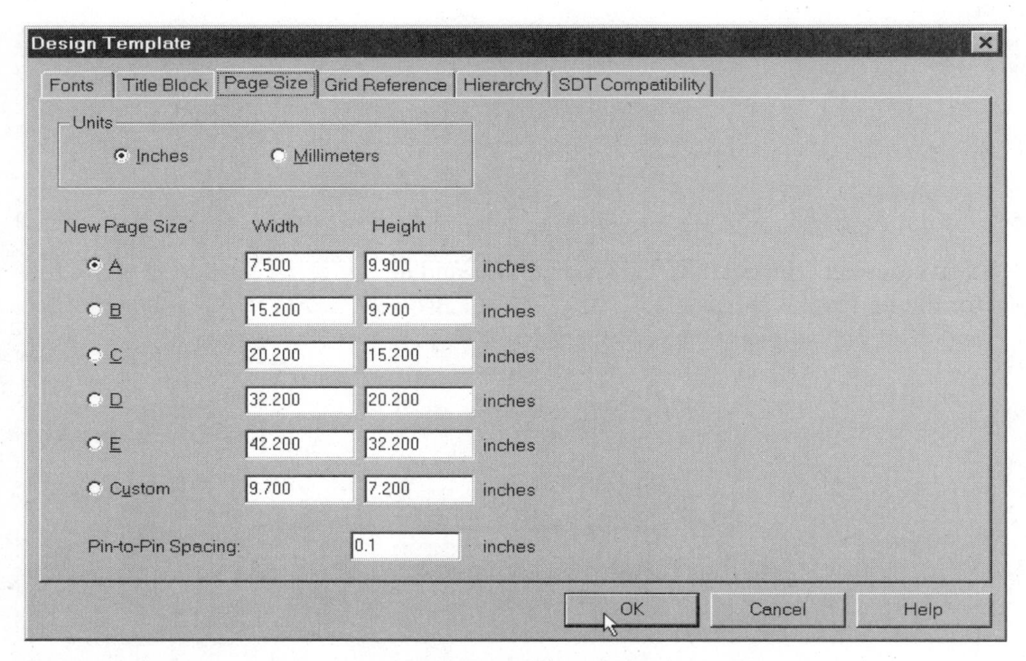

Figure 2-24 Design Template Page Size Settings

Standard ANSI A to E size page dimensions appear if inch units are selected. If you select metric units, ISO (International Standards Organization) A4 to A0 page dimensions appear instead. Note that the page size refers to the overall dimensions of the printed area. Most printers cannot print to the very edge of the paper, so a blank border area is required. The A size dimensions are set to be compatible with Hewlett-Packard laser printers using 8.5 × 11 inch paper.

If you require a nonstandard size, you can edit the standard ANSI or ISO page size dimensions or use the custom page size selection. Capture default settings are for landscape orientation (width greater than height). Note that the author has redefined the A size page as portrait orientation. Most companies prefer portrait orientation for 8.5× 11 inch technical documentation.

Pin-to-pin spacing refers to the grid spacing used on schematic pages. All electrical objects must be placed on a grid. The term *pin-to-pin spacing* sometimes confuses new users. It has no relation to the physical spacing of parts pins. Use of a .1 inch grid for schematic drafting is an established industry practice. Note that once a schematic page has been created, the grid spacing cannot be changed.

Design Template Grid Reference Settings

Click on the Grid Reference tab. This brings up the design template grid reference settings window shown in Figure 2-25. Grid references appear in a printed border area surrounding schematic pages. Most companies used alphanumeric ANSI grid references to identify drawing zones. For example, a technical manual might refer to a calibration trimpot located in zone C3.

You can select the count (number of zones) and alphabetic or numerical indexing for the horizontal and vertical axis. You can also set the width of the border area and printing and display options. For some reason, OrCAD writers chose to place the title block display and print options on the screen, even though they are not directly related to grid reference settings.

In most cases, the Capture defaults give good results and do not require modification.

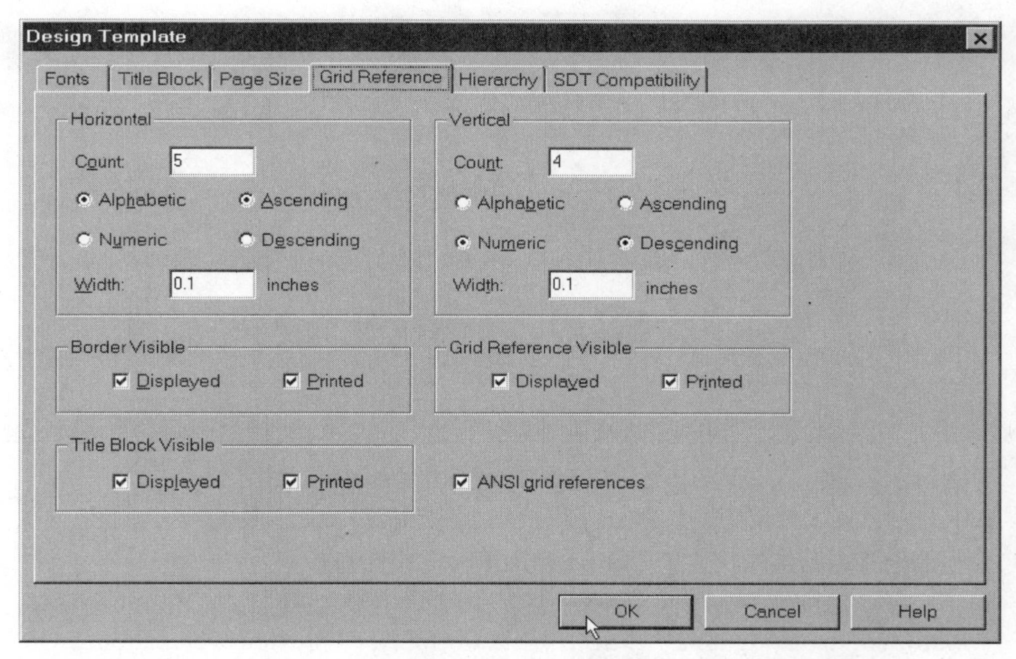

Figure 2-25 Design Template Grid Reference Settings

Design Template Hierarchy Settings

Click on the Hierarchy tab. This brings up the design template hierarchy settings window shown in Figure 2-26. Hierarchical blocks are used to represent circuit blocks in multi-sheet hierarchical schematics (see Chapter 1, page 31). Primitive hierarchical blocks and parts cannot have associated schematics. Nonprimitive objects can have associated schematics. In most cases, hierarchical blocks are nonprimitive. Parts, on the other hand, are generally primitive. All examples and exercises in this book use nonprimitive hierarchical blocks and primitive parts.

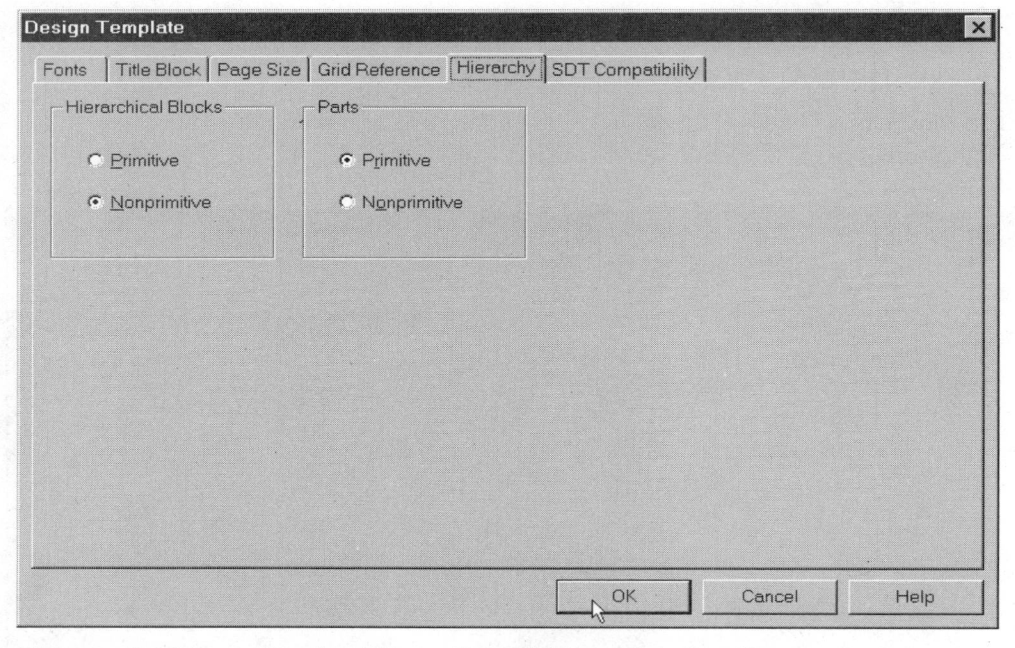

Figure 2-26 Design Template Hierarchy Settings

Design Template SDT Compatibility Settings

Click on the SDT Compatibility tab. This brings up the SDT compatibility settings window shown in Figure 2-27. SDT refers to OrCAD's previous generation, DOS-based schematic capture program. The settings in this window only affect backward translation from Capture to SDT. Since SDT software is obsolete, you are not likely to encounter a requirement for backward translation. You are much more likely to encounter the requirement for translation from SDT to Capture.

Translation from SDT to Capture is straightforward, since Capture directly opens most SDT files. Special translation considerations are discussed in Chapter 11.

For all practical purposes, you can ignore the SDT compatibility settings. The default values are probably acceptable if you should ever have a requirement for backward translation.

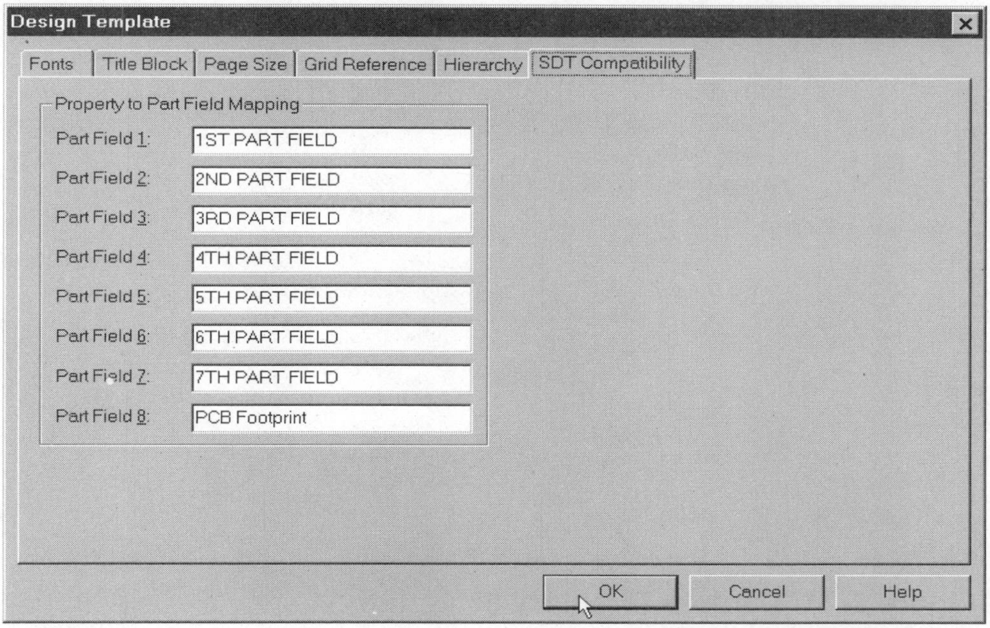

Figure 2-27 Design Template SDT Compatibility Settings

Completing the Configuration Process

If you have opened all the preferences and design template windows described on pages 52 to 65 and entered all the suggested settings, you have completed the initial configuration process. Exit Capture to save your settings.

Capture supports an additional level of configuration settings that affect only individual designs: design properties and schematic page properties. These are discussed in subsequent chapters.

Capture Directory Structure

Installation of Capture creates a number of new subdirectories on your hard drive. Specific subdirectories exist for designs, libraries, fonts, and third-party add-ons. An understanding of this directory structure helps when doing file maintenance, especially for design and library file backup.

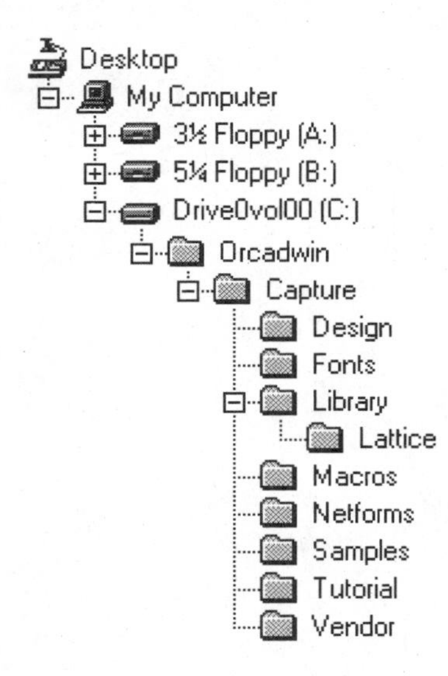

Figure 2-28 Capture Directory Structure

After initial installation of Capture, the directory structure appears as shown in Figure 2-28. All OrCAD Windows software is loaded into Orcadwin. All Capture-related files, including executables, are loaded into the Capture subdirectory. The Design subdirectory is reserved for designs. The Library subdirectory contains parts libraries, including any custom libraries that you might create. Lattice is a supplemental library. Netforms contains netlist postprocessing routines used for data transfer to PCB design systems. Third-party add-ons are loaded into the Vendor subdirectory.

You should maintain this directory structure to assure compatibility with future upgrades and product releases.

Files on the Disk Supplied with This Book

This book includes a disk that contains sample files and a custom parts library for the tutorial exercises covered in subsequent chapters. The disk also includes a special utility program for sorting bill of material files, a subject covered in Chapter 10. You can find more detailed information about the disk in Appendix A.

The sample files for the tutorial exercises are intended to help complete the exercises and to serve as examples of finished designs. In some cases, fine details on schematic figures within this book may be hard to make out because of the reduced size. You can use Capture to print out the sample schematic files for improved readability.

Use the Windows Explorer to copy the custom library file, Custom.olb, from the disk to your C:\Orcadwin\Capture\Library directory.

Backing up Design and Custom Library Data

To avoid disaster, always have three independent backups for any mission-critical computer data. Assume that data stored on a hard disk drive can disappear at any time without warning. Keep one backup copy on floppy disk and update this data after every computer session. Keep a second backup copy on a magnetic tape medium, such as a QIC-80 tape cartridge. This type of medium lends itself to backing up the entire disk. Make a current backup tape every week and store it offsite in case a disaster strikes. If you are working in corporate network environment, shared network drives are most likely backed up on a daily basis. You can copy design and library files onto a network drive and let your computer department handle the backup for you. The third backup can be a paper hardcopy.

With the Autorecovery feature enabled, an occasional abnormal program termination should not cause you excessive grief. However, if you are in the middle of an important project, regularly saving the design is still a good practice. Remember that Autorecovery does not save your data if you exit Capture.

Users upgrading from OrCAD SDT will appreciate the fact that Capture designs now include all parts data in a design cache that is stored in the design file, eliminating dependency on external libraries when design files are transferred or archived. However, users are still admonished from modifying any of the OrCAD-supplied libraries, as these modifications will be lost if subsequent Capture upgrades write over the libraries. Always keep new or modified parts in a custom library and keep a backup copy.

Conclusion

In this chapter you have learned how to install and configure OrCAD Capture. You are now ready to start the tutorial exercises, which will show you how to create and edit schematics with Capture.

Review Exercises

1. If you have access to the Internet, visit the OrCAD web site (www.orcad.com) Explore the Knowledge Base.
2. Examine the Capture libraries using Capture Library List as explained on page 50. Print out a listing of the DEVICE.OLB library. This library contains the most commonly used electronic part symbols.
3. Describe the difference between Capture preference and design template settings. What effect do these settings have on designs that are transferred to another user's system?
4. Describe two methods for setting snap to grid on and off.
5. What term does Capture use for grid spacing, and how do you set the grid spacing for new schematic pages?
6. Can you change the grid spacing on an existing schematic page?
7. What is the recommended grid spacing for electronic schematics?
8. Why is the design template page size setting 7.5 × 9.9 inch for an A size schematic printed on 8.5 × 11 inch paper?
9. Explain the difference between portrait and landscape orientation.
10. Use the Windows Explorer to examine the directory structure for your Capture installation.
11. What is the purpose of Capture's Autorecovery feature? Can you recover a design file if you exit without saving it?
12. When should you back up data and what media should you use?
13. What is the disadvantage of modifying Capture-supplied parts libraries?

3

Capture Basics

In the previous chapter, you learned how to install and configure Capture. The next several chapters are tutorial exercises intended to get you off to a fast start. In this chapter you will learn how to draw a basic single-sheet schematic. Your task will be to recreate the programmable power supply schematic shown in Figure 1-17. This figure will be the model for the exercise in this chapter. Take a moment and leaf back to page 30 in Chapter 1 and familiarize yourself with the model. You may want to run a copy of the figure and keep it handy for reference.

First Session – Introduction to Capture

Launch Capture from the Windows 95 desktop. The Capture session frame appears. Then click on File, New, and Design as shown in Figure 3-1 to create a new design. The screen shown in Figure 3-2 appears.

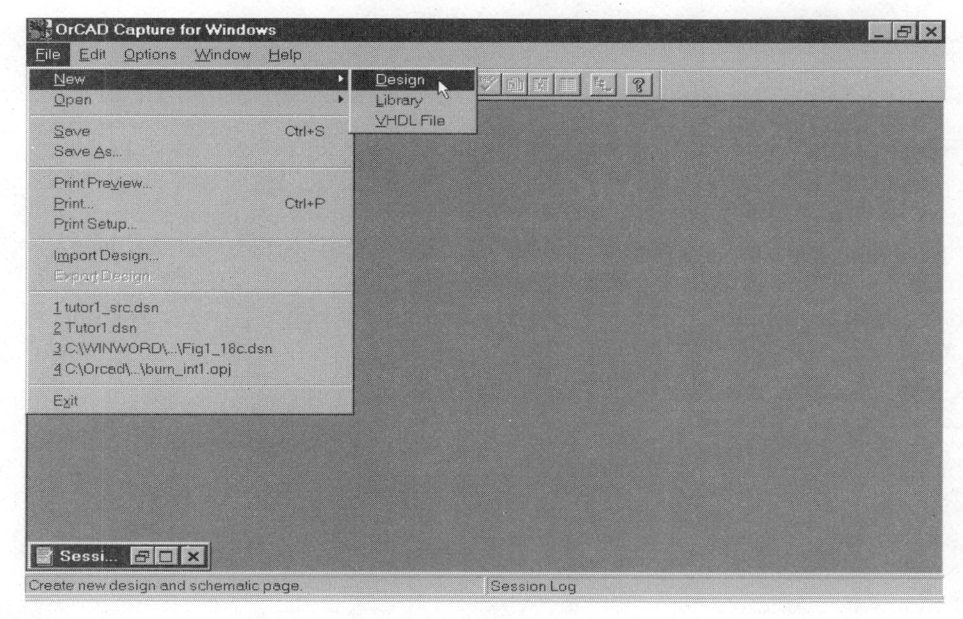

Figure 3-1 Creating a New Design

Note that Capture designs can contain one or more schematics. A schematic can consist of one or more pages (sheets). A Capture design includes a design cache. This cache contains all electronic part symbols used in the design. Once a part is retrieved from a library, it is stored in the design cache.

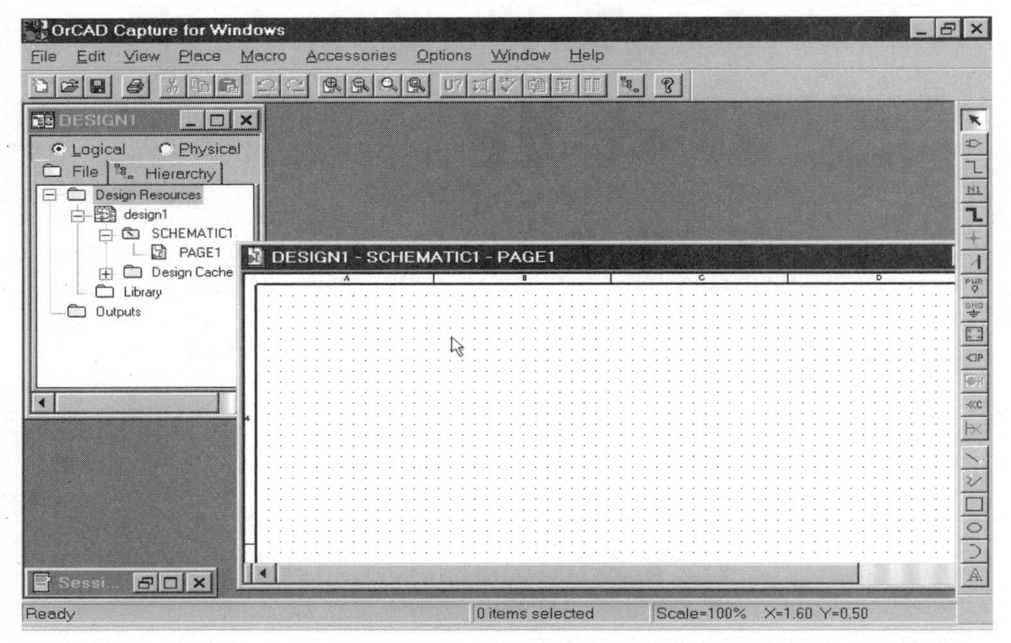

Figure 3-2 Project Manager and Schematic Editor Windows

The upper left window is called the project manager. The schematic editor window appears at the lower right of the screen. Each Capture design that you open has an associated project manager window. This window shows a graphical overview of the design structure and parts in the design cache. You will also use project manager to launch tools for postprocessing operations such as updating reference designators, generating bills of materials, checking for design rule violations, and creating netlists for PCB design.

You can navigate the project manager window as you do the Windows Explorer. You can expand or collapse the design structure by double clicking the left mouse button. Double clicking on a schematic in the design opens the schematic and causes icons to appear for each page in the schematic. Double clicking on a schematic page opens the page and causes it to appear in the schematic editor window. You can then click on the upper left corner of the schematic editor window to maximize the schematic window. Note that the project manager uses a file folder icon for schematics and a page icon for each page in the schematic.

The view menu area at the top of the project manager window allows you to select logical or physical views of a design. For the most part, you can ignore the physical view selection. The only exception is if you ever create a complex hierarchy. This is a design in which two or more hierarchical blocks refer to the same schematic. You might use a complex hierarchy when working with a gate array, in which a block representing an 8-bit register could occur in multiple locations in the design. In this case, selecting physical view would unfold the design and show all occurrences of the 8-bit register. In logical view, the schematic page representing the 8-bit register would only occur once.

Logical view is required for all normal drafting and editing operations. In physical view, only part properties such as reference designators can be edited.

Complex hierarchies are useful for PLD (programmable logic device) and gate array design but tend to confuse users drafting conventional schematics. None of the exercises in this book uses a complex hierarchy or requires selecting physical view.

Saving the New Design

Click on File and then Save. The first time you save a design, the file name dialog box shown in Figure 3-3 appears. This is a standard Windows file name dialog box. Navigate to the C:\Orcadwin\Capture\Design directory. Click on the Create New Folder icon and create a new Tutor1 directory. Then save the design using the name Pwr_Supply.dsn.

Introduction to the Schematic Editor

Click on the icon at the upper left corner of the schematic editor window and maximize this window. You can pan around the schematic editor window by using the scroll bars. You can also pan by holding the left mouse button down and dragging the mouse to the edges of the schematic area. If you pan to the bottom right of the schematic, you will see the title block area. Your screen will appear similar to Figure 3-4, except that your title block will have the default name and address values set as preferences when Capture was configured on your system.

The icons appearing across the top of the screen compose the main toolbar. The main toolbar includes tools for basic file operations, cut and paste, zoom, and postprocessing operations such as generating a bill of materials. Icons for postprocessing operations remain dimmed (grayed out). Postprocessing operations are global in scope (they affect the entire design) and can only be run from the project manager window.

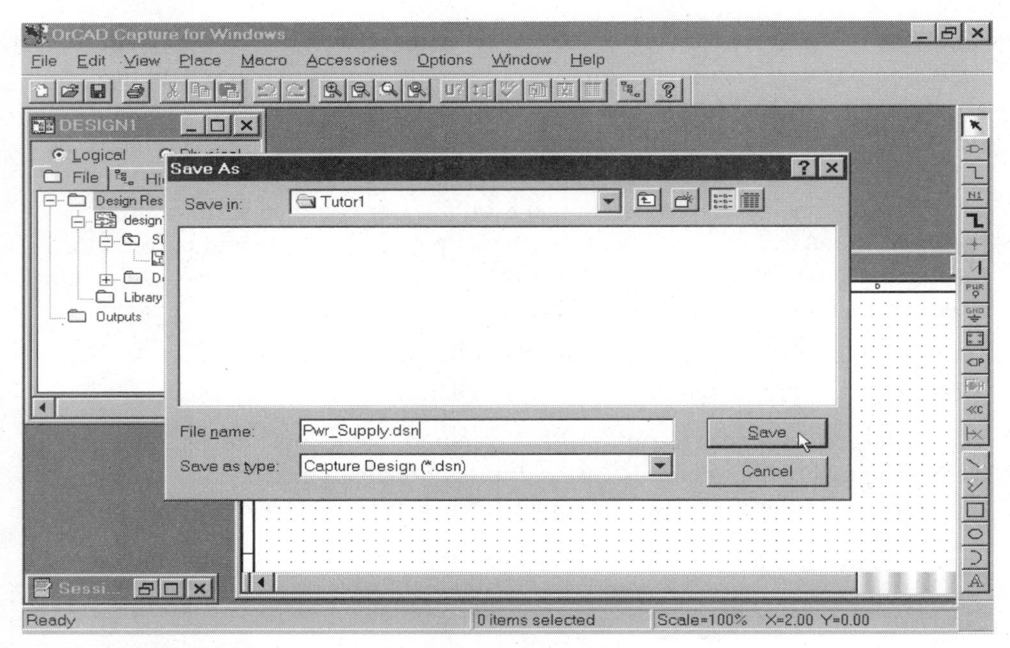

Figure 3-3 Saving the New Design

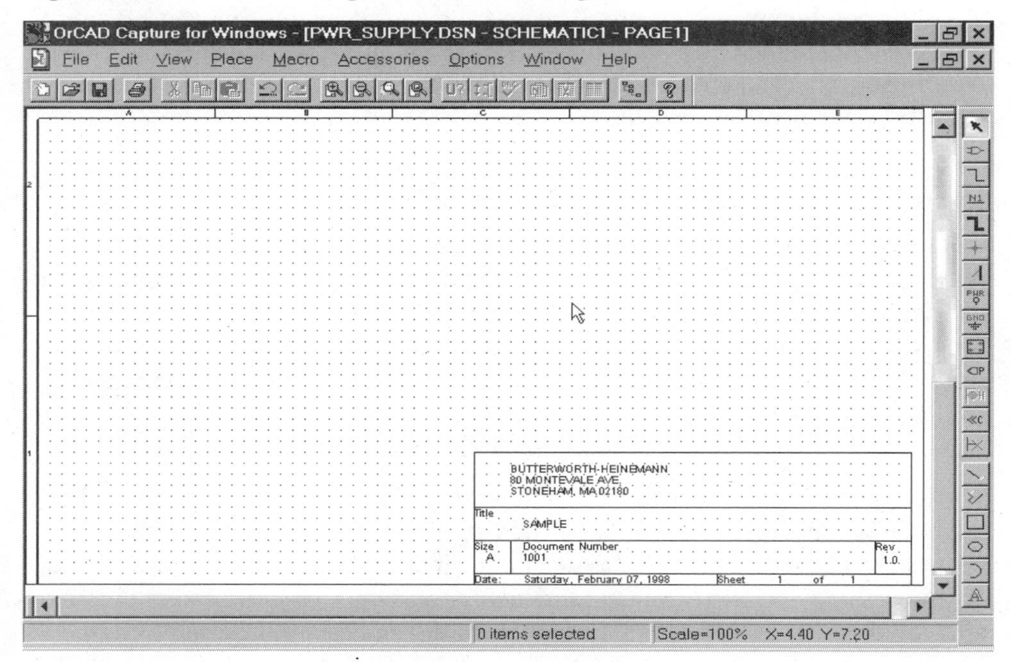

Figure 3-4 Schematic Editor Window

The icons appearing at the right side of the screen comprise the schematic editor tool palette. A different tool palette appears when running the parts editor. The schematic editor tool palette includes tools for placing parts, electrical symbols, and drafting objects.

As with most other Windows programs, the Capture main toolbar and tool palette are dockable. This means that you can click in the area surrounding the buttons and then drag the toolbar to a different location. You can also resize the toolbar.

When you click on some tool buttons, such as the text tool, a dialog box appears. You can also press the right mouse button to bring up a context-sensitive shortcut menu with additional command options. You can use the ESC key to cancel most commands.

The Main Toolbar

The main toolbar is displayed at all times when the Capture session frame or the project manager is active. In the schematic editor, you can click on View at the top menu bar and then click on Toolbar to toggle display of the tool bar on and off. The toolbar duplicates most of the commands on the top menu bar and associated pulldown menus. Most users find that clicking on the toolbar buttons is the quickest means of launching commands, but you can free up additional display area by turning off the toolbar display. Toolbar commands are summarized below.

ICON	TOOL	ACTION
	**New**	Creates a new document with the same properties as the active document. Similar action to the New command on the File menu.
	**Open**	Opens an existing design or parts library. Similar action to the Open command on the File menu.
	**Save**	Saves the active schematic or library part. Same action as the Save command on the File menu.
	**Print**	Prints the active schematic page or library part. Same action as the Print command on the File menu.
	**Cut**	Cuts selected objects from the active document and moves them onto the clipboard. Same action as the Cut command on the Edit menu.

	Copy	Copies selected objects from the active document and puts them onto the clipboard. Same action as the Copy command on the Edit menu.
	Paste	Pastes the clipboard contents into the active document at the pointer location. Same as the Paste command on the Edit menu.
	Undo	Undoes the action of the last command, if possible. Same as the Undo command on the Edit menu.
	Redo	Cancels the action of the last Undo command, if possible. Same as the Redo command on the Edit menu.
	Zoom In	Performs zoom-in operation in active document. Same as Zoom and In commands on the View menu.
	Zoom Out	Performs zoom-out operation in active document. Same as Zoom and Out commands on the View menu.
	Zoom Area	Uses mouse pointer to select a zoom area in active document. Same as Zoom and Area on the View menu.
	Zoom All	Performs zoom out that displays entire active document. Same as Zoom and All on the View menu.
U?	**Update Part References**	Assigns part references (reference designators) for parts in the selected schematic pages. Available from project manager only. Same as the Update Part References command on the Tools menu.
	Gate and Pin Swap	Used after PCB design process to back annotate parts references in the selected schematic pages. Available from project manager only. Same as the Gate and Pin Swap command on the Tools menu.
DRC	**Design Rules Check**	Runs check for electrical design rules violations in the selected schematic pages. Available from design project only. Same as the Design Rules Check command on the Tools menu.

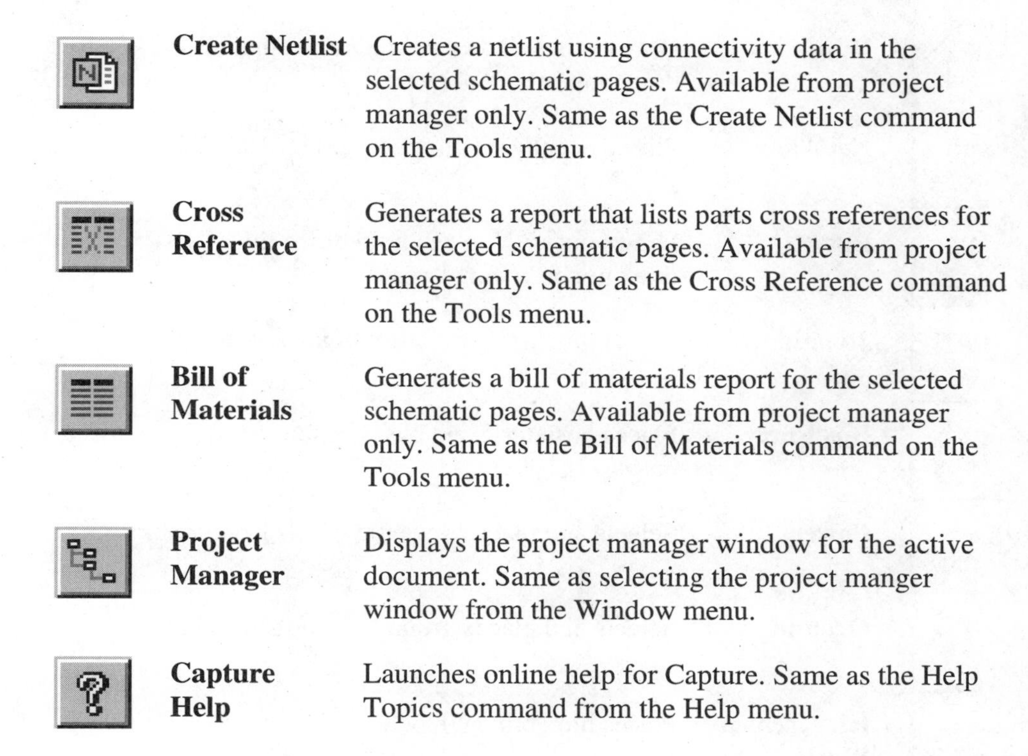

	Create Netlist	Creates a netlist using connectivity data in the selected schematic pages. Available from project manager only. Same as the Create Netlist command on the Tools menu.
	Cross Reference	Generates a report that lists parts cross references for the selected schematic pages. Available from project manager only. Same as the Cross Reference command on the Tools menu.
	Bill of Materials	Generates a bill of materials report for the selected schematic pages. Available from project manager only. Same as the Bill of Materials command on the Tools menu.
	Project Manager	Displays the project manager window for the active document. Same as selecting the project manger window from the Window menu.
	Capture Help	Launches online help for Capture. Same as the Help Topics command from the Help menu.

The Schematic Editor Tool Palette

The schematic editor tool palette is normally displayed when the Capture schematic editor is active. You can click on View at the top menu bar and then click on Tool Palette to toggle display of the tool palette on and off. The tool palette duplicates most of the commands on the top menu bar and associated pulldown menus. Most users find that clicking on the tool palette buttons is the quickest means of launching commands, but you can free up additional display area by turning off the tool palette display. Tool palette commands are summarized below.

ICON	TOOL	ACTION
	Selection Mode	Sets the normal editing mode and allows selection of objects using the mouse pointer.
	Part	Selects and places library parts.

	Wire	Draws wires. You can hold down the SHIFT key to draw nonorthogonal wires.
	Net Alias	Places net aliases on wires and buses.
	Bus	Draws buses. You can hold down the SHIFT key to draw nonorthogonal buses.
	Junction	Places junctions at intersections of wires.
	Bus Entry	Places bus entries on buses. Used to connect wires to buses.
	Power	Selects and places power symbols from a library.
	Ground	Selects and places ground symbols from a library.
	Hierarchical Block	Places hierarchical blocks.
	Hierarchical Port	Selects and places hierarchical port symbols from a library.
	Hierarchical Pin	Places hierarchical pins in the selected hierarchical block. Dimmed out unless a hierarchical block is selected.
	Off-Page Connector	Places off-page connectors. Use in multiple-sheet flat schematics (not hierarchical).
	No Connect	Places no-connect symbol on pins that are not connected to other electrical objects.
	Line	Draws lines. Note that lines are not electrical objects.

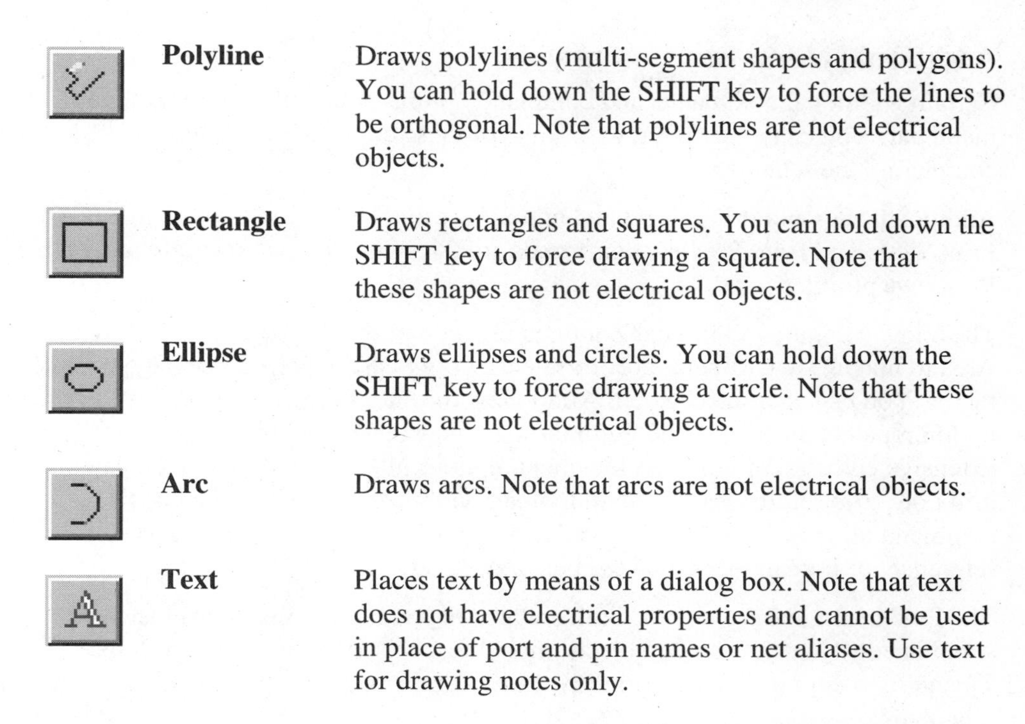

	Polyline	Draws polylines (multi-segment shapes and polygons). You can hold down the SHIFT key to force the lines to be orthogonal. Note that polylines are not electrical objects.
	Rectangle	Draws rectangles and squares. You can hold down the SHIFT key to force drawing a square. Note that these shapes are not electrical objects.
	Ellipse	Draws ellipses and circles. You can hold down the SHIFT key to force drawing a circle. Note that these shapes are not electrical objects.
	Arc	Draws arcs. Note that arcs are not electrical objects.
	Text	Places text by means of a dialog box. Note that text does not have electrical properties and cannot be used in place of port and pin names or net aliases. Use text for drawing notes only.

All of the tools listed correspond directly to identical commands available on the Place menu. The term *tool* is used if the command is launched from an icon. Several additional commands are available only from the Place menu, including the Place Title Block and Place Picture commands. As with most Windows programs, you can also use keyboard shortcuts for menu commands. For example, you can use the key sequence ALT, P, P to launch the Part tool.

Launching tools by means of clicking on buttons is more efficient and increases productivity. Typing commands is a somewhat of a throwback to the old DOS days. For modern EDA tools such as Capture running in a Windows environment, the mouse has become the primary means of command entry. Use of the keyboard has been relegated to typing in parts data and text notes. The only exceptions are certain commands such as Toggle Snap to Grid, in which a keyboard hotkey combination (CTRL+T) is more efficient than navigating through multiple levels of menus and dialog boxes with the mouse.

Tool palette commands are modal. This means that the command remains in effect until cancelled. For example, once you click on the Wire tool, you can draw multiple wires. After a command is launched, you can press the right mouse button to bring up a shortcut menu with an option to end the current command. You can also use the ESC key to end or cancel commands.

Additional Menu Bar Commands

Certain additional commands and command options are available from the top menu bar. You can experiment with some of these commands as you work on completing the schematic exercise.

Additional file- and printing-related options are available from the File menu. The Print Setup and Print Preview commands are the same as those found in most other Windows programs.

The View menu has additional Zoom command options. You can select Zoom Area to magnify a particular area or select Zoom Scale to set a percentage zoom factor. You can also select Zoom All to view the entire drawing area. The Redraw option repaints the screen and eliminates garbage that sometimes accumulates after extensive editing. Additional View menu options allow you to toggle display of grid dots, grid references, the main toolbar, and the tool palette. The Go To command allow you to shift the display to a particular X,Y location, grid reference, or bookmark (explained below).

The Find command from the Edit menu allows you to shift the display and simultaneously select named objects, including parts, nets, and text. Both the Go To and Find commands are primarily useful for navigating around large schematics.

The Bookmark command from the Place menu allows you to place a specially named bookmark symbol that appears as a small square. Bookmarks have some usefulness in conjunction with the Go To and Find commands for navigating around large schematics.

The Windows menu offers options for cascading and tiling screen windows and clicking on numbered windows that represent the session log, program manger, and open schematic pages. These options are fairly standard and should be familiar to most Microsoft Windows users.

Second Session – Starting the Single-Sheet Schematic

This exercise introduces the majority of the commands that you will routinely use in Capture for schematic drafting: selecting parts from the libraries and placing them on the page, interconnecting the parts, adding ground, power, junction, and text objects, and editing descriptions.

Keep a copy of Figure 1-17 handy for reference, because this is the model for the exercise. Start by orienting yourself on the page. Move the mouse cursor to pan to the lower right corner of the page, as shown in Figure 3-4.

Plan on spending several hours to complete the first exercise. If you run out of time and must stop, save your design as explained on page 71. You can launch the Pwr_Supply.dsn design again when you are ready to continue the exercise.

Selecting Libraries and Parts

All electrical parts are stored in libraries. Capture comes with a large number of libraries for various types of parts. You installed the libraries as part of the Capture installation in Chapter 2. When you first use the Part tool, you must configure the available libraries. If you were to make every Capture library available, scrolling through the resultant list of parts to select a particular part would become very time consuming. For most designs, you will require the Device and Custom libraries and perhaps one or two other additional libraries with specialized parts, such as TTL or CMOS.

Click on the Part tool. The dialog box shown in Figure 3-5 appears. Note that when Capture is first installed, the only configured library is Capsym. No parts appear because Capsym only contains power, ground, and hierarchical port symbols. This appears to be a minor idiosyncrasy of the installation process.

The next step is to add the Device and Custom libraries. Click on Add Library. This brings up the dialog box for browsing library files shown in Figure 3-6. Click on Device.olb or type in the library file name. Note that all Capture library files use the .olb file name extension. Then click on Open. The dialog box shown in Figure 3-5 reappears, and the Device library is now included in the libraries list. Repeat this process to add the Custom library.

Note that you can also delete a library by clicking on the library in the list box and then clicking on Remove Library.

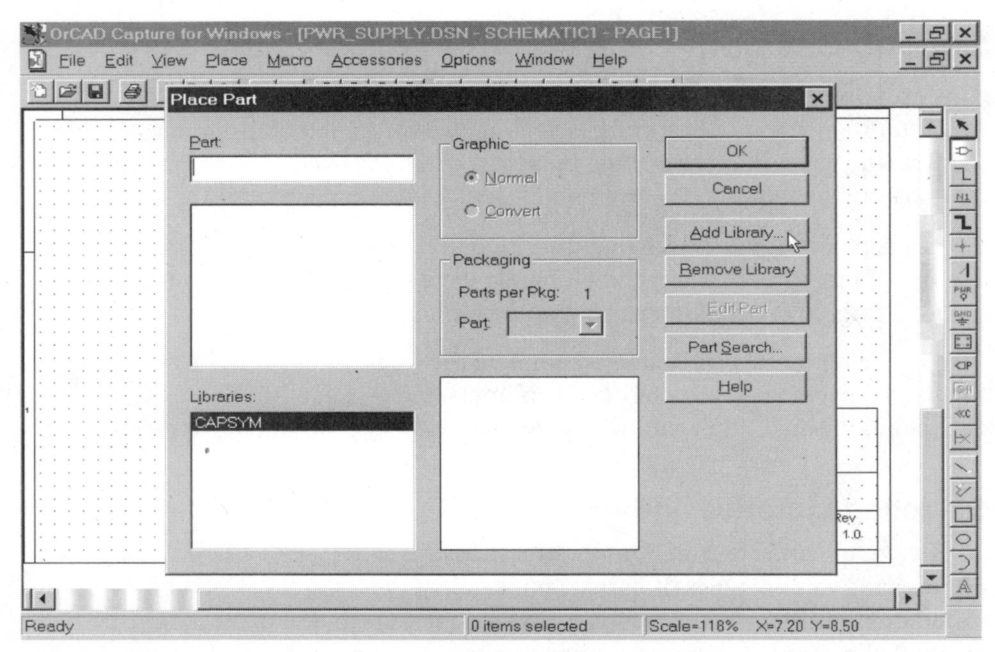

Figure 3-5 Dialog Box for the Place Part Tool

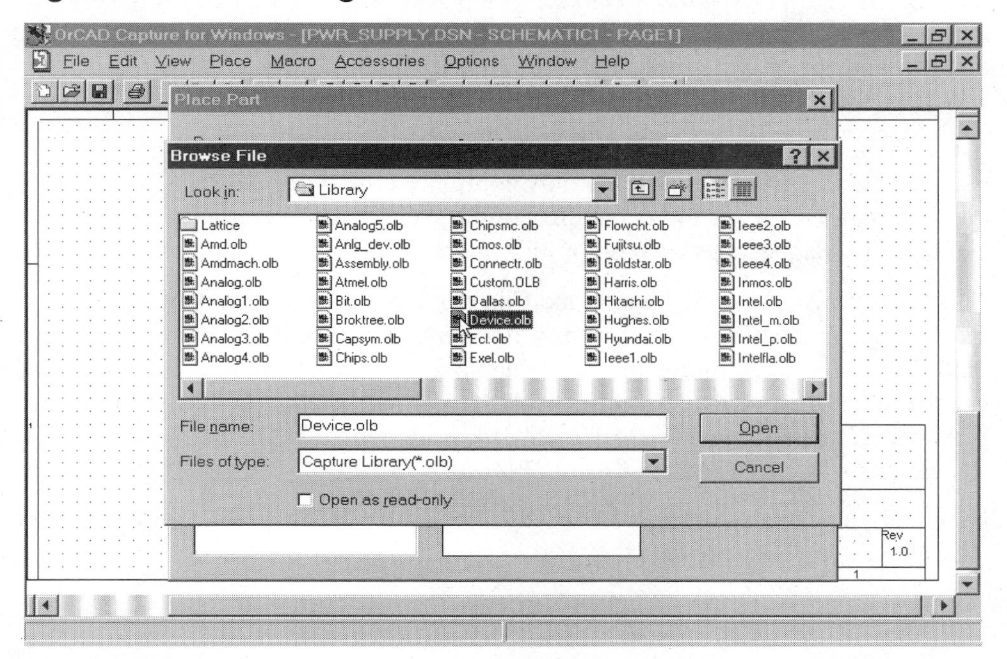

Figure 3-6 Dialog Box for Browsing Library Files

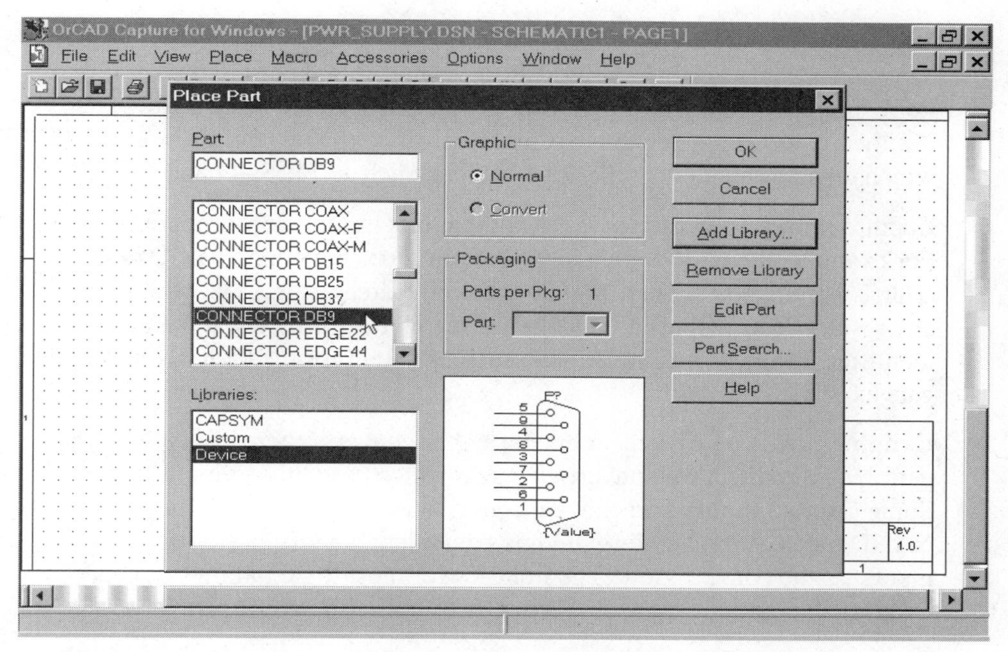

Figure 3-7 Selecting a Part from the Device Library

Once you have added the Custom and Device libraries, click on the Device library in the libraries list box. You can then use the parts scroll box to examine the library contents. Click on CONNECTOR DB9. The selected part appears in the part box, and a preview of the part is displayed as shown in Figure 3-7. Before you continue, let's examine a few aspects of library parts and the Place Part dialog box:

- Once the library list has been configured with the desired libraries, several means are available for finding and selecting library parts.
- If you are not certain about which libraries are required or you need to find the library that contains a particular part, you can type in the part name in the part box and click on Part Search. This tool searches through your entire library directory and reports which libraries contain the part. Note that in some cases, more than one library may contain different versions of the same part. You can use wildcard characters in the part name. Use an asterisk (*) to match multiple characters and a question mark (?) to match a single character.
- You can select one or more of the libraries on the list shown in the Place Part dialog box. To select multiple libraries, hold down the CTRL key while clicking on the desired libraries.

- You can enter the first several characters of the part name and a wildcard character (*) and then press ENTER to display all matches in the scroll box. For example * displays all parts and 74* displays all 74 series logic parts. You can then click on the desired part in the scroll box. You could also enter *244, which displays all "244" type octal buffers, such as 74LS244 and 74HC244.
- Capture libraries sometimes leave out manufacturers' alphanumeric prefixes. For example, the CMOS library (Cmos.olb) does not contain a CD4093 (Harris brand "4093" quad NAND gate). A search using CD* would not yield any results. You will usually get better search results by using the numeric identifier for the type of logic gate, such as *4093 for the CMOS quad NAND gate example.
- **Graphic box.** You can select the normal or convert (alternate) view of the part. Not all parts have convert views, in which case this option is dimmed out. Some Capture library parts have a convert view. For example, the 4093 CMOS NAND gate has the DeMorgan logic equivalent, an OR gate with inverted inputs, as the convert view. The libraries for most logic families include DeMorgan convert views for AND, OR, NAND, and NOR gates.
- **Packaging box**. This displays the number of parts in the package and allows you to select which part you will place. For example, the 4093 CMOS NAND gate has four parts with suffixes A, B, C, and D.
- **Edit Part tool**. You can click on this tool to launch the part editor to make changes to the selected part. The part editor is introduced in the next Chapter.

Placing Parts

Make sure you have selected the CONNECTOR DB9 part and then click on OK. Capture returns to the schematic editor and the part appears on the screen as a ghosted outline. This ghosted part outline moves with the mouse cursor. Hold down the left mouse button and move the mouse pointer against the upper edge of the schematic. Note that the display view pans up. Keep holding down the mouse button and pan up until the title block just disappears from view as shown in Figure 3-8). Then release the button to place the connector on the schematic.

Note that the Place Part command is modal. A ghosted connector continues to follow the pointer. Each time you press and release the left mouse button, an additional copy of the part is placed. To end the command, press the right mouse button. A shortcut menu appears as shown in Figure 3-9. Click on End Mode to end the Place Part command. Recall that you can also use the ESC key to end or cancel most commands.

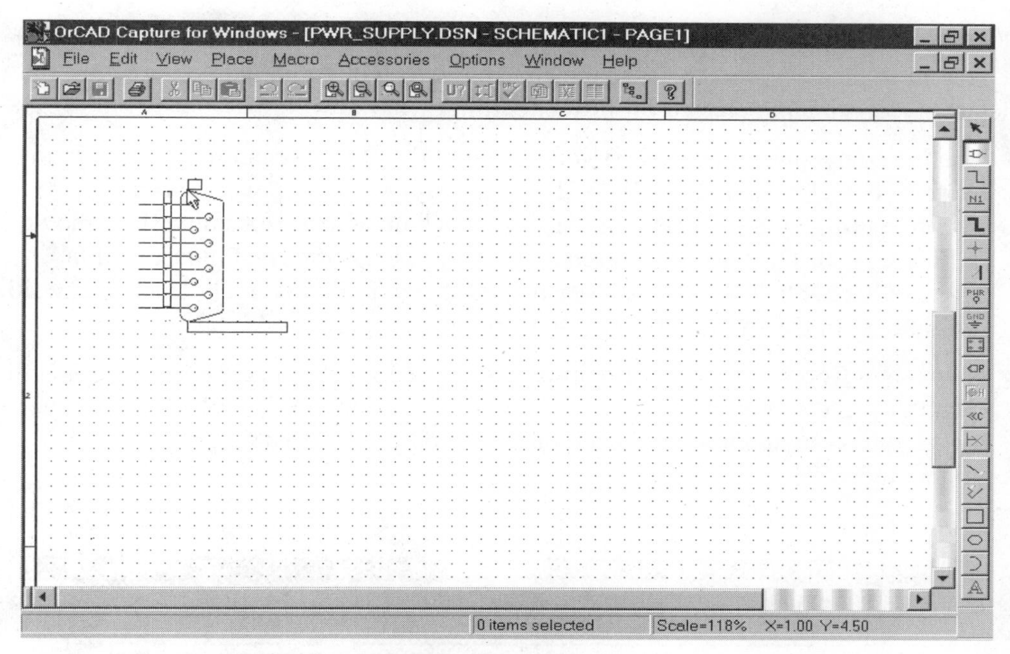

Figure 3-8 Placing the DB9 Connector

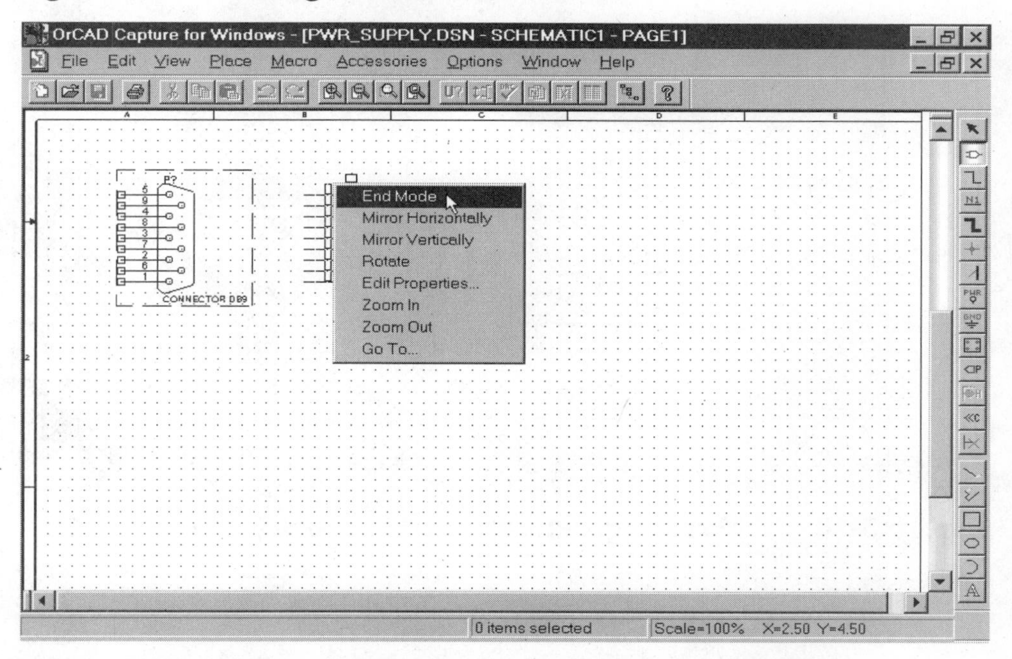

Figure 3-9 Ending the Place Part Command Mode

The connector that you placed now appears within a selection box. This is another feature of the Capture user interface. Whenever you place parts or draw wires, the last object remains selected for further editing.

You have now placed PL1. Next you must edit the part to orient it as shown in the model schematic (Figure 1-17). With the part still selected, press the right mouse button to bring up the shortcut menu. Then click on Mirror Horizontally as shown in Figure 3-10. The part is mirrored horizontally. Press the right mouse button again and then click on Mirror Vertically as shown in Figure 3-11. The part should now appear in the correct orientation as shown in Figure 3-12.

Note that a Rotate operation also appears on the shortcut menu. This operation will rotate the part counterclockwise in 90-degree increments. When placing parts, you use various combinations of mirror and rotate operations to obtain the desired orientation.

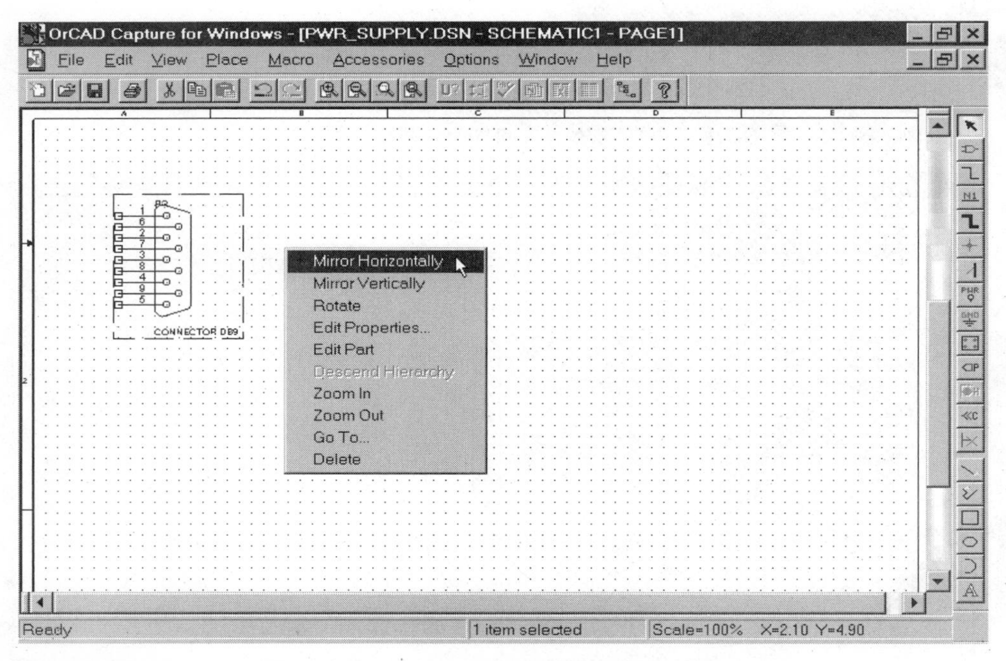

Figure 3-10 Mirroring the Part Horizontally

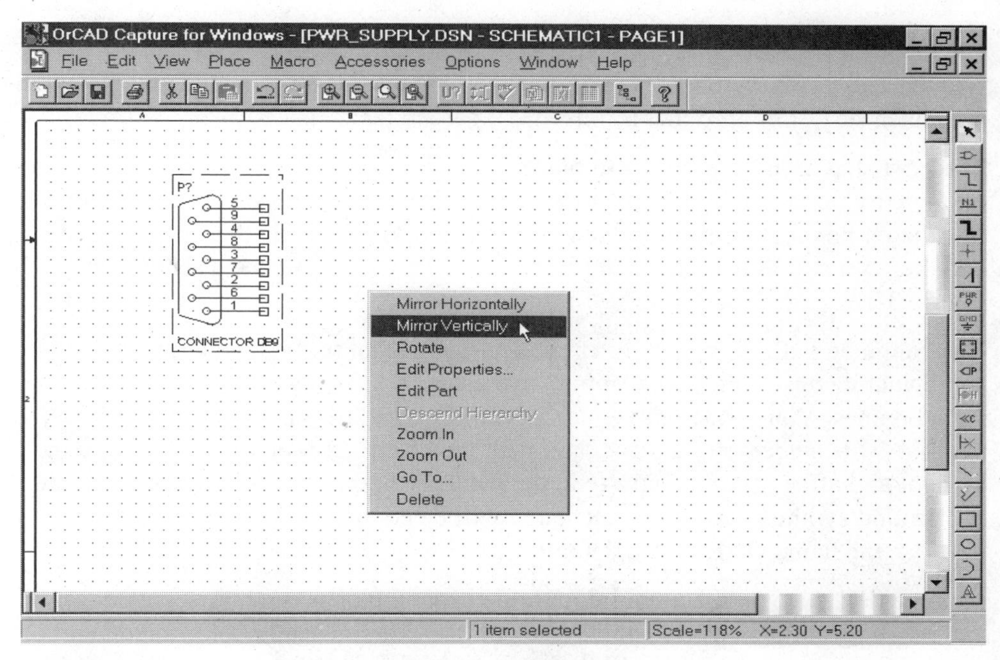

Figure 3-11 Mirroring the Part Vertically

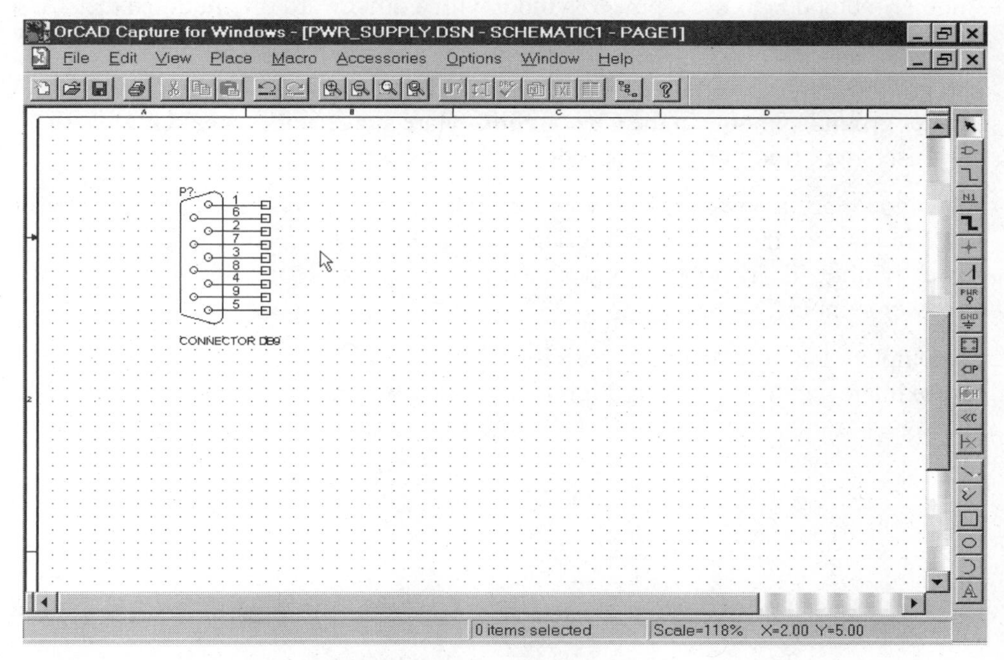

Figure 3-12 Completed Part Placement

Placing Ground and Power Symbols

Ground and power symbols (see Figure 1-3 on page 23) are selected and placed in a manner very similar to that for electrical parts.

The next step in the exercise is to place the ground symbol just below PL1. Click on the Ground tool. This brings up the Place Ground dialog box. Many aspects of this dialog box are identical to the Place Part dialog box. Add the Custom library and then select it. Select the GND symbol and then click on OK as shown in Figure 3-13.

The Name box allows you to enter an electrical signal name. The terminology is somewhat inconsistent since Capture also uses the term *net alias* for electrical signal names. You are actually entering the electrical net alias for the ground symbol. This name will be associated with the electrical properties of the ground symbol and affect its connectivity in the database and in any netlist you generate. All ground symbols with the same name will be connected together in the same net. Use the name GND as the default.

In some designs, you may require multiple ground planes, such as separate signal and power ground planes. Capture does not provide for display of ground symbol names. One method of differentiating ground planes is to use different symbols. Another method is to use a power symbol with a downward orientation, since these symbols have visible names.

The ground symbols supplied in the Capsym library are somewhat oversize. The smaller ground symbols in the Custom library are suggested as an alternative that most users will find more appropriate for dense schematics.

When you place a power object, you need to enter the signal name, such as +5V or +24V in place of the default value. The VCC_ARROW is recommended as a power symbol. To avoid a clumsy appearing schematic, do not mix different types of power symbols.

To complete this section of the exercise, place the ground symbol in the approximate location shown in Figure 3-14. Use the same technique as with placing a part.

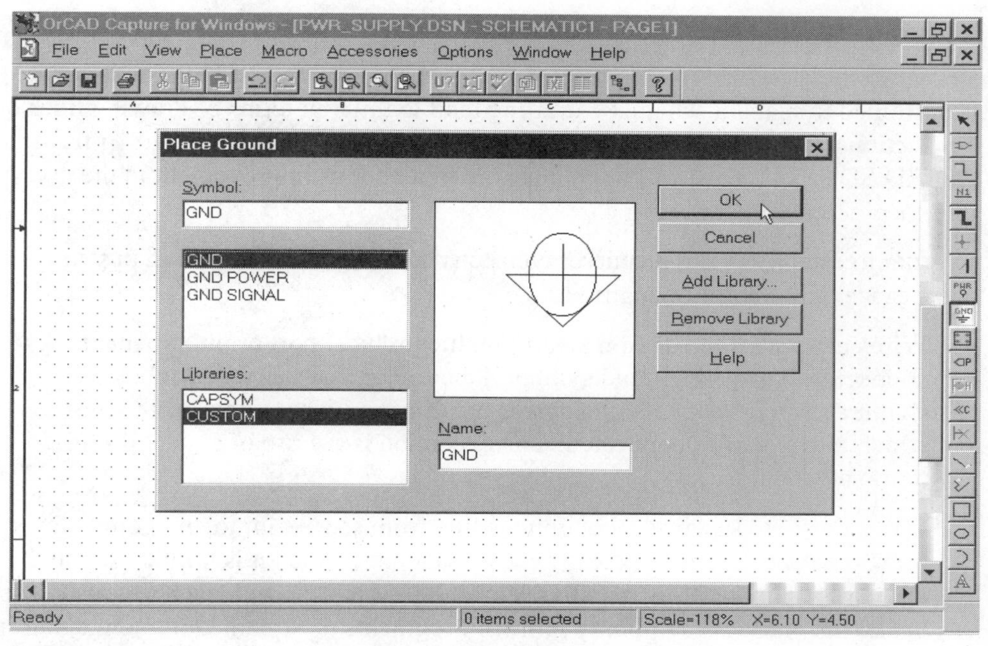

Figure 3-13 Selecting a Ground Symbol

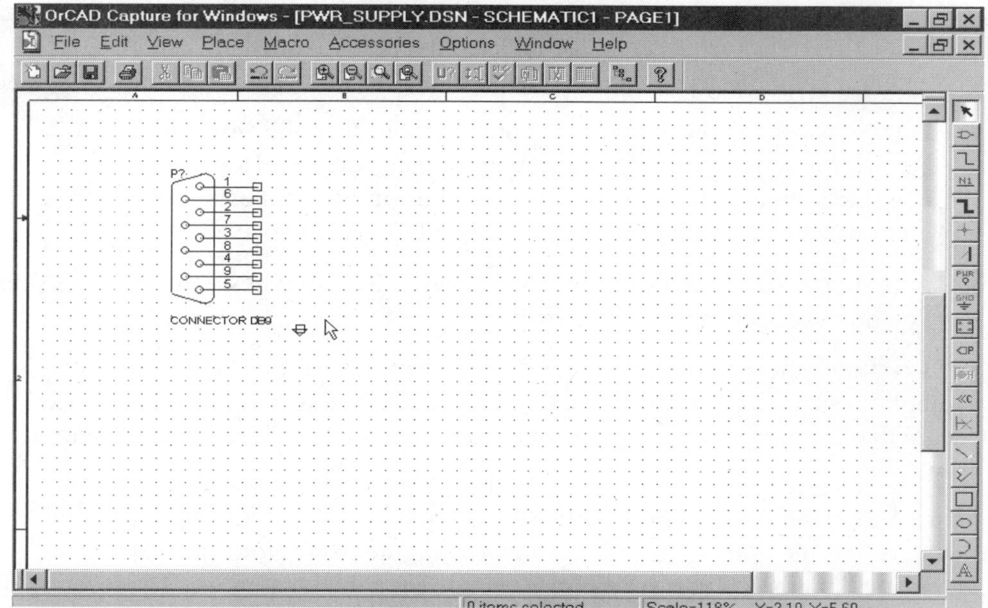

Figure 3-14 Placing the Ground Symbol

Placing Wires

Wires are the electrical interconnections between pins on parts. Individuals accustomed to manual schematic drafting tend to think in terms of drawing lines or connections. Capture uses the terminology *placing wires*. Users upgrading from OrCAD SDT to Capture will appreciate the tremendous improvement in the user interface in regard to drawing wires.

In order to establish valid electrical connectivity, wires must be drawn in strict accordance with the following rules:

- Wires must start and end on a part pin, hierarchical port or pin, off-page connector, power or ground symbol, junction on another wire, or bus entry. Capture displays connection squares on all unterminated objects and pins to which a wire can be connected. One exception is the use of a net alias to terminate a wire.
- Wires must be drawn as a continuous line from start point to end point. Never draw breaks where wires cross over one another. When it is not feasible to draw a continuous wire, net aliases must be associated with each segment. The use of net aliases to implicitly join wire segments is explained in more detail in subsequent exercises.
- Wires that cross over or intersect one another are not electrically joined unless a junction object is placed at the intersection point. Note that Capture automatically places a junction object if you "T" onto another wire.
- Do not overlap wires onto pins. Capture tries to snap the wire end to the connection square that appears on unterminated objects. However, it is possible accidentally to overlap a wire. The wire will not establish electrical connectivity, and the connection square will remain displayed.
- As with electrical parts and symbols, always place all wires on grid. Make sure you do not accidentally toggle snap to grid off.

Let's continue the exercise by drawing a wire between PL1 pin 1 and the ground symbol. Press the I key to zoom in. Click on the Wire tool. Click on the connection square that appears on pin 1 of the connector, as shown in Figure 3-15, to start the wire.

Move the cursor several grid locations to the right and then move it diagonally down toward the ground symbol as shown in Figure 3-16. Note that Capture automatically draws a corner whenever you make a diagonal move. The location of the corner depends on the initial direction of the move.

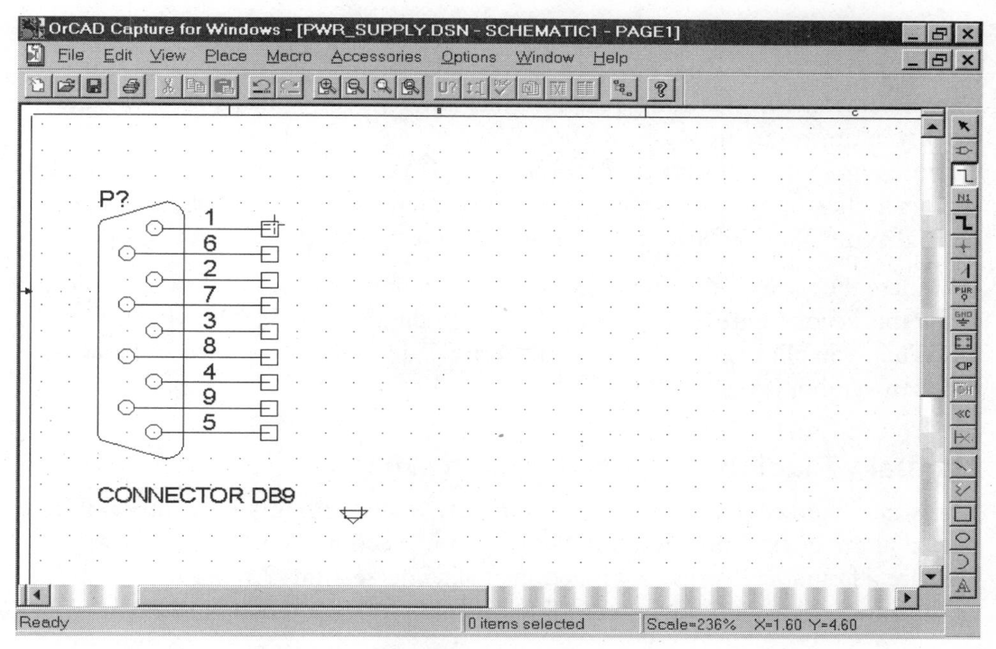

Figure 3-15 Starting a Wire

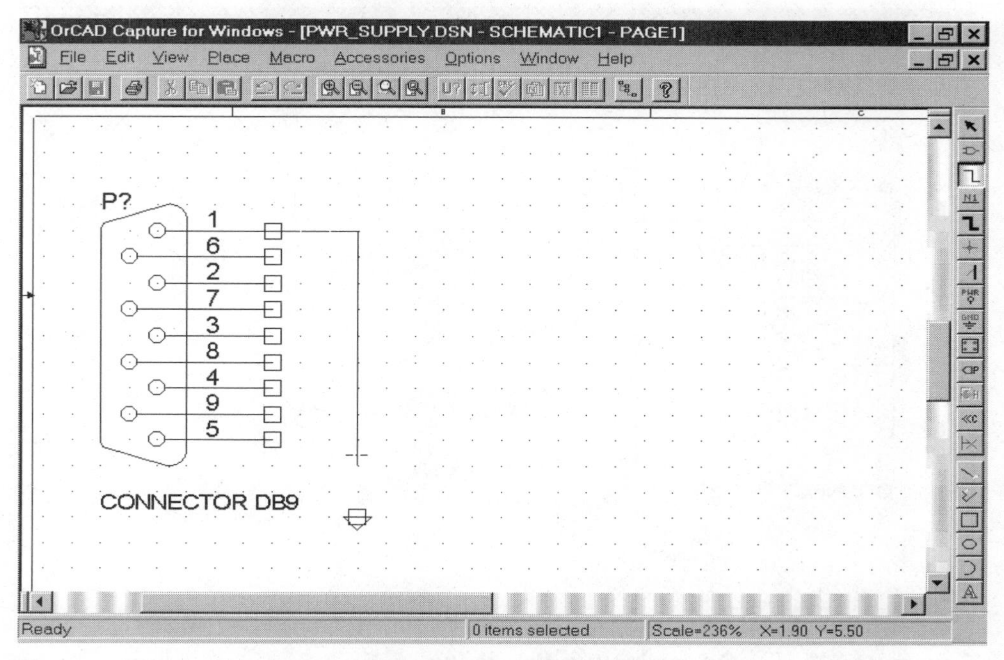

Figure 3-16 Automatically Drawing a Corner

Capture normally draws orthogonal wires and buses. To create diagonal segments, press the SHIFT key while drawing the wire. Whenever you click the cursor, Capture creates a corner point. Capture will automatically place the last corner when you draw a multi-segment wire. When you click on a connection square, Capture automatically starts or ends the wire. If you want to manually end a wire segment, click on the last corner point, press the right mouse button, and then click on End Wire. You can also use the ESC key to end a wire.

Complete the wire as shown in Figure 3-17. Then draw another wire from pin 2. Draw this second wire to a T intersection with the first wire as shown in Figure 3-18. When you click on the intersection point, Capture ends the wire and places a junction symbol (Figure 3-19).

Manually Placing Junctions to Join Wires

Wires that intersect are considered electrically joined only if a junction symbol is placed at the intersection point. In the preceding section, you explored how Capture automatically places a junction at a T intersection. To join wires that cross, you must use the Junction tool to manually place a junction symbol.

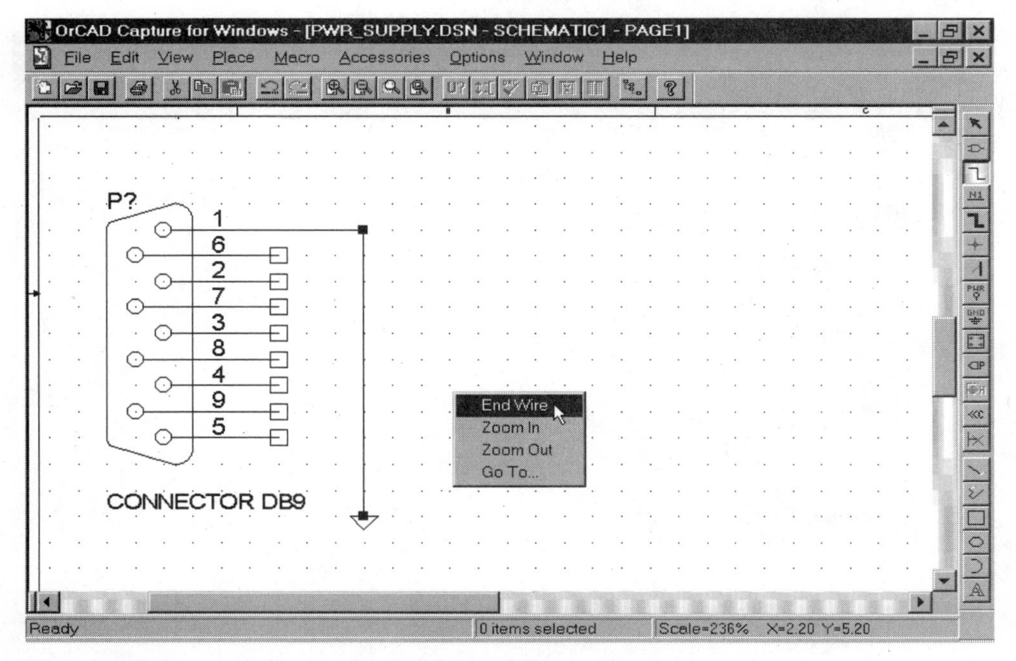

Figure 3-17 Completing the Wire

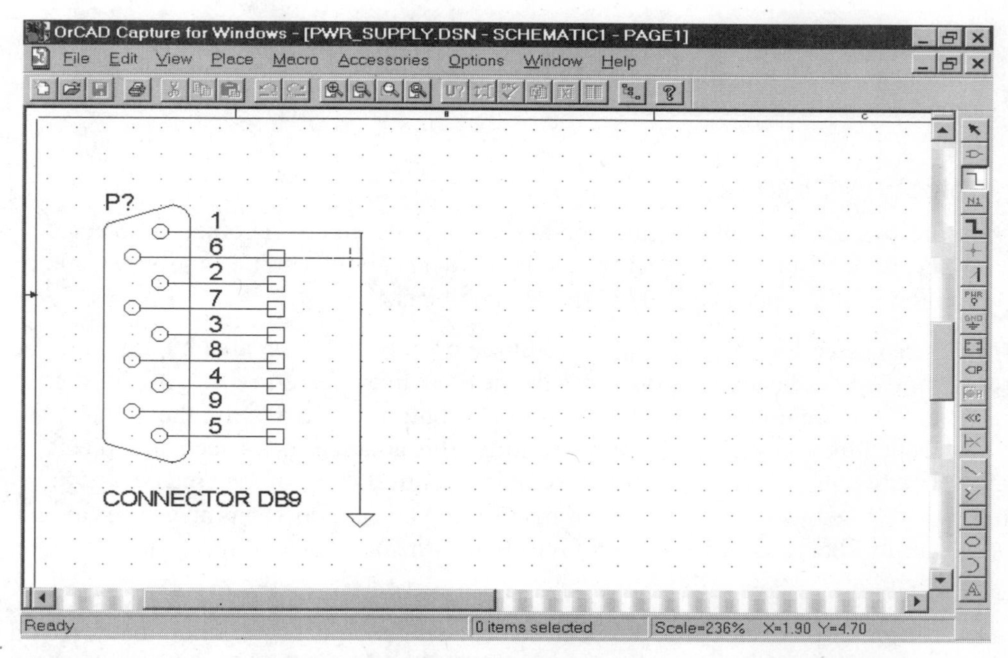

Figure 3-18 Drawing the Second Wire to a T Intersection

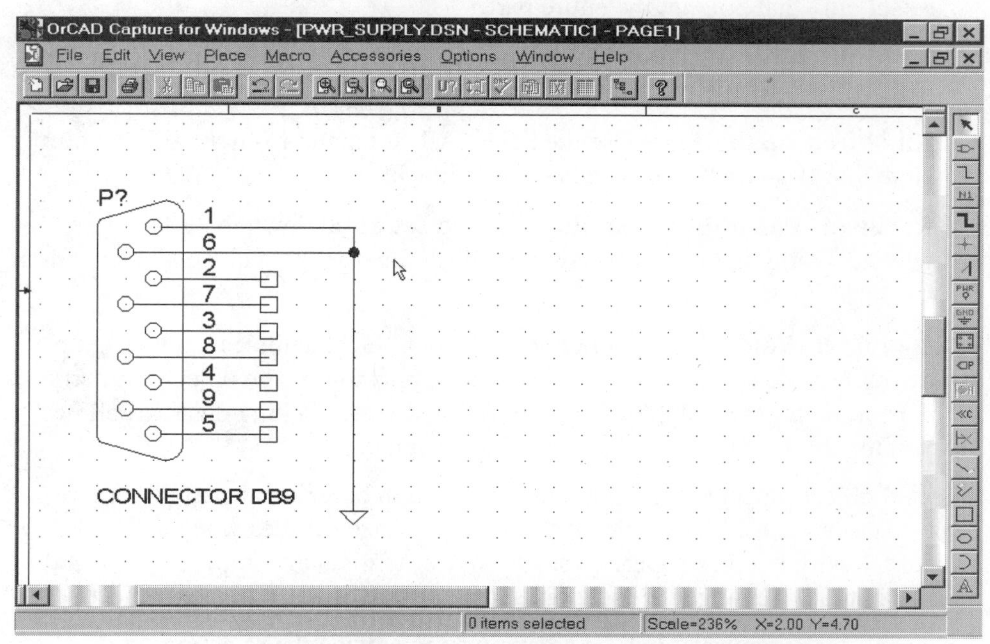

Figure 3-19 Completed T Intersection with Junction

If you've made it this far without making a mistake, you are doing very well. Before you go on to look at more drawing functions, let's take a quick look at some basic editing operations.

Selecting Objects

Before you can start any editing operation, you must select the objects. Selected objects become part of a selection set, which you can then edit as a group. Capture provides several means of selecting graphic objects:

- **Area selection**. You can select multiple objects within an area by dragging the mouse pointer diagonally across the desired area (the term *dragging* implies that you are holding down the left mouse button while moving the pointer). Depending on your preferences settings, this action either selects all objects intersecting the area or only those objects entirely within the area (see page 56). When using area selection, make sure you first position the pointer to an empty spot, otherwise you will only select the one object directly at the pointer location.

- **Single object selection.** You can select a single object by clicking on it. If you want to select a part, click on the part outline. If you click on a part pin, you select only that pin, not the entire part.

- **Multiple object selection.** Hold down the CTRL key while clicking on the objects.

- **All objects on the page**. Use the Select All command from the Edit menu. Note that Select All also selects the title block.

- **Removal of an object from the selection set**. Hold down the CTRL key while clicking on the object. Use the CTRL key to toggle objects in and out of the selection set.

- **Double clicking to edit properties**. You can select an object and bring up a dialog box for editing properties by double clicking on the object. Properties depend on the type of object. For example, double clicking on a part allows you to edit the part value and reference designator.

Selected objects appear highlighted in the selection color set as part of your Capture preferences. Many objects also display resize handles when they are selected. Resize handles appear as small boxes. As the name implies, you can drag a resize handle to change the location, shape, or size of an associated object.

To cancel a selection set, click on an empty area outside the selection set.

Practice selecting objects using area select. Zoom out. Drag the mouse pointer over the wires and ground symbol to define the selection area shown in Figure 3-20. After you release the left mouse button, the objects appear in the selection color. Resize handles appear at the corners and endpoints of the wires.

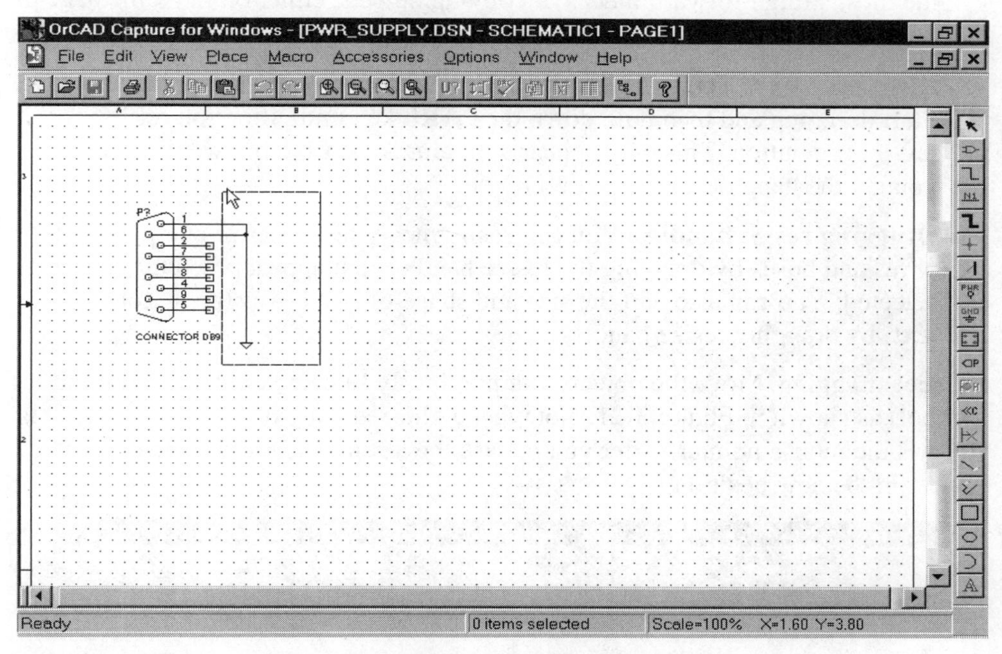

Figure 3-20 Selecting Objects Within an Area

Moving, Copying, and Resizing Objects

You can use several approaches for moving, copying, and resizing objects:

- **Dragging a single object**. Position the mouse pointer on the object (but not on a resize handle), hold down the left mouse button and drag the object. This action selects the object and then moves it to a new location in one operation. Note that connected wires will be dragged along with any part. When preparing to drag a part, do not position the pointer on a pin.
- **Dragging a selection set**. This works the same way as dragging a single object, except that all the objects move as a group. Remember not to position the mouse pointer on a resize handle or part pin.
- **Cut and paste**. You can use the Cut and Paste commands from the Edit menu to move objects. These work the same in Capture as you would expect from

other Windows programs. Cut and paste operations are very useful for moving objects between schematic pages.

- **Copy and paste**. Using the Copy and Paste commands from the Edit menu is the most straightforward means of copying objects, either on the same page or between pages.
- **Drag and copy**. You can use the CTRL key to copy an object or selection set. Start dragging and then hold down the CTRL key until after you release the left mouse button. This may appear a bit clumsy at first, but works well once you get the hang of it.
- **Dragging resize handles**. You can easily move the corner or end points of wires and buses by dragging the resize handles that appear when the object is selected. You can also edit the size and shape of text boxes (areas containing text) by dragging resize handles.

Practice dragging. Move the objects you previously selected (wires and ground symbol) as shown in Figure 3-21. Note that a ghosted copy of the selected objects follows the mouse pointer. After you release the left mouse button, the objects appear in the new position.

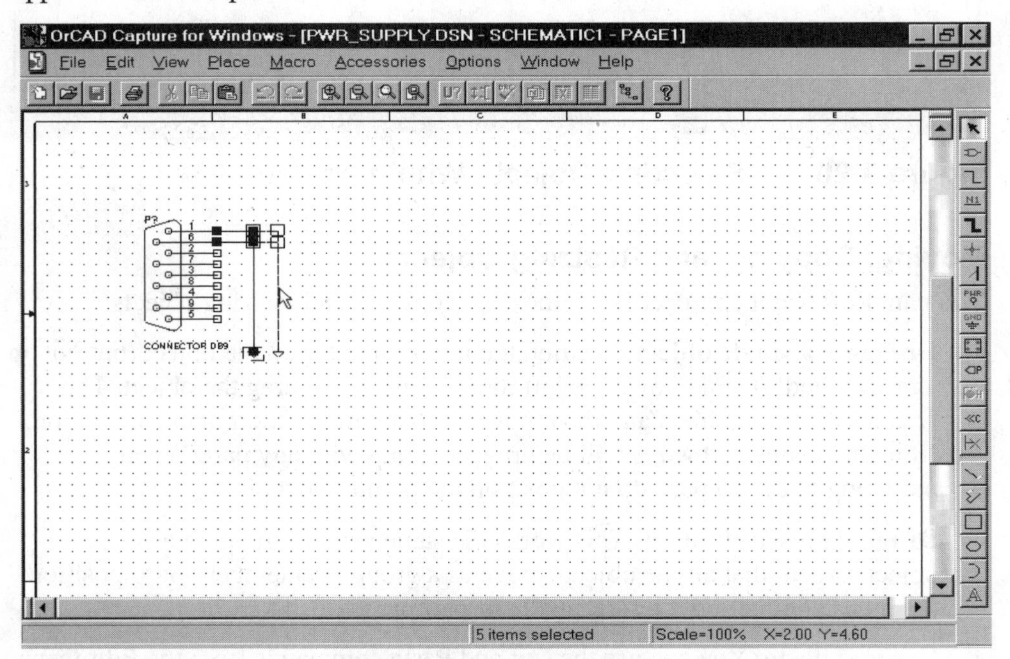

Figure 3-21 Dragging the Selected Objects

Deleting Objects

You can delete selected objects by two means:

1. Press the right mouse button to bring up the context-sensitive shortcut menu. Then click on the Delete option. Most users find that this approach is the easiest to use.
2. Use the Delete command from the Edit menu.

Both the Delete and Cut commands remove selected objects from the screen. However, using the Cut command in place of Delete is not a recommend practice, since the selection set is placed onto the Windows clipboard.

Practice deleting an object from the schematic. Click on the connector to select it. Then press the right mouse button and click on the Delete option as shown in Figure 3-22. The connector is deleted.

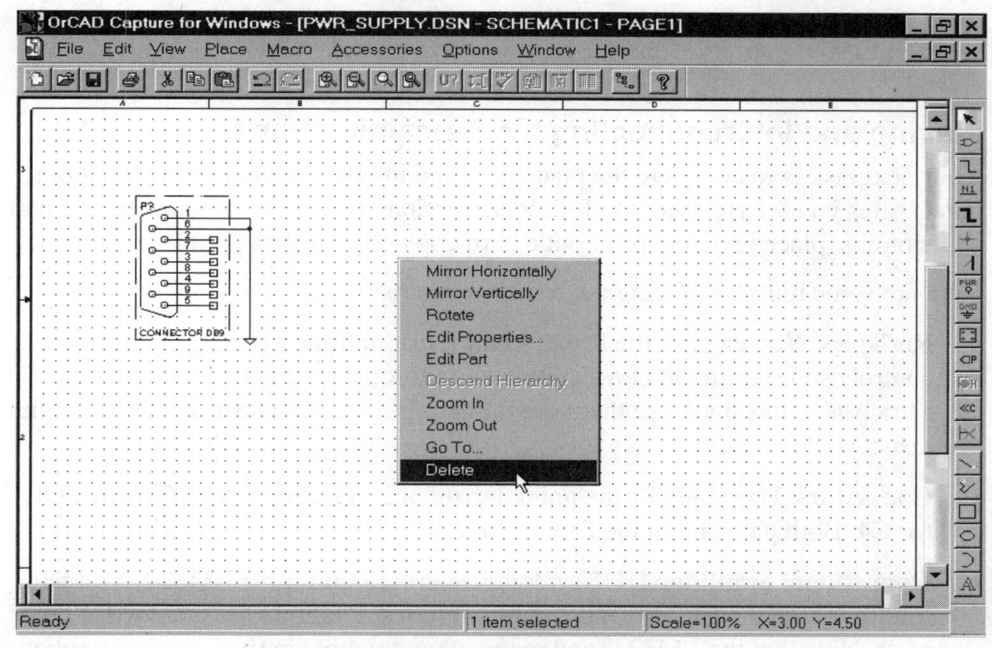

Figure 3-22 Deleting the Selected Object

Undo and Redo Tools

Recall that the Undo and Redo tool icons appear on the main tool bar at the top of the screen. You can use the Undo tool to correct many minor editing mistakes, such as inadvertently deleting the wrong part. If an action cannot be undone, the

Undo tool appears dimmed out. If you change your mind about an Undo operation, you can use the Redo tool. Redo is dimmed out if nothing has been undone. Note that Undo and Redo commands are also available from the Edit menu, but most users prefer clicking on the tool icons.

Unlike many other Windows programs, Undo in Capture is limited to a single step. This means that you cannot undo a series of commands or actions. Hopefully OrCAD will remove this limitation in a future release.

Practice using the Undo and Redo tools.

Third Session – Completing the Single-Sheet Schematic

You have now learned how to select parts from the libraries and place them on the sheet, draw wires and junctions, place power, and ground symbols and perform basic editing operations. Continue and complete the schematic.

Placing and Interconnecting the Remaining Parts

Your next task is to complete the placement and interconnection of the remaining parts. After you are finished, the following sections will show you how to edit part descriptions, place text, and print out a hard copy of your first Capture schematic.

Here are some hints on placing the remaining parts:

- **Resistors.** You can use the R part in the Custom library (European/industrial controls style) or in the Device library (traditional style). All examples and exercises in this book use the more modern European/industrial controls style resistors.
- **Capacitors**. Use CAP for the nonpolarized part, C3, and CAPACITOR POL for C4. Both parts are in the Device library.
- **Operational amplifier U1**. Try doing a part search for the LM324. You will find it in both the Custom and Analog libraries. Examine the graphic for both parts. Note that the LM324 (and many other opamps) in the Capture-supplied libraries is drawn with the noninverting input on top and long power pins. Recall that the preferred orientation for opamp circuits is with inverting input at the top. If you mirror the part vertically, the power pins wind up upside down (negative on top and positive on the bottom). Use the LM324 part in the Custom library. This part has the preferred orientation.

- **Molex connector PL2**. Use the CON5 part from the Device library. Note that most connector part names in this library start with CON.
- **Transistor Q1**. Use the NPN part from the Device library.
- **+24V power symbol**. Use the Place Power tool and the VCC_ARROW symbol with (signal) name +24V.

If you make a mistake, just use the Delete or Undo commands and start over. Initially place parts in approximate locations, then drag them as needed. Don't worry about precisely locating every part or wire in exactly the same place as the model in Figure 1-17. Use the model as a guideline and follow the general flow.

Depending on the sequence you use to draw wires, Capture will automatically place some of the required junction symbols. Place the rest manually.

Placing No Connect Symbols

Use the No Connect tool to place these symbols at the unused pins on PL1 and PL2. Click on the tool and then click on each pin that requires a no-connect symbol.

Editing Reference Designators and Part Values

At this point you have placed and interconnected all the parts on the schematic. Your screen should look like Figure 3-23. The screen appears somewhat messy because reference designator and part value text overlaps. The next step will be to annotate the schematic. Annotation refers to editing reference designators and part values (descriptions). The annotation process involves four steps for each part:

- **Edit the reference designator**. In most cases the alphanumeric prefix, such as R for a resistor, is correct and you only need to edit the numeric suffix. Capture reference designators default to a "?" suffix when the part is initially placed. Note that the five-pin Molex connector at the right of the sheet will require editing the prefix from J to PL. Capture provides a postprocessing tool that automatically numbers reference designator suffixes, on the basis of the order that the parts appear in the design. Use of this tool is covered in subsequent chapters.

- **Edit the parts value**. This involves entering a part number or other descriptive data. See Chapter 1 for further details. In this exercise, you will use only the part value field, which limits the part description to a single line. Capture allows multiple part description fields, including fields that are user defined. You will learn to use this feature in subsequent chapters.

- **Locate the reference designator**. Sometimes the reference designator can be left in the default location, but in most cases you will have to move it to improve legibility of the schematic. Group the reference designator with the part value near the part.
- **Locate the part value**. The general rule is that the reference designator should be located on top or to the left of the part value. For narrow parts such as resistors, an acceptable practice is to locate the reference designator directly above the part and the value directly below the part.

The reason for following the sequence just described is that the length and height occupied by part descriptions vary. You will find it much easier to locate the parts description last, when you know how much space is required. Neat and carefully placed annotations improve the readability of the schematic and reduce the possibility of confusion. This becomes an important issue with complex and crowded schematics.

To avoid potential problems during later postprocessing operations, capitalize all schematic annotations and text notes. You should also enter part values in a consistent manner to avoid problems when sorting the bill of materials. For example, if the design uses two 1% resistors, don't enter one part value as 10K 1% and another as 10.0K 1%.

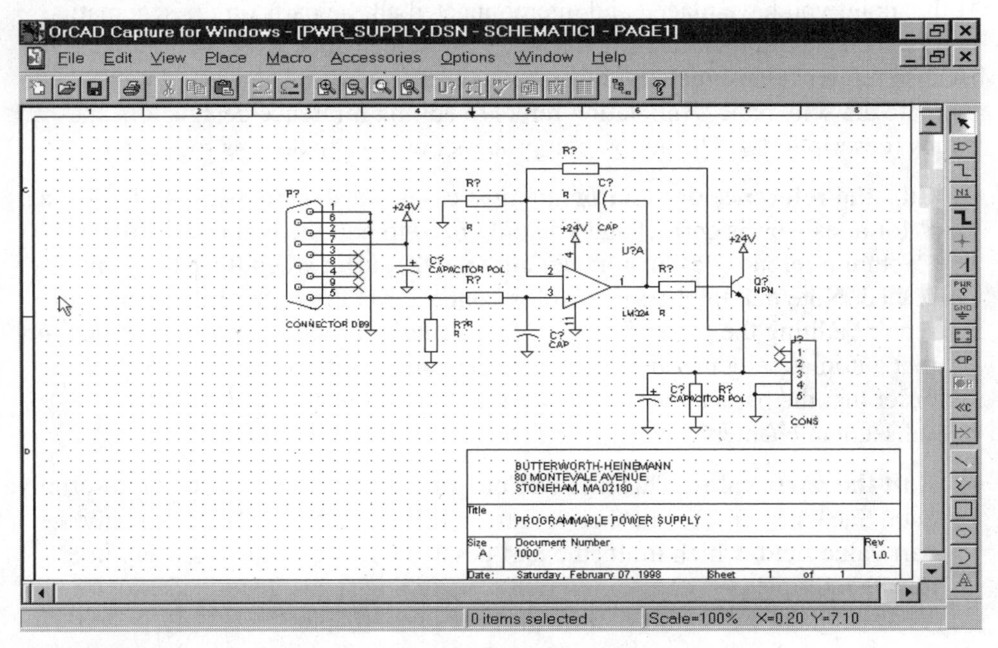

Figure 3-23 Schematic before Annotation

To edit a reference designator and part value, double click on the part. You can also click on the part to select it, press the left mouse button to bring up the shortcut menu and then click on Edit Properties.

Let's start with R4 (feedback resistor on top of the opamp). Zoom in on this area and then double click on the resistor to bring up the Edit Part dialog box as shown in Figure 3-24.

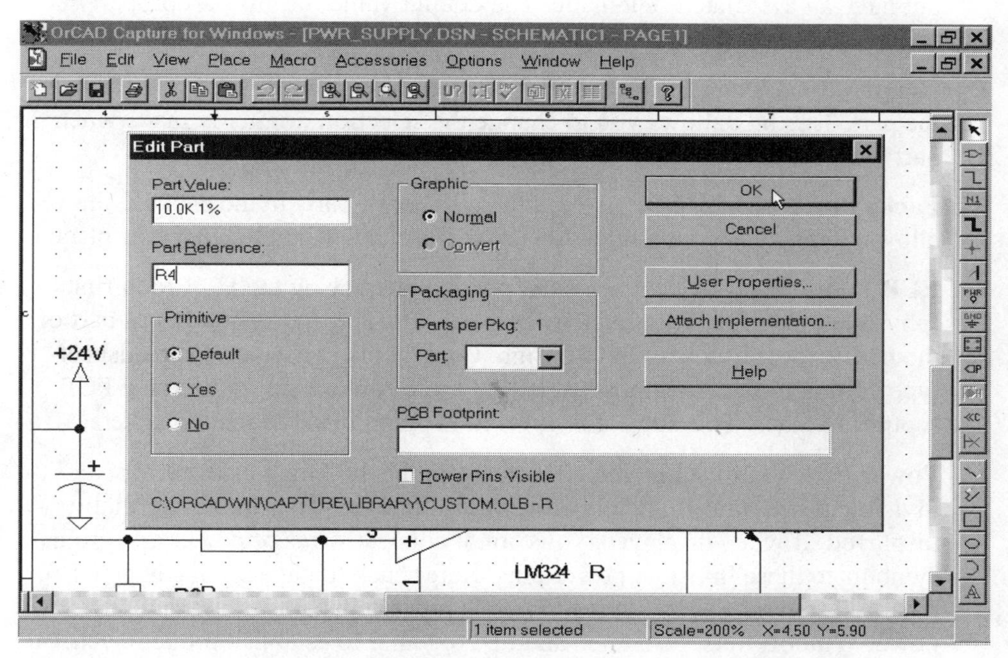

Figure 3-24 Dialog Box for Editing Part Properties

The Part value box will initially display the library part name. Backspace to delete the name and enter 10.0K 1% as the new part value. The Part Reference box displays the reference designator. This box will initially show the default reference designator prefix (R in this case) followed by a question mark (?) which indicates the reference designator has not yet been annotated. Enter the values shown and then click on OK.

For complex schematics, Capture provides the Update Part References tool that automatically annotates reference designators. For simple schematics such as the one in this exercise, you can manually annotate the reference designators at the same time that you enter the part values. In some cases, reference designators for parts, such as connectors or front-panel controls, may require certain preassigned values. The remaining reference designators can be left unassigned and then automatically annotated with the Update Part References tool. These considerations will be covered in subsequent chapters.

The Edit Part dialog box also allows you to enter or change certain other properties:

- **Primitive box**. In most cases parts are primitive. This means that the part does not have an associated schematic. The default value for this part property is one of the design template settings (see page 64 in Chapter 2).
- **Graphic box**. You can select either the normal or convert (alternative) view of the part. This box allows you to change the selection originally made when the part was placed.
- **Packaging box**. This box displays the number of parts in the package and allows you to change the selection originally made when the part was placed.
- **PCB Footprint box**. You can enter a value identifying the PCB footprint (physical type of package such as 14-pin DIP). This field is output as part of the netlist for subsequent PCB design. Capture also provides a spreadsheet-type editing tool, which most users find more convenient for entering PCB footprint values. This subject is covered in detail in subsequent chapters.
- **Power Pins Visible checkbox**. Clicking on this box toggles display of invisible power pins. Invisible power pins on logic gates are not normally displayed. If when displayed, you cannot connect wires or power and ground symbols to these invisible power pins. Refer back to page 36 in Chapter 1 for details. Temporarily displaying invisible power pins aids in determining how power symbols must be named and tied together to establish proper electrical connectivity to the invisible power pins.
- **User Properties option**. Clicking on this option brings up a detailed list of properties. You can also add user defined properties that can be extracted by postprocessing routines. This subject is covered in subsequent chapters.
- **Attach Implementation option**. If the part is non-primitive you can click on this option to attach a schematic. If you have defined certain user properties, you can add additional information such as an EDIF or VHDL file that describes a PLD implementation. This subject is beyond the scope of the book. Refer to the OrCAD documentation for further details.
- **Source Library**. A text line at the very bottom of the window indicates the path and name of the library where the part originated. Note that once a part has been placed, it is stored in the design cache. If the external library is edited or deleted, the part is still available in the design.

The next step is to drag the reference designator and part value into the proper position. Try to match the positions on the model schematic. Click on the reference designator as shown in Figure 3-25 and then drag slightly to the left so that it appears in the position shown in Figure 3-26.

Next, drag the part value to the position shown in Figure 3-26. Note that because snap to grid is enabled, you cannot precisely align the part value with the reference designator. This is a minor but annoying flaw. Use the snap to grid toggle hotkey combo CTRL+T. You can now precisely align the part value as shown in Figure 3-27. Remember to toggle snap to grid on again when you are finished. Several notes of caution apply:

1. Capture does not display a status line that indicates whether snap to grid is enabled. All electrical objects must be placed and edited with snap to grid enabled. Only part properties text, text notes, and nonelectrical drafting objects may be placed off grid.
2. If an object (such as the part value in this example) is initially offset from grid and then precisely aligned with snap to grid disabled, the object will snap back to its original offset position if it is later dragged with snap to grid is enabled. This can be extremely annoying if you accidentally click on the object.

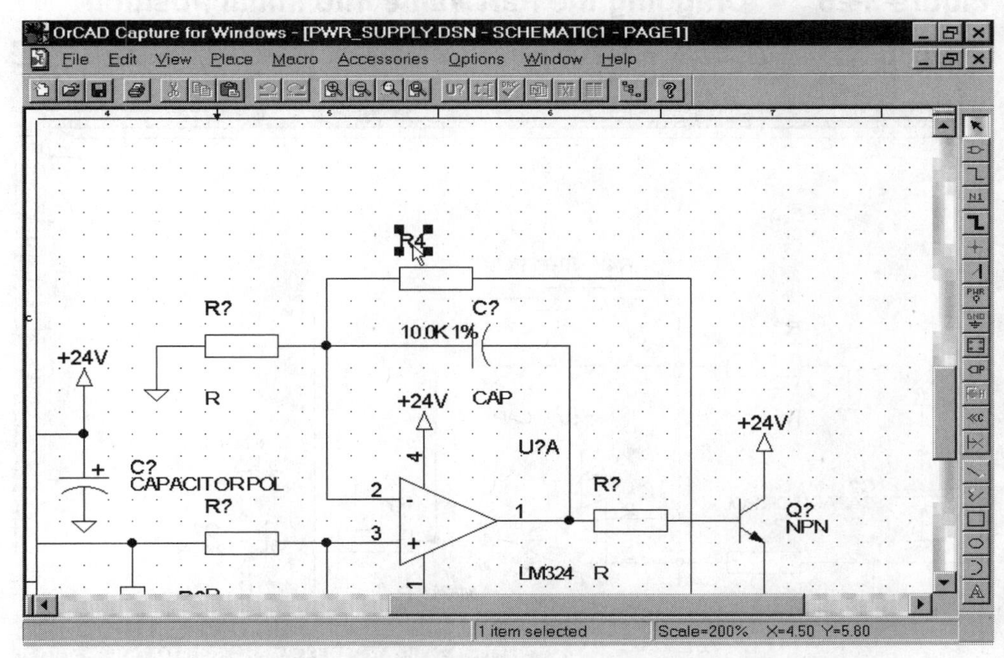

Figure 3-25 Dragging the Reference Designator

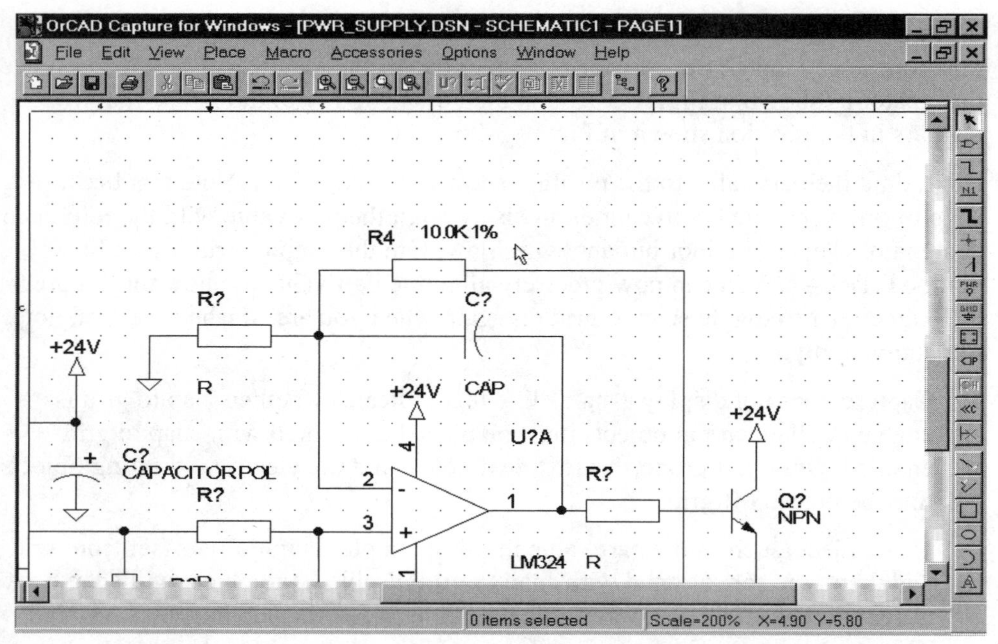

Figure 3-26 Dragging the Part Value into Initial Position

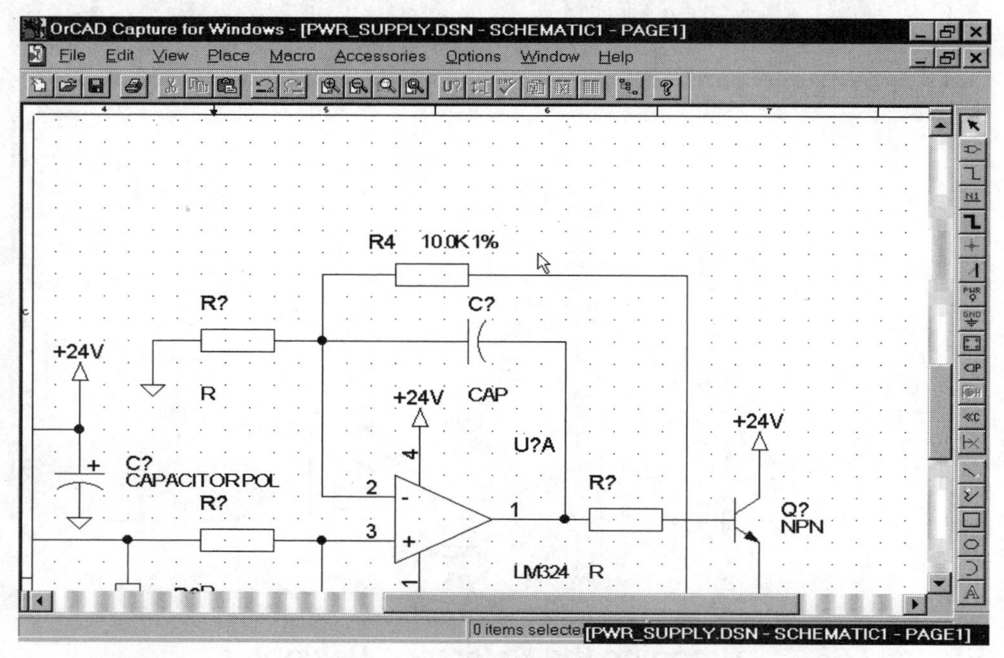

Figure 3-27 Precisely Aligning the Part Value

Finish annotating your schematic. You can use the model as a guideline, but don't be overly concerned about precise positioning. Use your own judgment.

Editing Power and Ground Symbols

Unlike the capability to edit parts, the capability to edit power and ground symbols with Capture is limited. Once these symbols have been placed onto the schematic page, you can only rename them.

Double click on one of the +24V power symbols to bring up the dialog box shown in Figure 3-28. If you correctly name power and ground symbols when you first place them, you will rarely need to rename them. A word of caution! You must use precise names. For example, if a design has a +12V power plane, power objects named +12, 12V, +12VOLTS, or +12 V would be not be associated with the +12V plane.

A dialog box similar to the one shown in Figure 3-28 appears when you double click on a ground symbol.

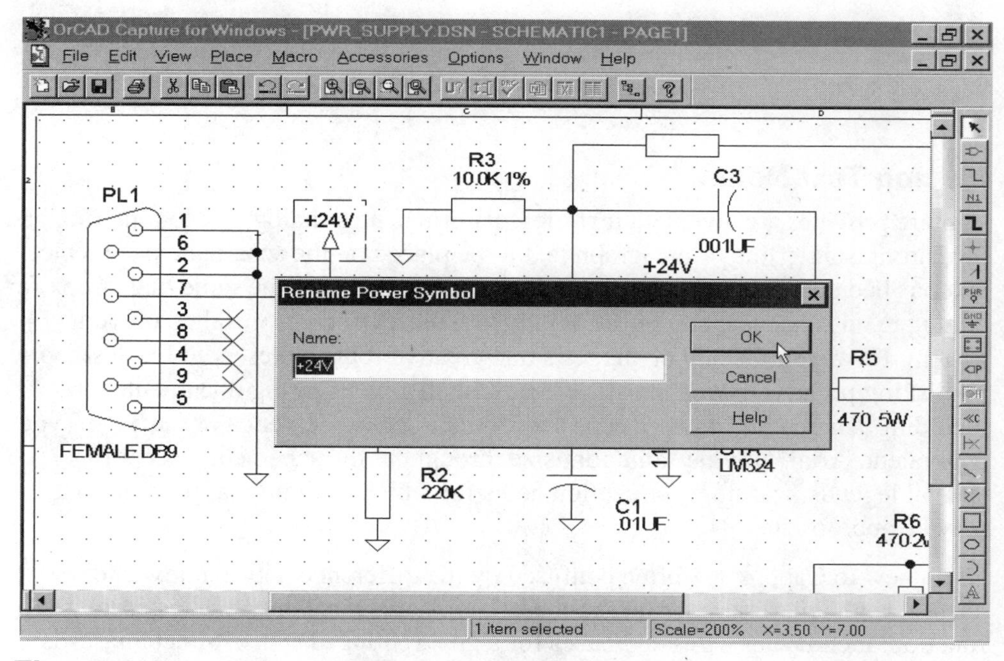

Figure 3-28 Renaming a Power Symbol

Your first schematic is now almost complete, and it should appear as shown in Figure 3-29. All the parts have been annotated, and all that remains is placing some text notes and editing the title block.

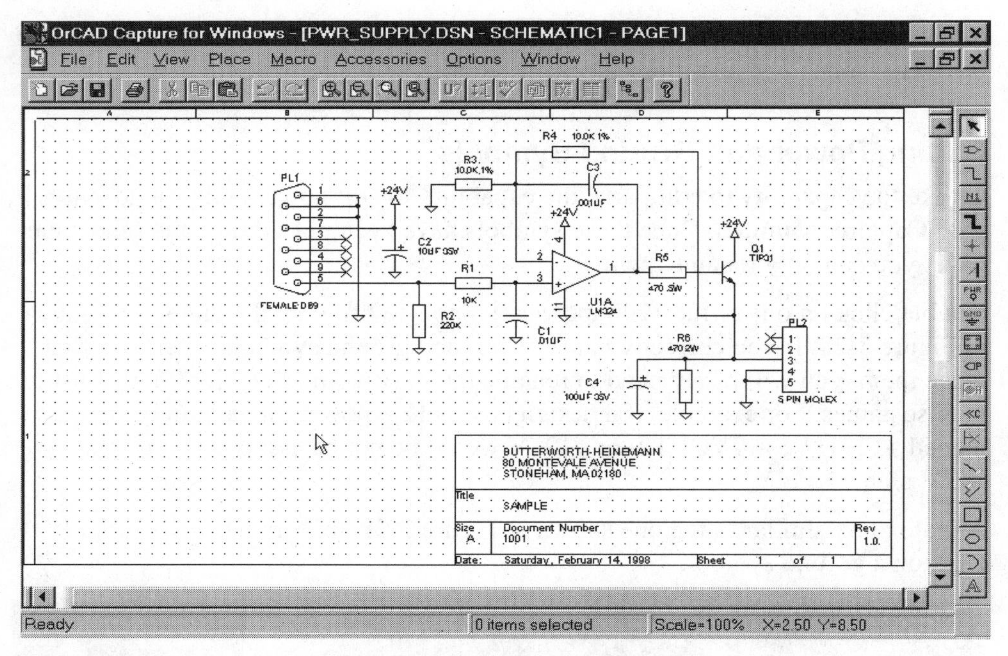

Figure 3-29 Schematic with All Parts Annotated

Placing Text Notes

Capture provides a convenient text tool that brings up a dialog box for entering and formatting single-line or paragraph text to be placed on the schematic page. Once text has been placed, you can double click on it to bring up the same dialog box for editing. You can select any of the Windows True Type fonts installed on your system. However, the author suggests that creative fonts be reserved for desktop publishing projects, not schematics. Traditionally, engineering documentation including schematics has used sans serif fonts, such as Helvetica or the True Type equivalent, Arial. Use the same font size throughout the schematic except for special legends, warnings, or caution notes for which a slightly larger bold font may be appropriate.

Users new to Capture are often confused by the difference between text and net aliases. Both appear similar on the sheet. Net aliases have only one specific function: identifying wires and buses for signal routing. The use of net aliases is explored in the next chapter. Text, on the other hand, is used for general purpose descriptions and notes. In the model used for this exercise, text is used to annotate the signal names at the two connectors.

Use the Text tool as shown in Figure 3-30 to enter and place the text note that appears in the lower left corner of the model schematic. Some special considerations related to the Text tool are as follows:

- **Single line of text**. Text is entered into the text dialog box. If you press the ENTER key at the end of a line of text, you immediately exit the Text tool. The line of text you entered appears as a box that follows the mouse pointer. When you have positioned the text box and click the left mouse button, the text appears on the screen.
- **Multiple lines of text**. Hold down the CTRL key while pressing the ENTER key to advance to the next line. For long paragraphs, use the scroll bar to navigate between lines.
- **Windows clipboard text**. You can import clipboard text from other applications into the text dialog box by using the CTRL+V key combination.
- **Color and font boxes**. Default values are established with your text preferences settings (see pages 53 and 60 in Chapter 2). You can click on these boxes to override the default settings.
- **Rotation**. You can select text rotations in 90-degree increments.

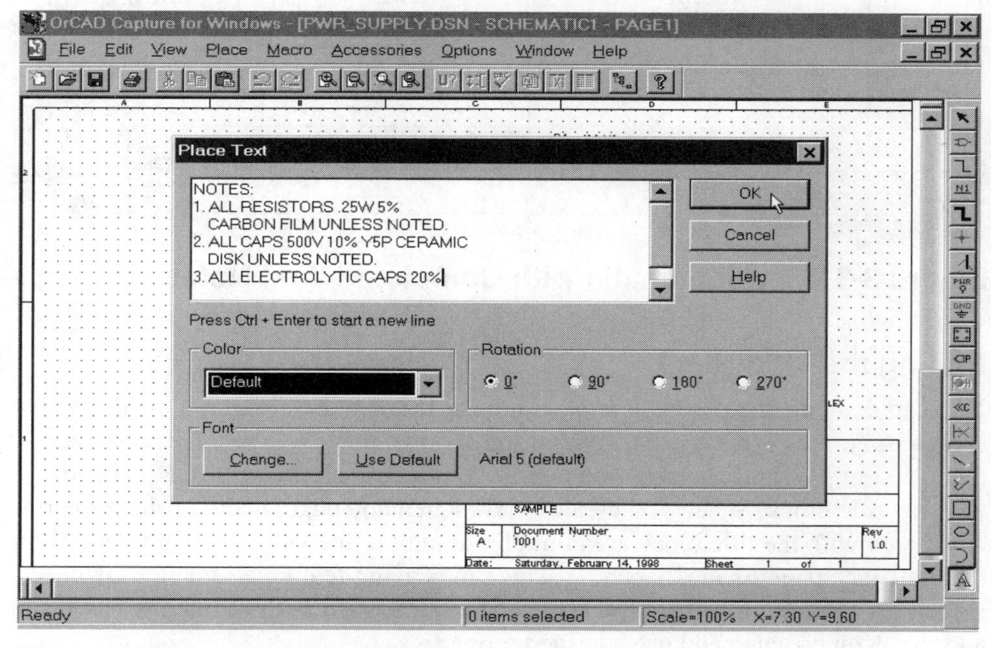

Figure 3-30 Dialog Box for the Text Tool

Note that if you edit an existing paragraph and add additional lines, you may need to resize the text box on the screen to fit the text that you added. Click on the text box and then drag the resize handles until the text appears properly formatted.

If you need to reposition text, click on it and drag it as you would with other objects.

Enter the remaining text notes on your schematic so that it appears as shown in Figure 3-31.

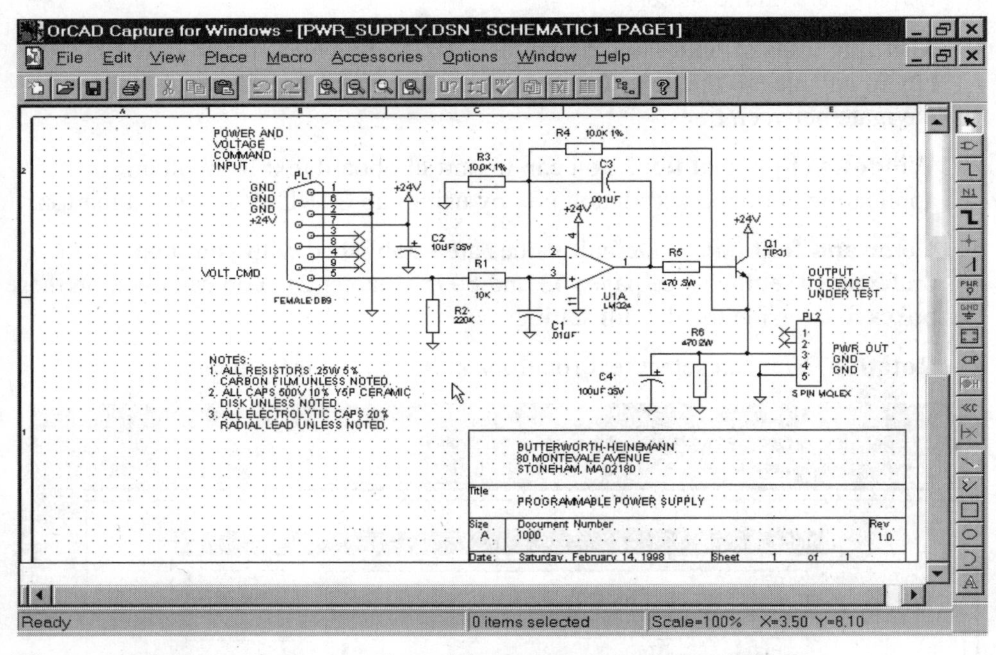

Figure 3-31 Schematic with Completed Text Notes

Editing the Title Block

You can double click on blank area within the title block to bring up the dialog box shown in Figures 3-32 to 3-34. As you scroll down the list, it becomes apparent that Capture provides a substantial number of predefined properties associated with the title block. When you create a new schematic page, many of the properties associated with the title block area initially default to the values set on your design template (see page 61 in Chapter 2). This time-saving feature eliminates the need repeatedly to enter information that stays the same, such as the company name and address. You can also add user-defined properties.

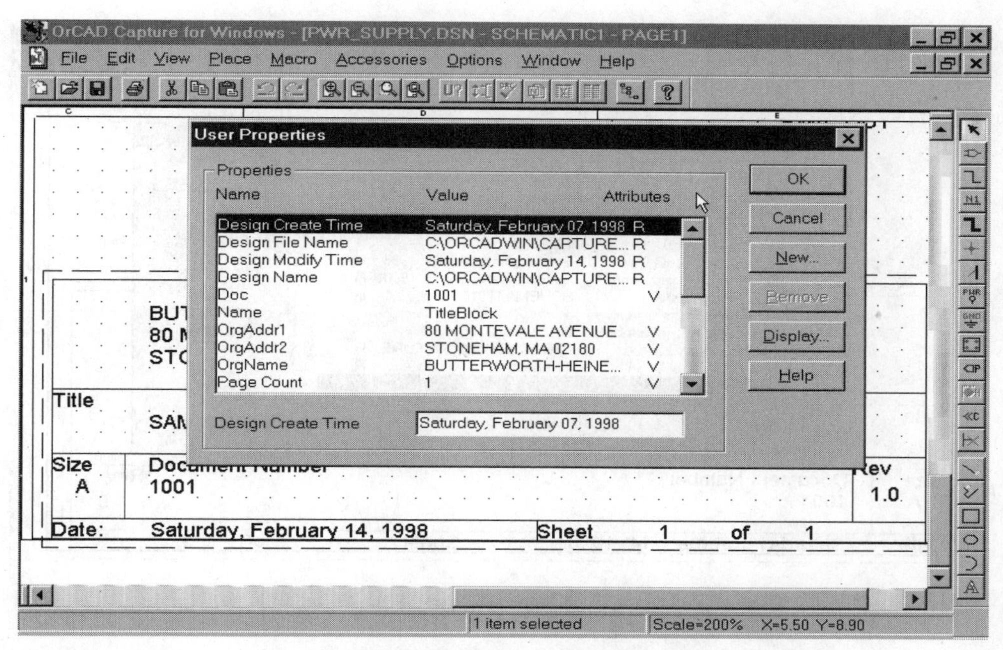

Figure 3-32 Dialog Box for Title Block Properties

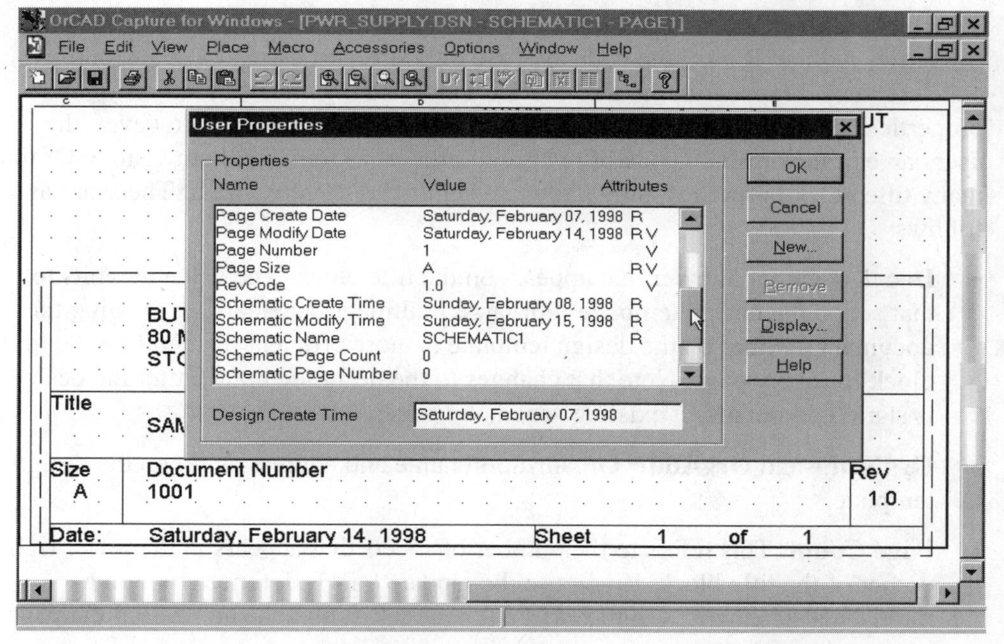

Figure 3-33 Dialog Box for Title Block Properties (Cont'd)

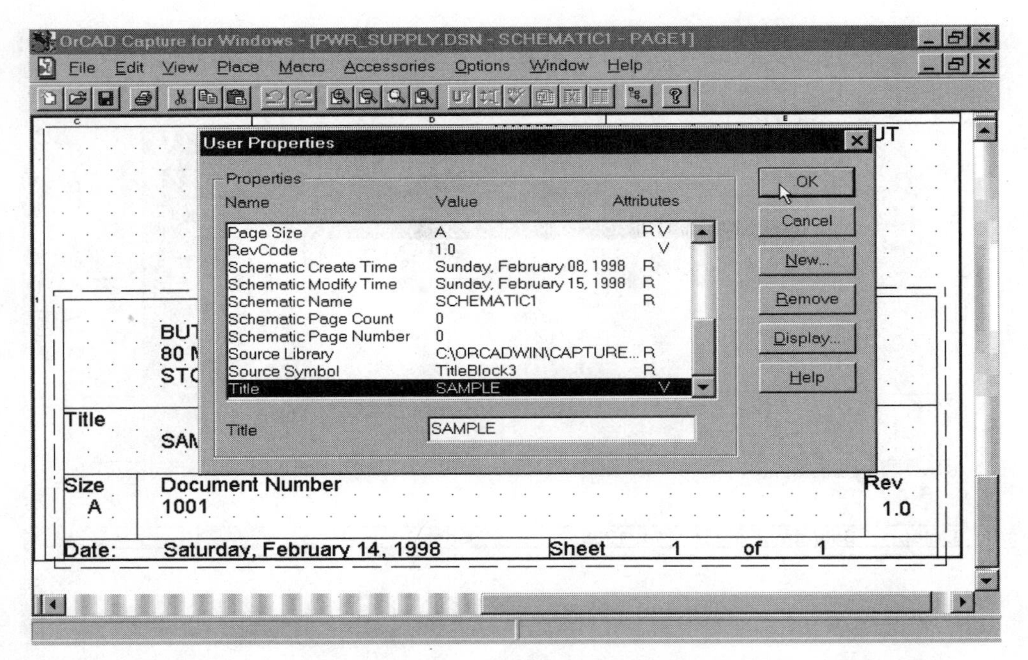

Figure 3-34 Dialog Box for Title Block Properties (Cont'd)

Title block properties have an identifying name, value, and attribute codes. Properties with an R attribute code are read only. Properties with a V attribute code are visible (appear in the title block). The system generates read-only properties. The user cannot edit the value of read-only properties. However, the user can edit the attribute code of read-only properties to make them visible. Of the many title block properties, only a few are significant to most users. These are as follows:

- **Doc**. Document number that appears on the title block. The value defaults to that set on the design template. For large multipage schematics, changing the document number on the design template is more efficient than editing the title block for every page. Note that changes to the design template after the design is started do not affect existing schematic pages.
- **OrgName** and **OrgAddr**. Organization name and address as set on the design template.
- **Page Count**. This refers to the number of sheets that appears in the lower right corner of the title block. If you use the Update Part References tool on a multisheet schematic, Capture will automatically calculate and assign the page count value.

- **Page Modify Date**. The date that appears in the lower left corner of the title block. The system updates this read only whenever you make an edit on the schematic page.
- **Page Number**. The sheet number that appears in the lower right corner of the title block. If you use the Update Part References tool on a multisheet schematic, you can select an option that automatically assigns page numbers. However, most users prefer to manually assign page numbers in some logical sequence.
- **Page Size**. The value set on the design template appears in this field.
- **Rev Code**. The revision code or number that appears on the title block. The value defaults to that set on the design template. The user generally edits the revision code or number whenever a change is made to the schematic.
- **Title**. The name of the schematic. The value defaults to that set on the design template. The author recommends that schematic titles include a common overall descriptive name and an extension (modifier) that identifies particular functional circuit blocks represented on individual pages. For example, POWER SUPPLY, LINE INTERFACE and POWER SUPPLY, REGULATOR. The overall name could be set on the design template before the design is started. As each page is created, the title value is edited to add the extension.

To edit a particular property, scroll down to the desired property and double click on it. Or you can click once on the property and then click on the Display option.

To complete the title block for the schematic, edit the title and document number. Double click on the title property to bring up the dialog box shown in Figure 3-35. Note that this is very similar to the Text tool dialog box, except that you can only enter a single line of text. You can use the mouse or keyboard cursor keys to scroll along the text line. You can click on the visibility check box to toggle the visibility attribute.

You can also edit visible title block properties directly by double clicking on them. If you double click on the title text, a dialog box the same as the one shown in Figure 3-35 will appear.

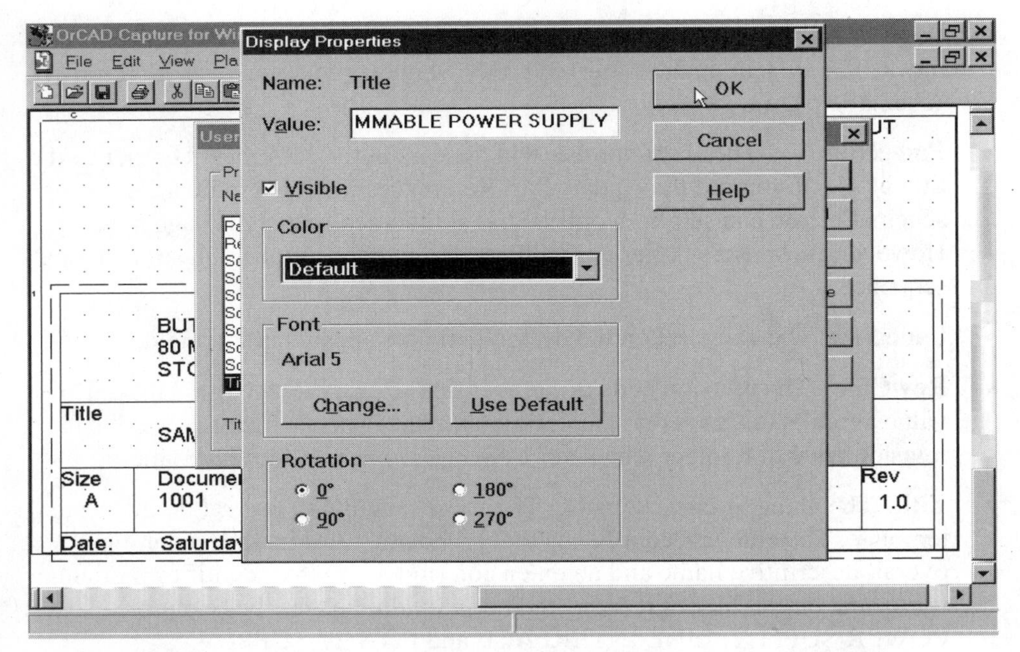

Figure 3-35 Dialog Box for Editing the Schematic Title

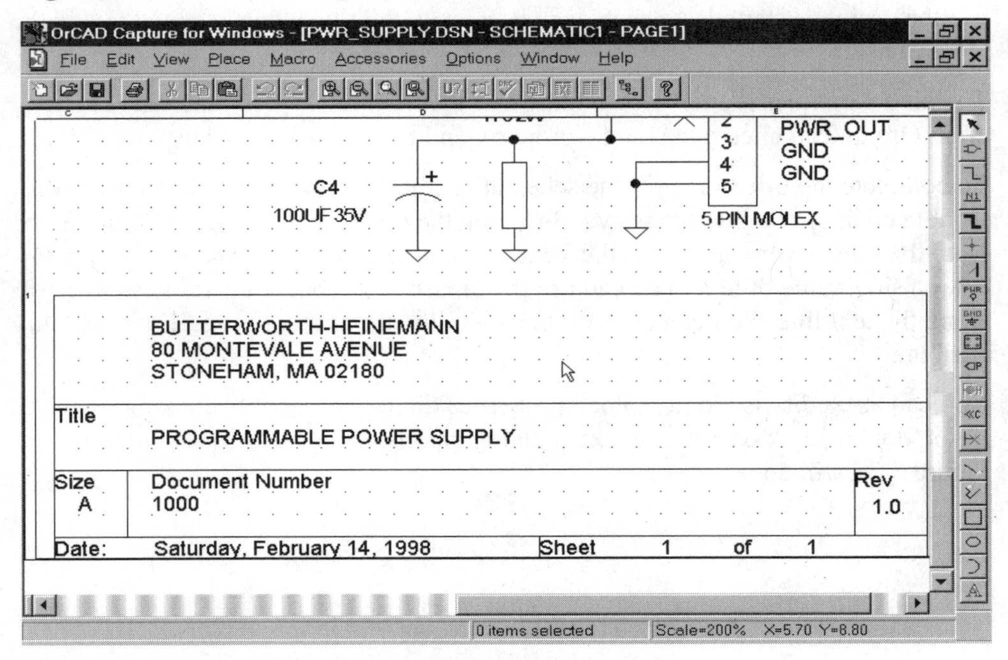

Figure 3-36 Completed Title Block

When you have completed the edits, your title block should appear as shown in Figure 3-36. Congratulations! You have finished drafting your first schematic. You have now learned most of the Capture commands used for everyday schematic drafting. To conclude the single-sheet schematic exercise, you will print out a hardcopy and then save the design to disk.

Printing the Schematic

To print out a hardcopy of your first schematic, you will require access to a laser or ink jet printer.

Printing from Capture is similar to printing from other Windows programs. Previous DOS versions of OrCAD SDT were not very forgiving of printer errors and would often lock up if the printer ran out of paper. Luckily, these problems have been corrected. Any standard hardcopy devices for which Windows drivers have been installed are supported, including networked devices. Printer errors (such as out of paper or offline) are handled by Windows and automatically recovered.

Capture provides two means of printing schematics. For multipage schematics, you can use Project Manager to select pages for printing. This operation will be covered in the next chapter, which deals with multipage schematics. You can also print any open schematic by using the Print tool or Print commands from the File menu. The Print tool is less flexible because it does not offer any setup options. From the File menu, you can access the Print Setup, Print Preview, and Print commands. Print Setup allows you to select the Windows printer. If you ever use a custom page size, you will find that the Print Preview command is very useful for checking fit and orientation. The Print command from the File menu gives you setup options, including scaling, offsets, and print quality. Although most new laser printers support 600 DPI resolution, the author has found that 300 DPI (or the medium resolution option on some print drivers) gives outstanding results and saves time.

Print out your schematic. Use the Print Setup command to verify that the correct printer has been selected. Experiment with the Print Preview command and try printing at various resolutions to see which gives the best results with your system.

Saving the Design

Recall that a Capture design consists of one or more schematics each containing one or more pages and a design cache that contains a copy of all parts used in the design. You normally work on one or more open schematic pages. Good practice dictates saving open schematic pages whenever major changes are completed. This reduces the likelihood of loosing work because of operator errors, power failures, or computer crashes. You can save an open schematic page by using the Save tool or Save command from the File menu. You can also use the Close command from the File menu. The Close command will prompt you with an option to save any changes if you have not already done so.

Once all the schematic pages have been saved and closed, Capture returns to the project manager screen. You can then use the Save tool or Save command from the File menu to save the design. If you want to save the design or make a copy under a new name, use the Save As command. The Close Project command from the File menu closes an open project. The Exit command quits Capture. Both commands will prompt you with options to save any changes if you have not already done so.

If you use the Windows Explorer to examine your design directory, you may find the files with the following extensions:

- **.DSN**. This is the main design file that contains all schematic pages and the design cache.
- **.DBK**. Backup design file.
- **.OPJ**. Project file that contains information about additional design resources displayed in project manager.
- **.OLB** and **.OBK**. Library file and backup library file normally stored in the parts library directory.
- **.SCH** and **.LIB**. Schematic and library file written out in the older OrCAD SDT format for backward compatibility.

As explained in the section on backing up data in Chapter 2, good practice is to keep each design and all its associated files in a separate subdirectory. You can then easily make a backup to floppy disk by using the Windows Explorer to copy the entire design subdirectory.

At this point, save your design, exit Capture and practice making a backup copy of the design onto a floppy disk.

Conclusion

You have now completed the single-sheet schematic exercise and are on your way to learning the ins and outs of Capture. In this chapter, you have learned to use all the basic tools required to create and print out a simple one-page schematic. In the next chapter, you will build on this knowledge and learn how to create a multipage hierarchical schematic structure.

Review Exercises

1. Explain the difference between the project manager and schematic editor windows.
2. Project Manager allows selection of logical or physical view. Which type of view is most appropriate for creating and editing simple schematics?
3. List the tools available on the main toolbar. Which tools can be run only from Project Manager?
4. List the tools available on the schematic editor tool palette.
5. Capture normally draws orthogonal wires and buses. What key must you hold down to draw nonorthogonal wires and buses?
6. What action occurs if you hold down the SHIFT key while drawing polylines, rectangles, and ellipses?
7. Keyboard shortcuts can be used in place of clicking on most tool buttons. What is the keyboard shortcut for the Part tool?
8. What keyboard hotkey combination can be used to toggle snap to grid?
9. What is the action of the ESC key?
10. Describe the difference between the normal and convert view of a part. Draw out the DeMorgan equivalents of AND, OR, NAND, and NOR gates.
11. What is the default name for ground symbols? How would circuit connectivity be affected if you changed the name of one ground symbol in your design?
12. Describe the basic rules for making electrical interconnections with wires.
13. Describe several means of selecting objects. What key is used in combination with mouse clicks to add or delete objects from a selection set?
14. How do you bring up a dialog box to edit the properties of any object displayed on the screen?
15. Describe basic operations involved in moving, copying, and resizing objects. What is meant by the term *dragging*?
16. Outline the basic procedure for editing reference designator and part value properties.

17. The Text tool allows you to enter paragraphs with multiple lines of text. What key combination must be used to advance to a new line of text? How can you import Windows clipboard text?
18. The title block has a number of important properties. List the read only properties that are usually visible. List additional visible properties that may require editing.
19. List the file extensions for Capture design-related files.

4

Hierarchical Design

In the previous chapter, you learned how to create and print out a simple schematic. The mechanics of creating this schematic were similar to manual drafting techniques. In this chapter, you will build your skill level beyond the basics and you will start to unlock the power of schematic capture. This will entail creating a hierarchical schematic and then using postprocessing tools to generate a bill of materials.

Your task will be to recreate the four-sheet hierarchical schematic shown in Figures 1-18A through 1-18D. These figures represent the schematic for a simple motorcycle ignition system, which is used as the model for the exercise in this chapter. You should make a copy of the figures and keep them handy for reference. Before starting, review the material in Chapter 1 on hierarchical schematics (pages 31-35), especially the OrCAD Capture terminology. You may want to briefly review the material in Chapter 3. Prerequisites for the exercise in this chapter include familiarity with the basics of creating designs, placing and editing objects, and saving the design.

First Session – Creating a Hierarchical Design

The model for the exercise is a single-level hierarchy. The first schematic sheet shown in Figure 1-18A on page 32 is a block diagram that shows interconnections between additional sheets (circuit blocks) in the next level down. Capture refers to this first schematic as the *root schematic*. Project Manager displays a backslash symbol in the file folder representing the root schematic. Capture also refers to this root schematic as the *parent*. Schematics in the next level down are referred to as *child* schematics.

Launch Capture from the Windows 95 desktop. The Capture session frame appears. Click on File and then New Design. Note that the name shown on the title bar at the top of the screen initially appears as DESIGN1 – SCHEMATIC1 – PAGE1. Click on File and then Save. Create a new folder called Tutor2. Then save the design using the name Ignition_Sys.dsn. Note the underline character in the name.

The first task is to create the root schematic. Start by editing the title block. Edit the schematic title in the title block area. Use the title IGNITION SYSTEM as shown in Figure 4-1.

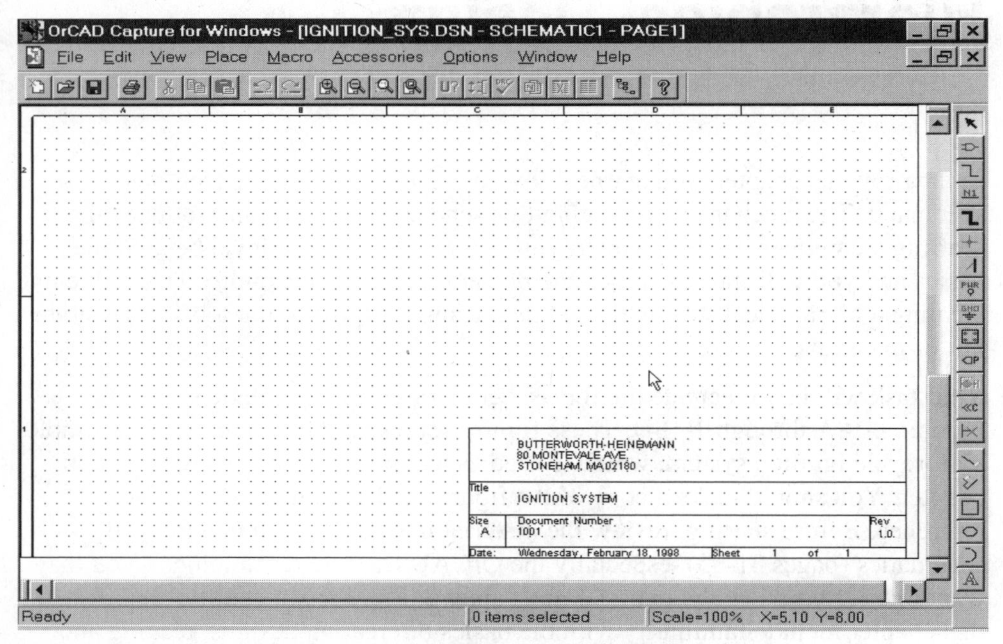

Figure 4-1 Starting on the Root Schematic

Placing Hierarchical Blocks

The next step is to create the hierarchy. This entails placing three hierarchical blocks, adding hierarchical pins for the signals, and interconnecting these pins with wires.

Use the model as a guideline for object locations. You can initially place objects in approximate locations and then make any required adjustments by dragging the objects.

Start with the MICROCONTROLLER hierarchical block. Click on the Hierarchical Block tool on the tool palette. The dialog box shown in Figure 4-2 appears. This dialog box allows you to set the following properties:

- **Reference**. This value appears on top of the hierarchical block. It identifies the block for the reader. You should use a plain text description that summarizes the overall function of the block.

- **Implementation Type**. You can select from several options that determine the implementation of the block: schematic view, VHDL, EDIF, project. If the block will represent a schematic, select schematic view. VHDL and EDIF are industry standard file formats. VHDL, EDIF and project implementation selections are typically used to attach files when the block represents a PLD device.
- **Implementation name**. If the implementation is a schematic, this specifies the name of the schematic folder in project manager. The name appears at the bottom of the hierarchical block. Use a short name or mnemonic, similar to a filename, that you can easily associate with the circuit block.
- **Path and filename**. Primarily applicable to VHDL, EDIF and project implementations in which information is in a separate file. Leave blank for schematic implementations.
- **Primitive**. In most cases, you can use the default setting that is taken from the Design Template. If you followed the configuration instructions in Chapter 2, hierarchical blocks will default to nonprimitive. This means you can descend into the block.
- **User Properties option**. Clicking on this option brings up a detailed list of properties. You can also add user-defined properties.

Enter the property values as shown in Figure 4-2. Then click on OK. A small crosshair-style cursor appears on the screen. Position the crosshair at the desired location for the upper left-hand corner of the block. Then hold down the left mouse button and drag to form the block. The block remains selected as shown in Figure 4-3 with resize handles visible to allow editing size and position. When you have finished editing, click on the background outside the block.

Use the process you just learned to place the remaining two blocks as shown in Figure 4-4. Use the same reference and implementation names as in the model. Approximately position the three blocks in the locations shown. Note that you can click on a block and then drag it.

If you make a mistake or want to change block properties, double click on the block. You can also press the right mouse button to bring up the shortcut menu and then click on Edit Properties. Either action will bring up the dialog box shown in Figure 4-2.

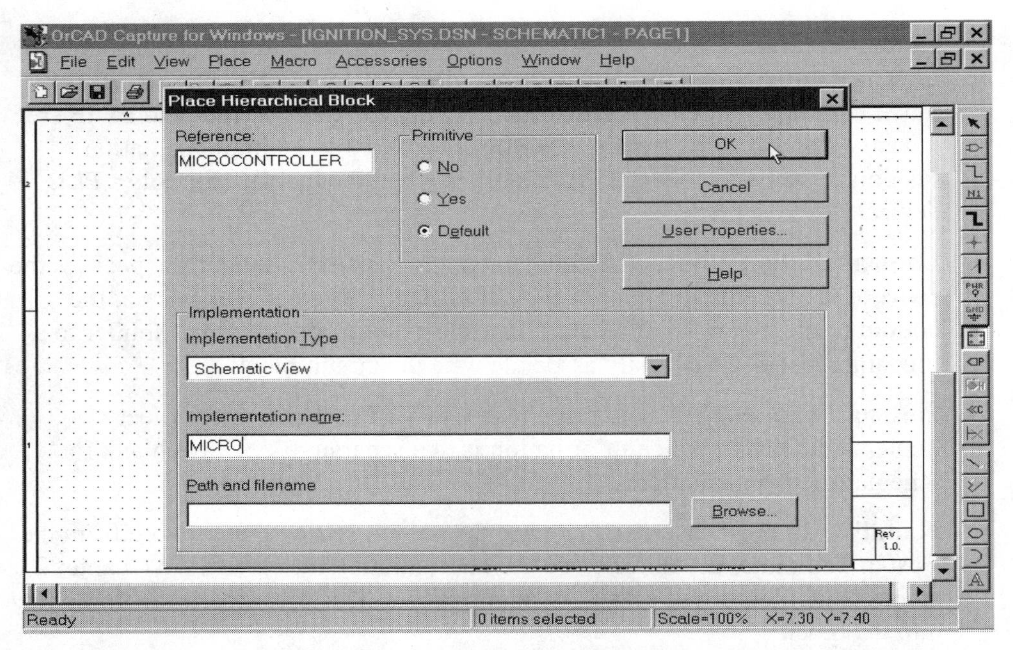

Figure 4-2 Dialog Box for the Place Hierarchical Block Tool

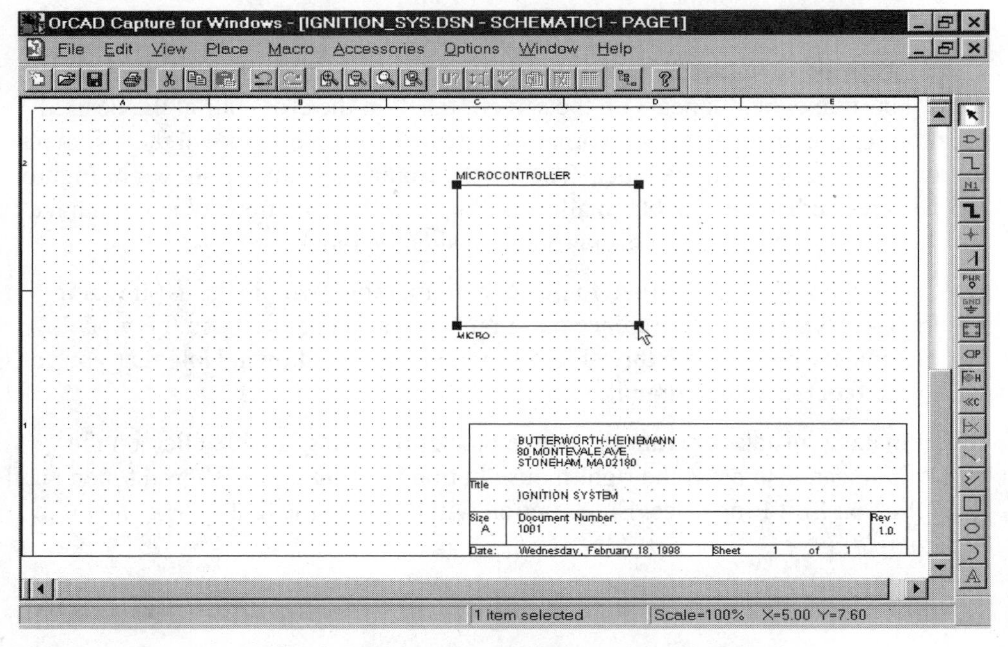

Figure 4-3 Placing the First Hierarchical Block

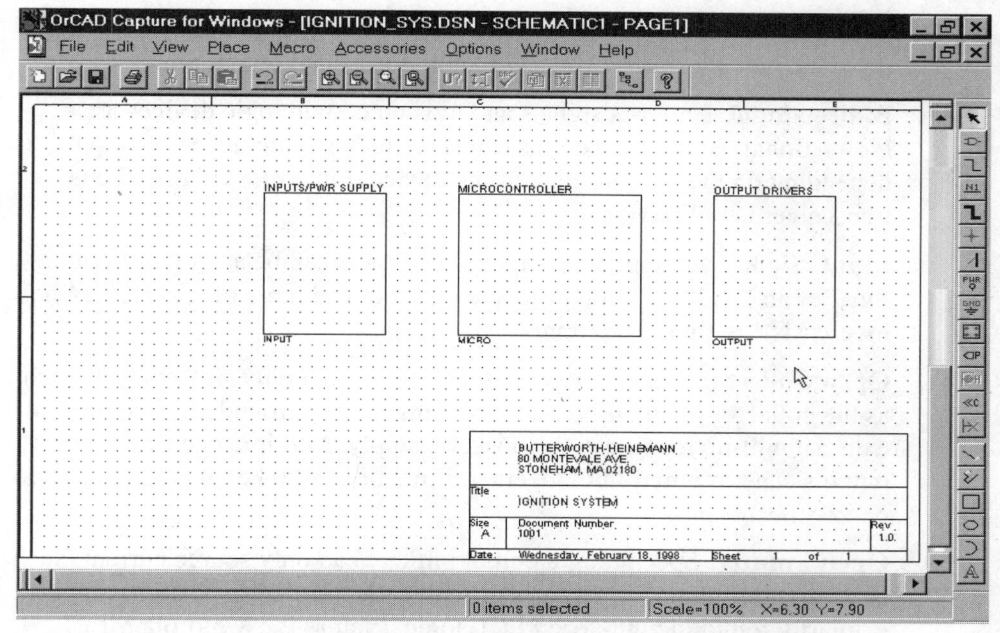

Figure 4-4 Root Schematic with Completed Blocks

Placing Hierarchical Pins

Hierarchical pins are the entry and exit points for signals that run between hierarchical blocks. Recall that power and ground are automatically interconnected between all schematics in a hierarchy, unless measures are taken to isolate them.

The Place Hierarchical Pin tool on the tool palette is used to place pins. Before you can use this tool, you must click on the block to which you plan to add pins. Start with the INPUT block. Click on this block to select it. Then click on the Place Hierarchical Pin tool. The dialog box shown in Figure 4-5 appears. This dialog box allows you to set several pin properties:

- **Name**. The electrical signal name that appears next to the pin. This name becomes the net alias for the pin.
- **Type**. You can select the electrical properties of the signal associated with the pin. This information is used for electrical design rules checking. Capture supports the following electrical signal types:

  3 State. Special logic output signal often used when multiple devices appear on a common bus. The three signal states are: active low, active high, or off (high impedance). In the off state, the output appears as an open circuit that does not contend with other devices that may be trying to drive the bus. Pin 2 of a 74HC373 latch is a 3-state output pin.

Bidirectional. Data bus signals and pins on certain logic devices are bidirectional. Bidirectional pins can serve either as inputs or as outputs, depending on the internal state of the device. Pin 2 of a 74HC245 bus transceiver is bidirectional.

Input. Signals are applied to input pins. Inputs can be analog or digital. Pin 2 of an LM324 opamp is an analog input and pin 1 of a 74HCT14 hex inverter is a digital logic input.

Open Collector. Special logic output pin that can only sink current (such as the collector of an NPN transistor with emitter grounded). Requires an external pull-up resistor to the positive supply. Multiple open collector outputs can be "wired-OR" together. Pin 2 of a 7406 hex inverter is an open collector pin.

Open Emitter. Special logic output pin that can only source current (such as the emitter of an NPN transistor with collector tied to positive supply). Primarily found in high-speed ECL logic, such as the Motorola MECL 10K/10H logic series. Requires an external resistor network for proper termination.

Output. Signals are generated at output pins. Outputs can be analog or digital. Pin 1 of an LM324 opamp is an analog output and pin 2 of a 74HC14 hex inverter is a digital logic output.

Passive. Typically found on passive devices such as resistors or discrete semiconductors without external power connections. You can also use passive pins for visible power connections on ICs.

Power. Supplies power to the part and is automatically connected to ground or one of the supply rails. Power pins are usually invisible. The name of the power pin, such as GND, VCC, VSS, or VDD, determines to which power or ground plane the pin is connected.

- **Width**. Determines whether or not the pin represents a single signal (scalar) or multiple signals (bus). Connect wires to scalar pins and buses to bus pins.
- **User Properties option**. Clicking on this option brings up a detailed list of properties. You can also add user-defined properties.

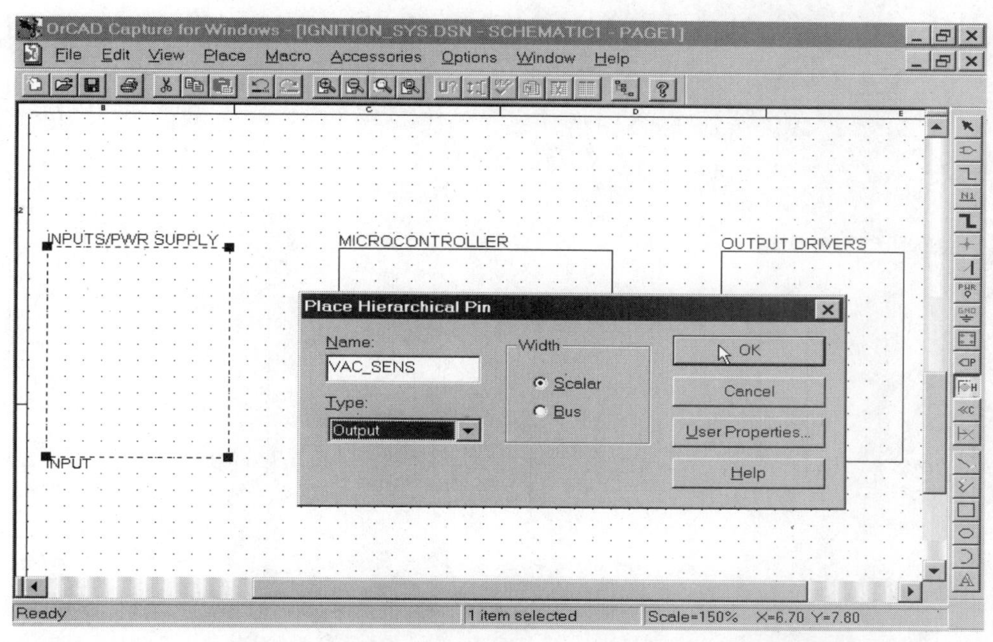

Figure 4-5 Dialog Box for the Place Hierarchical Pin Tool

The hierarchical pin type used on a block should match the type of device to which the pin connects within the block. An output pin drives circuitry outside the block. An input pin receives signals into the block. An output pin on one block or part generally drives an input pin on another block or part.

Enter VAC_SENS as the pin name and select an output type. Then click on OK. The hierarchical pin appears on the periphery of the selected block as shown in Figure 4-6. You can move the pin around the periphery with the mouse pointer. When you click the left mouse button, the pin is placed and the name appears next to it.

The Place Hierarchical Pin tool remains modal (active) and you can continue to place additional pins on the same block. To place another pin, press the right mouse button to bring up the shortcut menu shown in Figure 4-7 and then click on Edit Properties. Enter the information for the new pin. Note that when placing successive pins, you cannot switch to another block without first exiting the tool.

Place the remaining hierarchical pins. When you are finished, your root schematic should appear as shown in Figure 4-8.

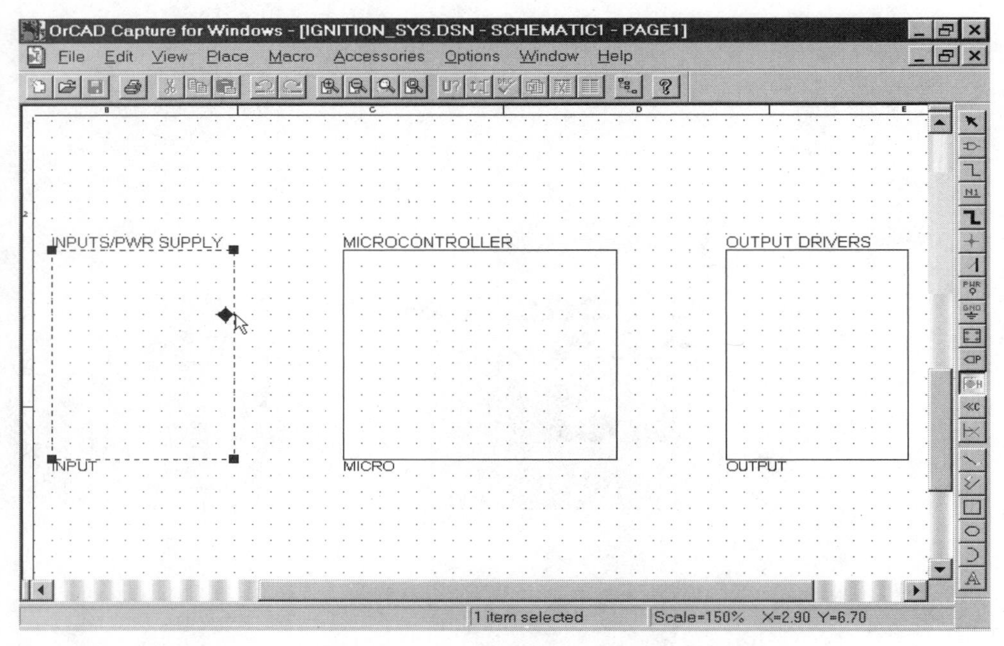

Figure 4-6 Placing the First Hierarchical Pin

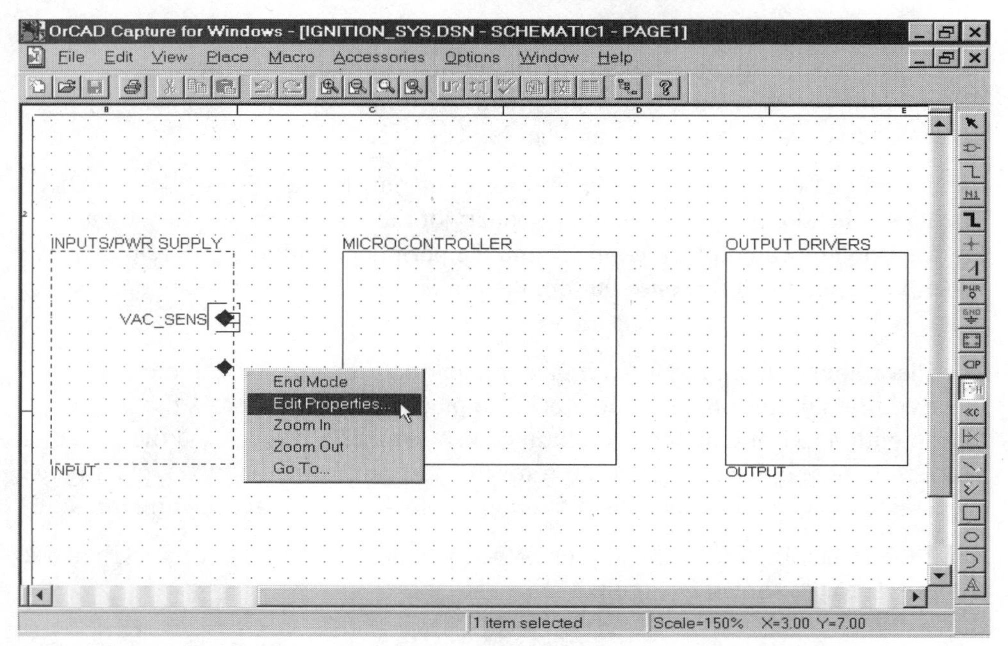

Figure 4-7 Editing Properties before Placing Another Pin

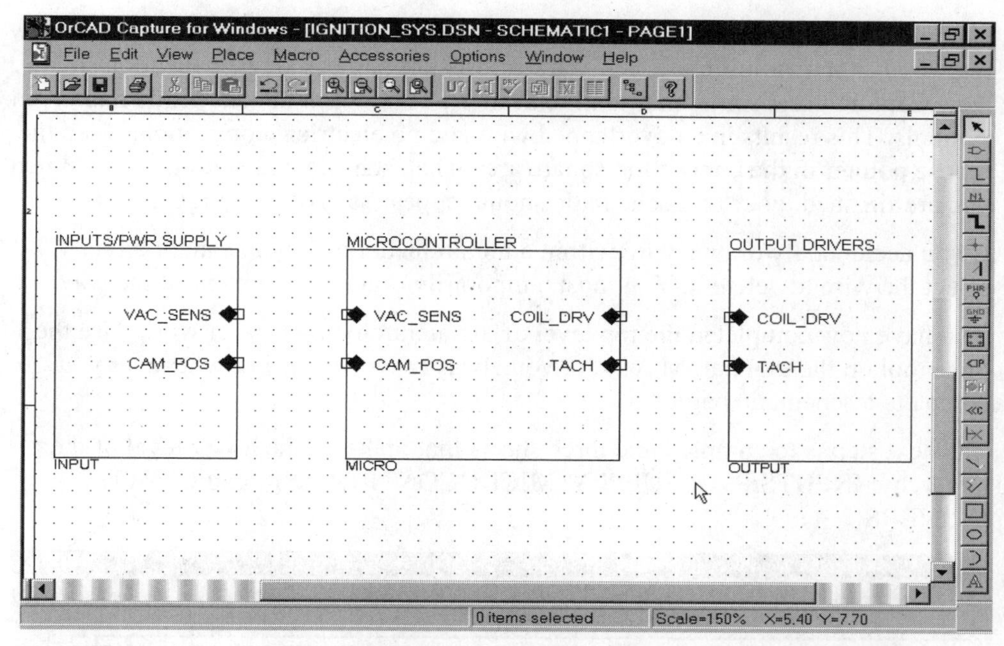

Figure 4-8 Root Schematic with Hierarchical Pins Placed

If you need to move or edit the properties of a hierarchical pin, you can directly select the pin by clicking on the diamond-shaped pin symbol. You do not need to click on the block first. If you first click on the block or click outside the diamond, you will not be able to select the pin.

Interconnecting Hierarchical Blocks

An important fact to remember is that the hierarchical pin name defines a signal name (net alias) only within a given hierarchical block. Two hierarchical pins with the same name on different blocks are not automatically interconnected. Conversely, you can interconnect hierarchical pins with different names. The only exceptions are power and ground, which are not considered signals. Power and ground are always common to an entire schematic unless special measures are taken to isolate a section of the circuitry.

Hierarchical blocks are interconnected via wires and/or buses to hierarchical pins. The same rules apply as for any other type of part that has pins. Wires must run to the connection squares on the pins.

Complete the root schematic by drawing wires between the hierarchical pins. Draw the wires as shown in Figure 4-9. Wires must start and end on a connection square. New Capture users sometimes tend to draw wires between the diamond-shaped pin symbols. This results in an overlapped wire and no electrical connection. Keep the mouse pointer in the connection square area when starting and ending wires. When you are finished, your root schematic should appear as shown in Figure 4-10.

If you accidentally draw a wire within a hierarchical block, you cannot directly select the wire to delete it. You must temporarily drag the block out of the way.

You have now completed the top level of the hierarchy. Save your work. Use the Save tool on the tool bar. Make sure you always save your work whenever you complete a schematic page.

The next step is to complete the three sheets that make up the lower level of the hierarchy: INPUTS/PWR SUPPLY, MICROCONTROLLER, and OUTPUT DRIVERS.

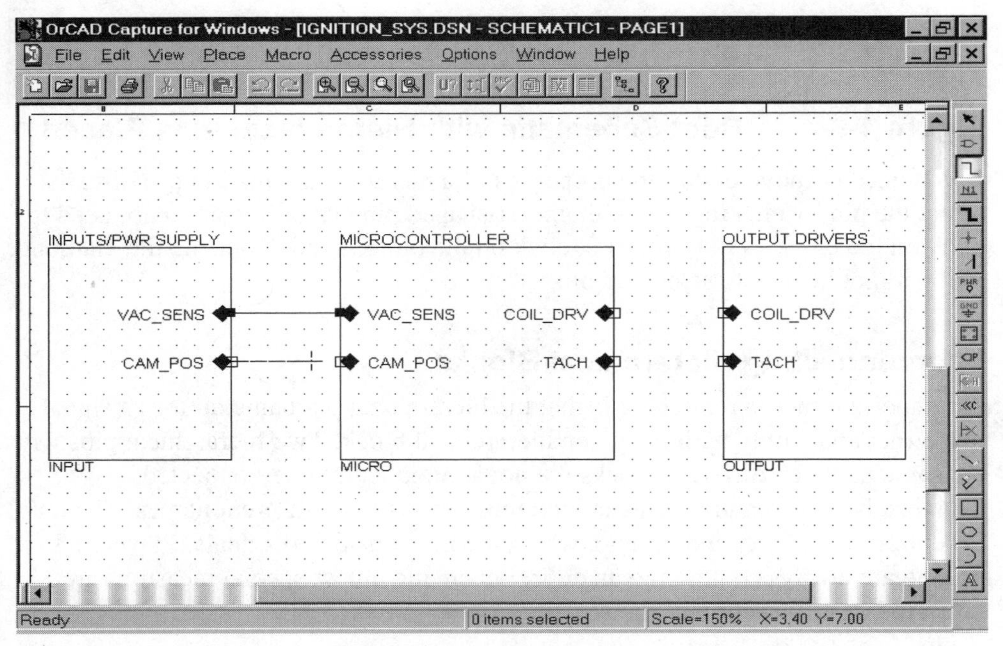

Figure 4-9 Drawing Wires between Hierarchical Pins

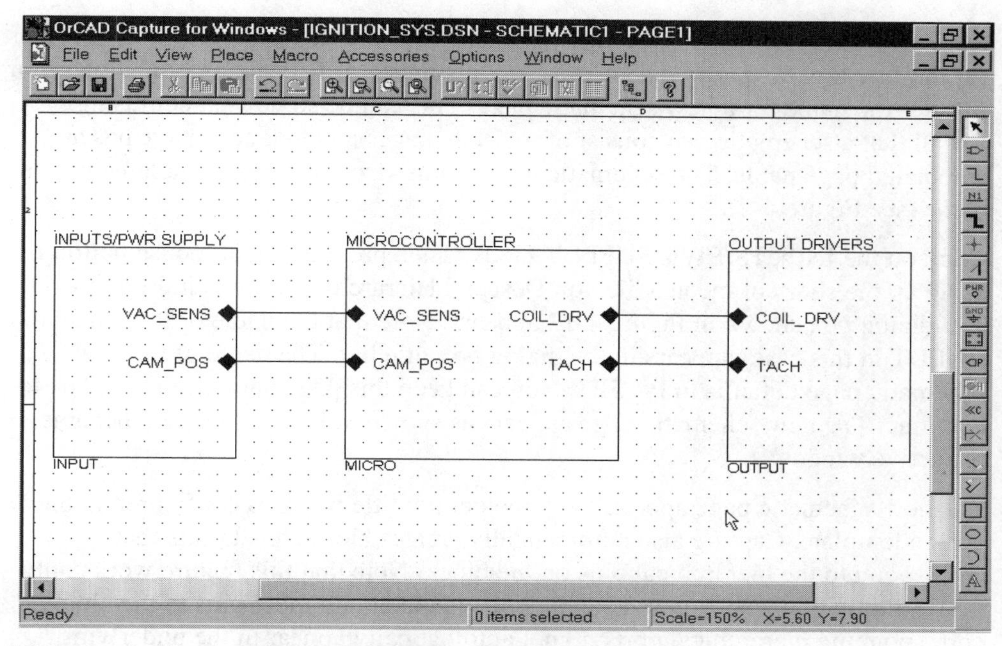

Figure 4-10 Completed Root Schematic

Navigating throughout a Hierarchical Design

Working on a hierarchical design usually requires a considerable amount of navigation between the various levels and schematics. Capture provides several means of navigating throughout a design. You can open a particular schematic page by double clicking on it in Project Manager. Multiple schematic pages can be opened simultaneously with each page occupying a separate window. This works similar to other Windows programs that allow multiple open documents. You can click on a schematic window that is in the background to make it active. You can also resize the schematic windows.

Although OrCAD Capture running under Windows 95 is relatively stable compared with some other EDA programs, abnormal terminations can occur. Whenever you are working with multiple open schematic windows, you should frequently save your work. If you have made edits in the active window, good practice dictates using the Save command from the File menu before you switch windows or navigate the hierarchy.

Many users prefer navigating throughout the design by means of the root schematic. Navigating from the root schematic to a lower level is called *descending* the hierarchy. Navigating the opposite direction to return to a higher level is called *ascending* the hierarchy. When you descend into a hierarchical block for the first time in a new design, a dialog box appears. This dialog box allows you to enter the name for a new schematic page. This can be somewhat confusing. Recall that a schematic can consist of multiple pages and that each page has an associated page name. The schematic name comes from the block implementation name (see Figure 4-2).

Click on the INPUTS/PWR SUPPLY block. Then press the right mouse button to bring up the shortcut menu. Click on Descend Hierarchy as shown in Figure 4-11. The dialog box shown in Figure 4-12 appears. Note that the name of the schematic (INPUT in this case) appears in the dialog box title bar. The name of the new schematic page defaults to PAGE1. You can keep this page name. Click on OK to continue. The new schematic page appears as shown in Figure 4-13 and becomes the active window.

The new schematic page appears with the default title block as configured in the design template. Capture also automatically creates hierarchical ports that correspond to the hierarchical pins on the block. Note that this feature works only the first time you open a new schematic. If you later add more pins to the block, corresponding hierarchical ports do not automatically appear in the underlying schematic.

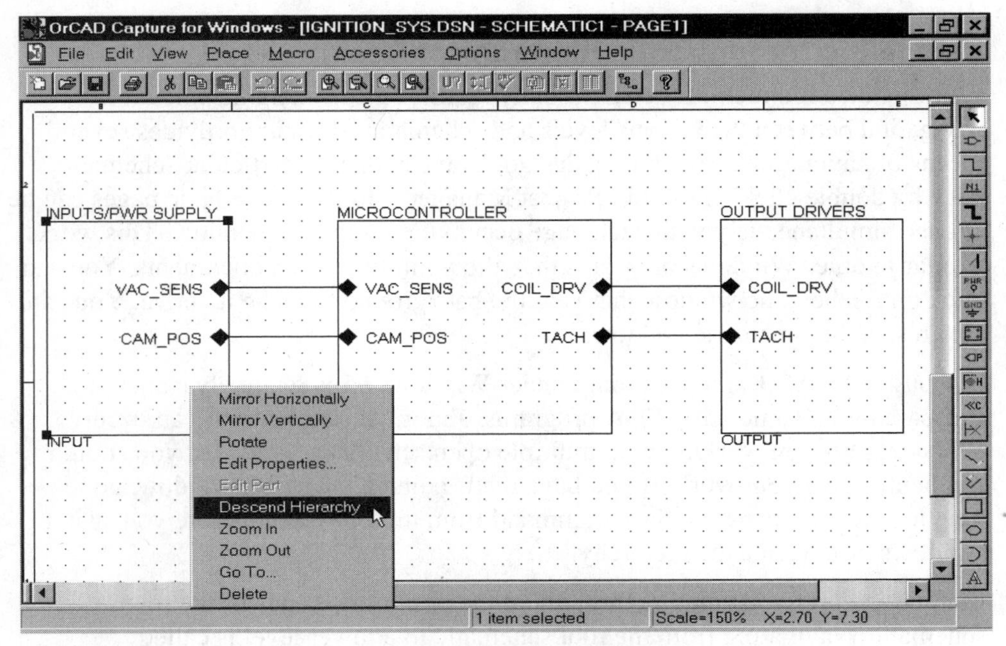

Figure 4-11 Descending into the Hierarchy

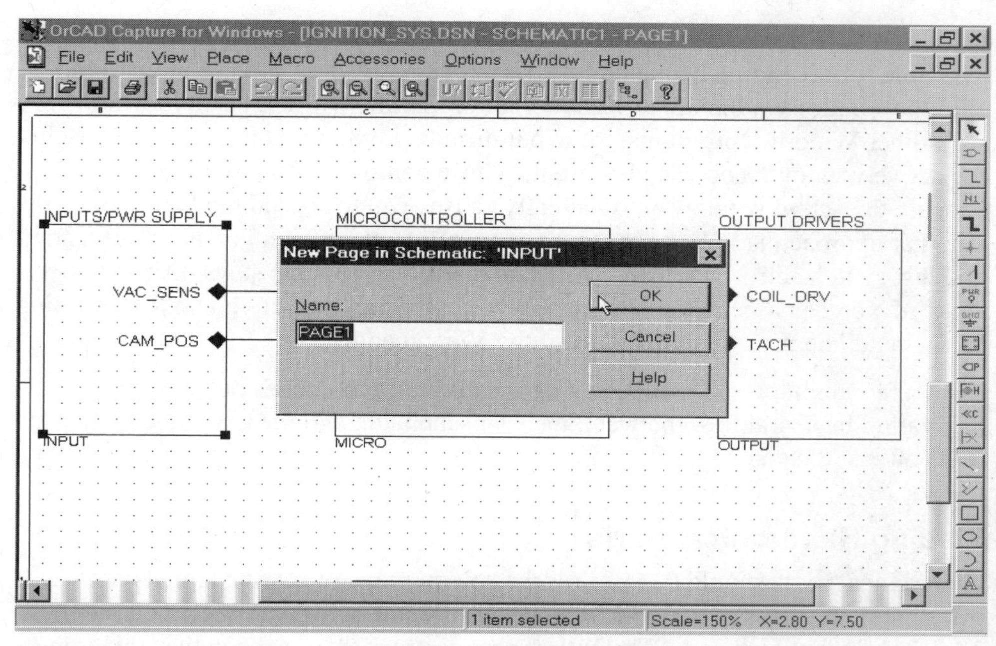

Figure 4-12 Opening a New Schematic Page

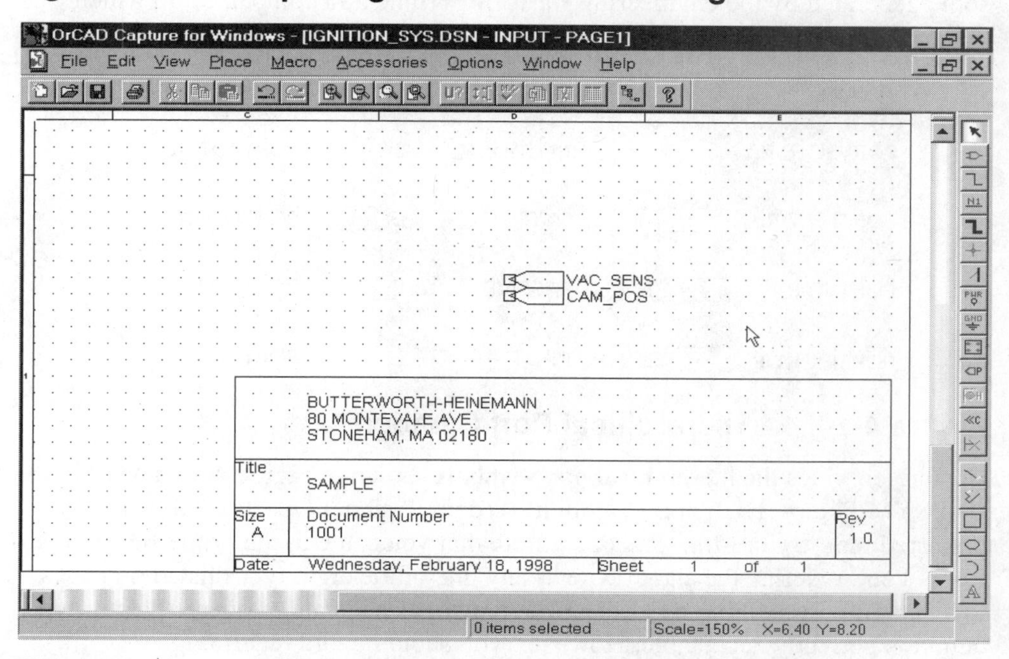

Figure 4-13 Initial View of the New Schematic Page

Once you have descended down into the hierarchy, several approaches are available for ascending back up to the root schematic. You can use the familiar Windows buttons at the upper-right corner of the title bar in the active window. As with other Windows programs, these buttons allow you to reduce, enlarge, or close the active window. You can also use the Close command from the File menu. Closing the active window automatically causes you to ascend the hierarchy. Another approach is to leave open any schematics into which you have descended and just switch active windows. Note that the root schematic stays open unless you close it. You can ascend the hierarchy without closing the active window by using the Ascend Hierarchy command from the View menu.

Take a few minutes and practice using the various approaches outlined for navigating back and forth between the root schematic and the new schematic page that you just created.

Placing Hierarchical Ports

Hierarchical ports should be used for the sole purpose of routing signals onto and off schematic pages in a hierarchical design. Never use hierarchical port symbols in place of connector parts. The author has seen numerous schematics, including some generated by the engineering staffs of Fortune 500 companies, in which hierarchical port symbols were misused.

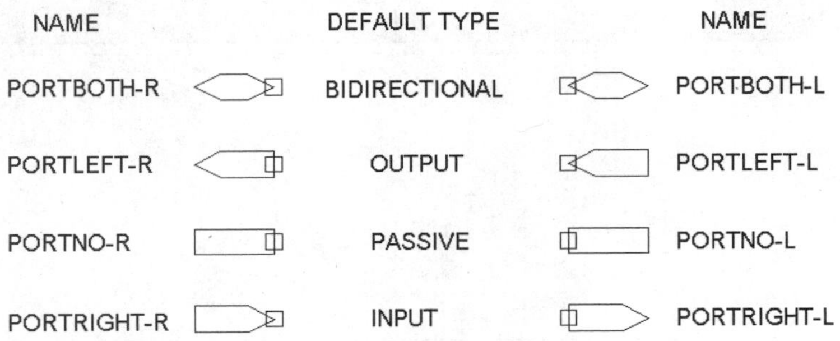

NAME	DEFAULT TYPE	NAME
PORTBOTH-R	BIDIRECTIONAL	PORTBOTH-L
PORTLEFT-R	OUTPUT	PORTLEFT-L
PORTNO-R	PASSIVE	PORTNO-L
PORTRIGHT-R	INPUT	PORTRIGHT-L

Figure 4-14 Hierarchical Port Symbols

Capture provides the hierarchical port symbols shown in Figure 4-14 as part of the CAPSYM library. Each port symbol has a default signal type as established by preferred industry drafting practices. Note that you can edit the properties of a port symbol and redefine the signal type as any one of the eight types listed on pages 119 and 120. Redefining a PORTLEFT symbol as a 3 state or open collector type output is perfectly acceptable. However, you should avoid redefining a

PORTRIGHT symbol as one of the output signal types, since this will cause confusion. You can use the PORTNO symbol for passive and power signal types.

A one-to-one correspondence must exist between the hierarchical ports shown on the schematic page and the hierarchical pins shown on the block representing the schematic. For each pin shown on the INPUTS/PWR SUPPLY block in the root schematic (Figure 4-10), a port with the same name and type (input, output, and so forth) must be shown on the INPUTS/PWR SUPPLY schematic page. The ports can be located anywhere on the page; location and order are irrelevant. Only the name and type must match the hierarchical pin. General practice is to place input ports on the left and output ports on the right. Maintaining some spatial relation between the order of ports on the schematic page and pins on the block improves readability but may not be possible on complex and crowded designs.

Recall that Capture automatically places appropriate ports when you first create a new schematic page in a hierarchical design. To add ports at a later time, you must use the Hierarchical Port tool on the tool palette.

Practice using the Hierarchical Port tool. Descend into the INPUTS/PWR SUPPLY schematic page (Figure 4-13). Select and delete the existing VAC_SENS port. Then click on the Hierarchical Port tool. The dialog box shown in Figure 4-15 appears. This dialog box is very similar to the box that appears when placing power or ground symbols (see page 87). You can select a symbol from the library list and enter an electrical signal name.

Select the PORTLEFT-L symbol and enter the signal name VAC_SENS. Then click on OK to place the port.

The Hierarchical Port dialog box does not provide a means of selecting the signal type. Ports are initially placed with the default signal types listed in Figure 4-14. In most cases, you can accept these default types. If you need to change the signal type, double click on the port symbol. The dialog box shown in Figure 4-16 appears. You can then change the signal name and signal type. If you want to change only the port name, double click on the name instead.

Note that once you place a port symbol, you cannot change the symbol type, i.e., from PORTLEFT to PORTNO. You must first delete the incorrect symbol and then place a new symbol. You can select a PORTLEFT or PORTRIGHT symbol, click the right mouse button to bring up the shortcut menu, and then horizontally mirror the port, but this is not a good practice. Mirroring the port does not change the signal type and thus can lead to errors.

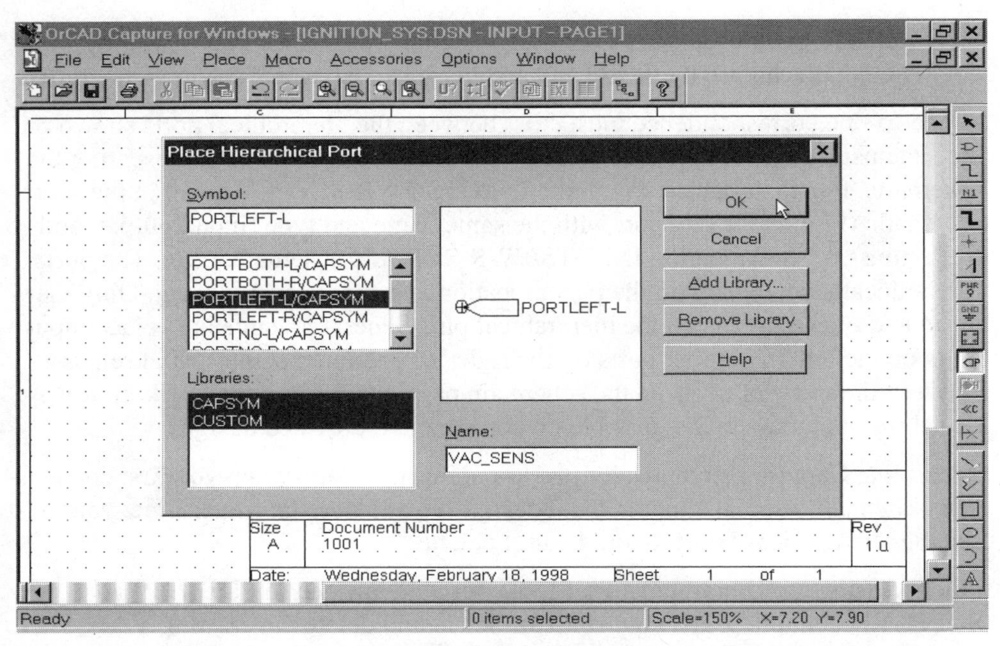

Figure 4-15 Dialog Box for the Hierarchical Port Command

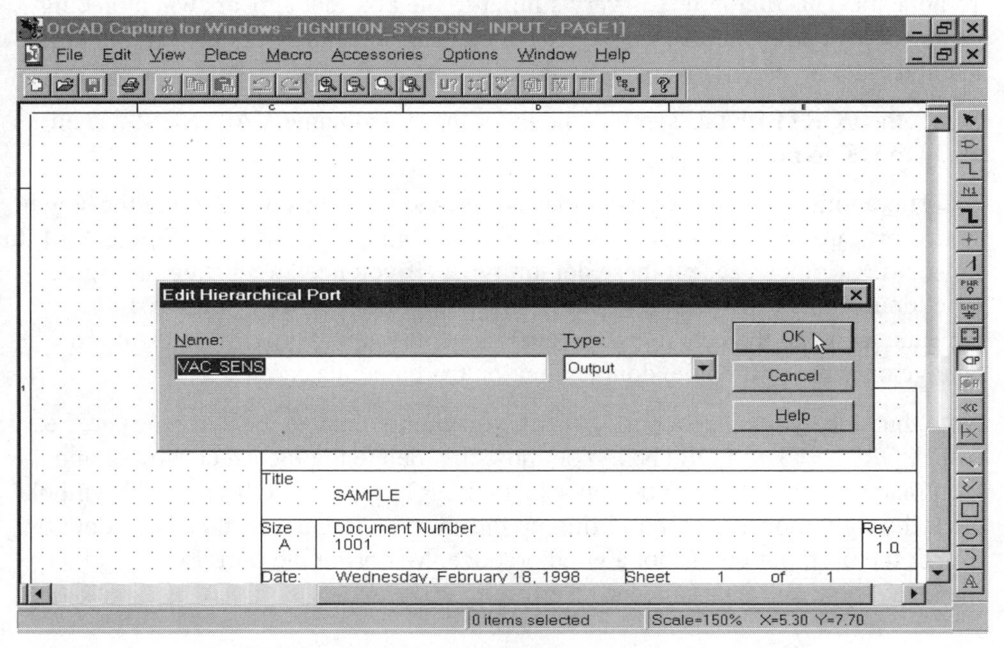

Figure 4-16 Editing Hierarchical Port Properties

Completing the Second Sheet

The completed second sheet is shown in Figure 1-18B (page 33 in Chapter 1). By now you should be able to complete this sheet on your own using the skills learned in Chapter 3. The design example is based on an ignition system for a Harley-Davidson motorcycle. The upper section of sheet 2 is the power supply that uses an automotive-type voltage regulator, the Micrel MIC2951. U1 is a Senisys HA-640 Hall Effect position sensor that generates a timing reference pulse based on camshaft position. A vacuum sensor signal is input on PL3. The third sheet shows a Microchip PIC16C71 RISC microcontroller with an on-chip four channel A/D converter. The two trimpots R5 and R6 are used to set the rpm limit and adjust the timing advance curve. The last sheet shows two output drivers. Q1 drives an electronic tachometer with a +12V square wave. Q3 is a special Darlington transistor, manufactured by Fuji, that can drive an ignition coil.

This is a real-world circuit. Some of the parts are unique and may not be familiar. Listed below are some hints on getting these parts from the Capture libraries. In each case, the part name and library are given.

- **Connectors PL1 - PL3**: HEADER 1 from the Custom library. Note the space between HEADER and the character 1.
- **Hall Effect Sensor U1**: HA-640 from the Custom library.
- **Voltage regulator U2**: MIC2951 from the Custom library. Horizontally mirror.
- **Varistor RV1**: VARISTOR from the Device library.

Completing the Third Sheet

The third sheet involves many of the same considerations as the second sheet. Note that both input and output ports are used on sheet three. Listed below are the special parts:

- **Trimpots R5 and R6**: POT from the Custom library. Rotate 90 degrees and horizontally mirror.
- **Microcontroller U3**: PIC16C71_ALT from the Custom library. Horizontally mirror.
- **Resonator Y1**: RESONATOR from the Custom library. Rotate 90 degrees and horizontally mirror.

Placing Net Aliases on Wires

You must often route wires from one end to another on a crowded schematic. The resulting maze of wires reduces clarity. This can cause confusion and errors in tracing signal paths. On sheet 3 the CAM_POS signal is an input and the corresponding port appears on the lower left side. The signal must run to pin 6 on U3, which is on the other side of the sheet. The signal connection is made with two wire segments that are implicitly joined by means of net aliases. The Capture connectivity database considers these two wire segments the same as a continuous wire.

In order for a net alias to become associated with a wire, the lower left corner of the net alias must touch the same grid point as the wire. Imagine the character "C" enclosed in a box. The lower left corner of this box must be on the wire. Capture automatically snaps net alias objects into position on wires. Capture will not allow you to incorrectly place a net alias or drag it away from a wire. If a net alias appears on the screen, you can safely assume that it is positioned correctly. Note that a net alias is also one of the few objects that you cannot cut and paste. Presumably, Capture imposes this limitation to prevent connectivity database errors.

Complete sheet 3 to the point that you are ready to place the net aliases. Select the wire going to pin 6 on U3. Note that it is sometimes difficult to double click on a wire. Often the best approach to selecting a wire is to drag the mouse pointer across it. Click on Edit Properties from the shortcut menu to check the net association of the wire. You will see a Capture-assigned net name such as N00117 (your system may assign a different name).

Next, place net aliases for CAM_POS on the two wire segments. Click on the Net Alias tool from the tool palette. The dialog box shown in Figure 4-17 appears. You can enter the alias name, rotation, and font properties. In most cases you can accept the default font properties that were configured on the Design Template. Figure 4-18 shows the completed sheet 3 with CAM_POS net aliases in position.

Check the net association for the wire going to Pin 6 on U3. The net name should now be CAM_POS.

Try editing both CAM_POS net aliases. Change the name to DUMMY. Now check the net association of the wire going to Pin 6 again. Note that the net name is still CAM_POS. The reason is that the name of the hierarchical port overrides any net alias.

Change the net aliases back to CAM_POS and complete sheet 3.

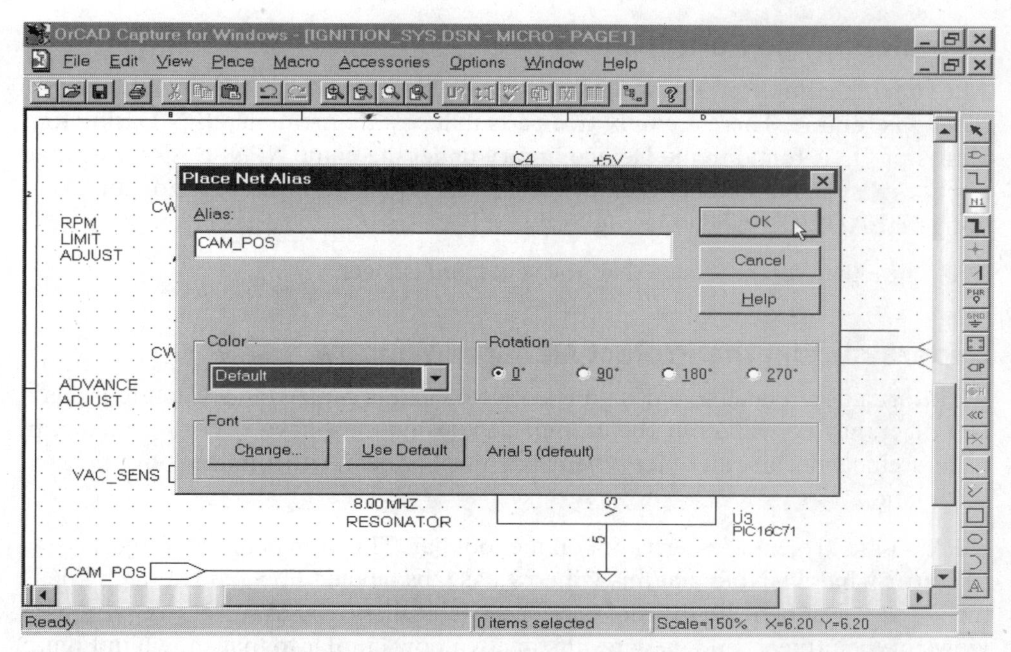

Figure 4-17 Dialog Box for the Place Net Alias Command

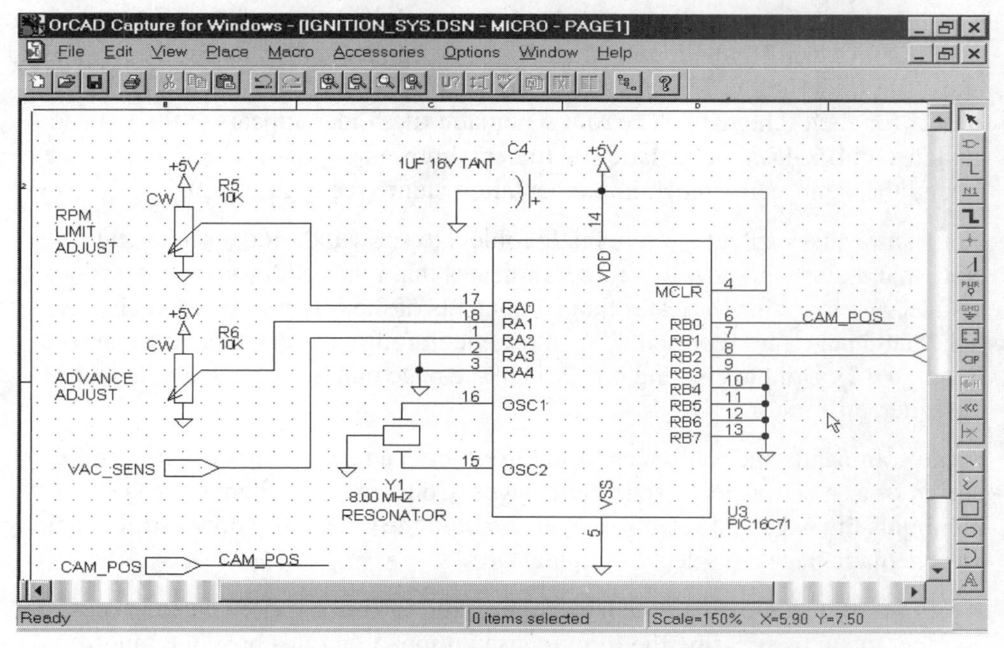

Figure 4-18 Sheet Three with Completed Net Aliases

Completing the Fourth Sheet

After having completed the previous three sheets, you should find that drawing sheet 4 is routine. There are only two parts that require mention. NPN Darlington transistor Q3 is found in the Device library under the name NPN DAR. Note that a space must be entered between the characters NPN and DAR. Horizontally mirror HEADER 1 when placing PL4 and PL5.

Make sure that you have saved all the schematic sheets.

Understanding the Project Manager Window

All information related to a design is collected and organized in a Capture project. A project may contain only one design. The design must have a root schematic. In a hierarchical design, all other schematics must branch out from the root schematic.

Click on the Project Manager tool on the tool bar. This brings up the Project Manager window. Note that this window has tabs labeled File and Hierarchy on the left side just beneath the tool bar. These tabs allow you to select two possible views of the project. File view results in a window similar to that shown in Figure 4-19. The file view window shows the design structure and contents. A "circuit" icon represents the design with the design name preceded by a backslash, i.e., \ignition_sys.dsn. If you double click on the design icon, it expands or collapses like a file folder in the Windows Explorer. If you expand the design, file folders appear for each schematic. The root schematic file folder appears at the top and contains a backslash. Schematic file folders representing hierarchical blocks are named the same as the block implementation name (see Figure 4-2).

You can further expand the schematic folders to show individual schematic pages. Note that a plus sign next to a folder indicates that it has additional contents not yet visible. A minus sign indicates that all contents of the folder have been displayed. Schematic pages are represented by a page icon followed by the page name (see Figure 4-12). Double clicking on a schematic page opens the page and launches the schematic editor.

Some commands and tools such as the Print tool can be run either on the entire design or a selection set of schematic pages. Clicking on a schematic page highlights the page and selects it. You can hold the CTRL key down while clicking on schematic pages to select multiple pages.

If you select a single object such as a schematic folder or page, you can access a shortcut menu by pressing the right mouse button. You must have the mouse pointer on the object while pressing the right button. Note that the shortcut menu is not available if you select multiple objects. Depending on the object you select, the shortcut menu includes commands to add files, add schematics or pages, edit pages or properties, rename, and save. These shortcut commands are similar to commands found on the File and Design menus.

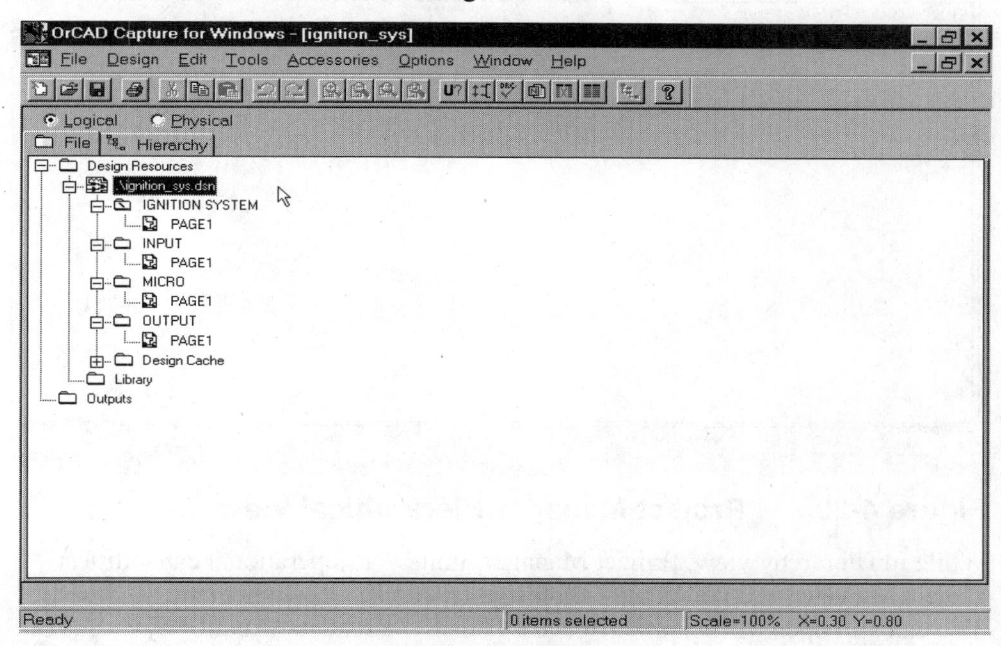

Figure 4-19 Project Manager File View

You can use the Rename command to rename schematic folders and pages. When you first start a new design, the root schematic will initially be named SCHEMATIC1. At this point in the exercise, your root schematic will still have the initial name. Click (not double click) on the root schematic and then use the Rename command to change its name to IGNITION SYSTEM as shown in Figure 4-19. All the schematics in your hierarchy now have concise and recognizable names.

If you click on the Hierarchy tab, the Project Manager window provides an overview of the hierarchical structure of the design. The window will look like Figure 4-20. A logic symbol icon now represents all objects. If you double click on the root schematic, the lower schematics appear. If you double click on one these, parts within the schematic appear.

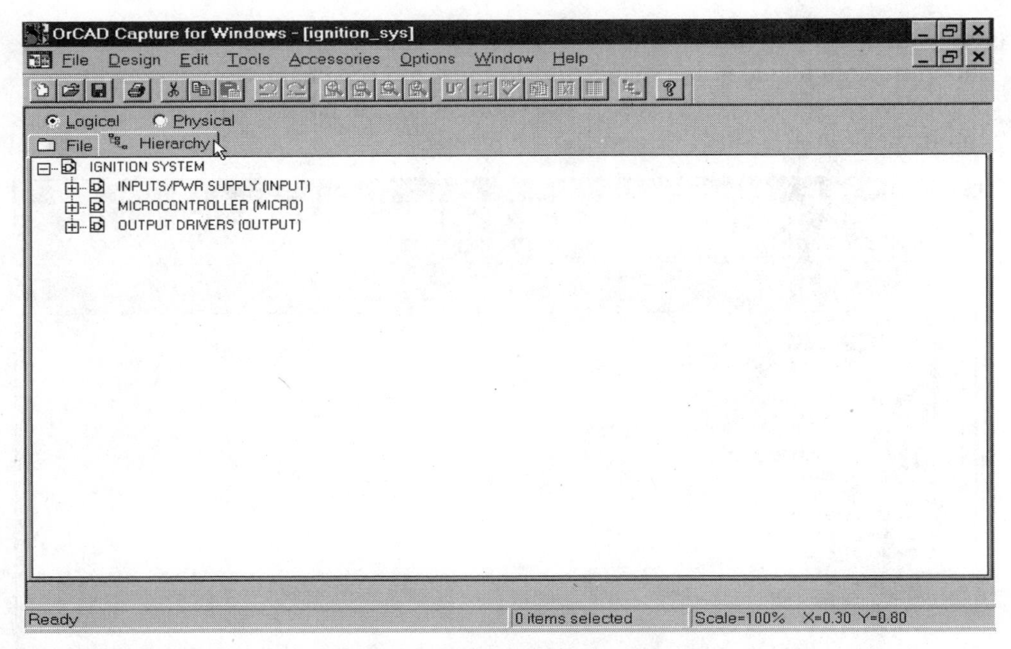

Figure 4-20 Project Manager Hierarchical View

While in Hierarchy view, Project Manager limits your operational capabilities. Project Manager will not display all design resources, only schematics and parts. You can still open a schematic page and launch the schematic editor by means of the shortcut menu, but double clicking does not work.

Printing the Design

In Chapter 3 you learned how to print individual schematic pages using the Print tool. Using this approach for a multisheet design would be very time consuming. Fortunately, Capture provides an easy means of printing multiple schematic pages or an entire design. Just select the appropriate objects in Project Manager and then click on the Print tool or use the Print command from the File menu. If you select the design object, Project Manager will print all schematics. If you select one or more schematic folders, Project Manager will print all pages within those folders. You can also select individual schematic pages for printing.

Print out all the pages in your design. You can also experiment with selecting schematic folders or pages for printing.

Second Session – Introduction to Postprocessing

The remainder of the exercise in this chapter introduces two Capture postprocessing tools: Cross Reference Parts and Bill of Materials. You will learn to run the tools and examine their output files.

Running the Parts Cross Reference Report

The Cross Reference Parts tool generates a parts cross reference report. Most users will find this report helpful in locating certain errors in parts usage and reference designator assignments. The report can also list unused parts (gates) in multiple-part packages. The report consists of an ASCII text file that contains a list of all parts in the design. Each part appears on a separate line, which includes an item number, reference designator, part description, schematic name, sheet number, and library of origin.

The information in the cross reference report is particularly useful for catching errors involving duplicate reference designators. Duplicate reference designators can easily occur when parts or sections of circuitry are copied and pasted during editing operations. Good schematic drafting practice also dictates showing all unused gates. The cross reference report provides a convenient means of identifying unused gates.

You must run postprocessing tools such as Cross Reference Parts from the Project Manager screen. Click on the design icon to select the entire design. Then click on the Cross Reference Parts tool on the tool bar. The dialog box shown in Figure 4-21 appears. The following options are available:

- **Scope**. You can select whether the tool processes the entire design or just the selected schematic folders or pages. In most cases, running the Cross Reference Parts tool makes sense only if you process the entire design.
- **Sorting**. You can select the sort order either by part value or by reference designator. Most users prefer sorting by reference designator, which generates a report similar to a bill of materials.
- **Report options**. Reporting of X and Y coordinates may help locate parts on very large schematics but unnecessarily clutters the report for most applications. You should select the option to report unused parts (gates).
- **Report file**. If you select the option to view the output, Capture launches the Windows Notepad with the report file. The report file path defaults to the current design directory and the file name defaults to the design name with an .XRF extension. You can change these defaults if required.

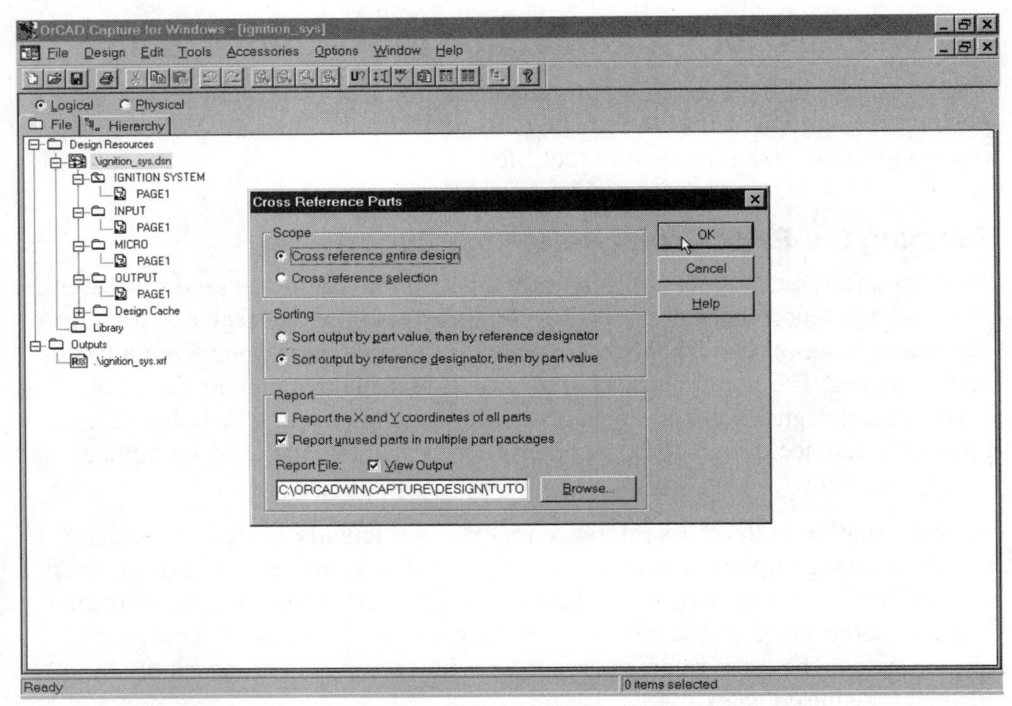

Figure 4-21 Dialog Box for the Cross Reference Parts Tool

Make sure you select the View Output option and then click on OK to run the Cross Reference Parts tool. The tool requires only a few seconds to process a small design such as the four-sheet schematic in this exercise. When the tool has completed processing the report, the Windows Notepad is automatically launched and appears as shown in Figure 4-22.

The body of the report is tab delimited, and the columns do not line up perfectly. If you want to print a neat copy, you must add tab characters on some lines. The identification header at the top of the report contains properties taken from the title block on the root schematic. Although recognizable, the header information appears somewhat disorganized. Unfortunately, the problems with tab-delimited data and disorganized-looking headers are typical of all Capture reports. Considerable editing is usually required.

If you are not familiar with the Windows Notepad, use the Help command or refer to your Windows documentation (a suggested Microsoft reference is listed on page 41 in Chapter 2). Try printing the report file from Notepad. Use the Print command from the File menu in Notepad.

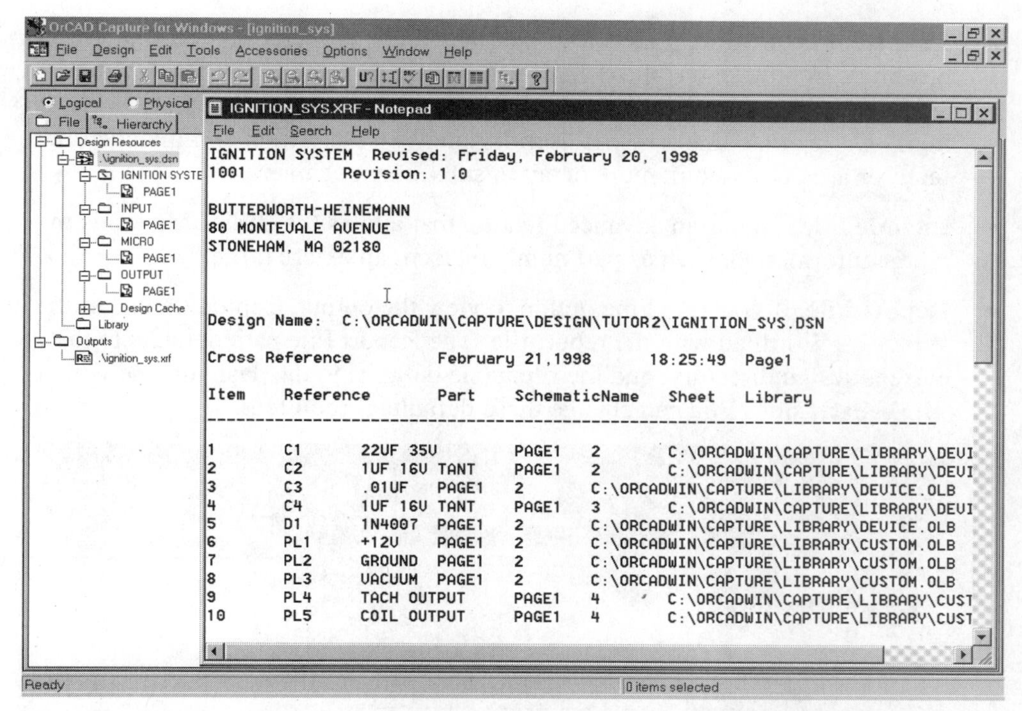

Figure 4-22 Cross Reference Report in Windows Notepad

Creating a Bill of Materials Report

The Bill of Materials tool generates a bill of materials report. A bill of materials report is one of the basic documents required to manufacture a product. The bill of materials report is an ASCII text file that includes an identification header and a parts list. The identification header contains properties taken from the title block of the root schematic. The parts used in the design are partially sorted and organized into a parts list.

You must run the Bill of Materials tool from the Project Manager screen. Click on the design icon to select the entire design. Then click on the Bill of Materials tool on the tool bar. The dialog box shown in Figure 4-23 appears. The following options are available:

- **Scope**. You can select whether the tool processes the entire design or just the selected schematic folders or pages. In most cases, running the Bill of Materials report makes sense only if you process the entire design.

- **Line Item Definition.** You can format the parts list, including the column headings. In most cases, the default values give proper results. For more information, click on Help within the dialog box. The option to place each item on a separate line rarely is useful. Line Item Definition and Include File features are discussed in more detail in subsequent chapters.
- **Include File.** This is an advanced feature that allows the Bill of Materials to merge information, such as part numbers, from an external file.
- **Report File**. If you select the option to view the output, Capture launches the Windows WordPad with the report file. The Report File path defaults to the current design directory, and the file name defaults to the design name with a .BOM extension. You can change these defaults if required.

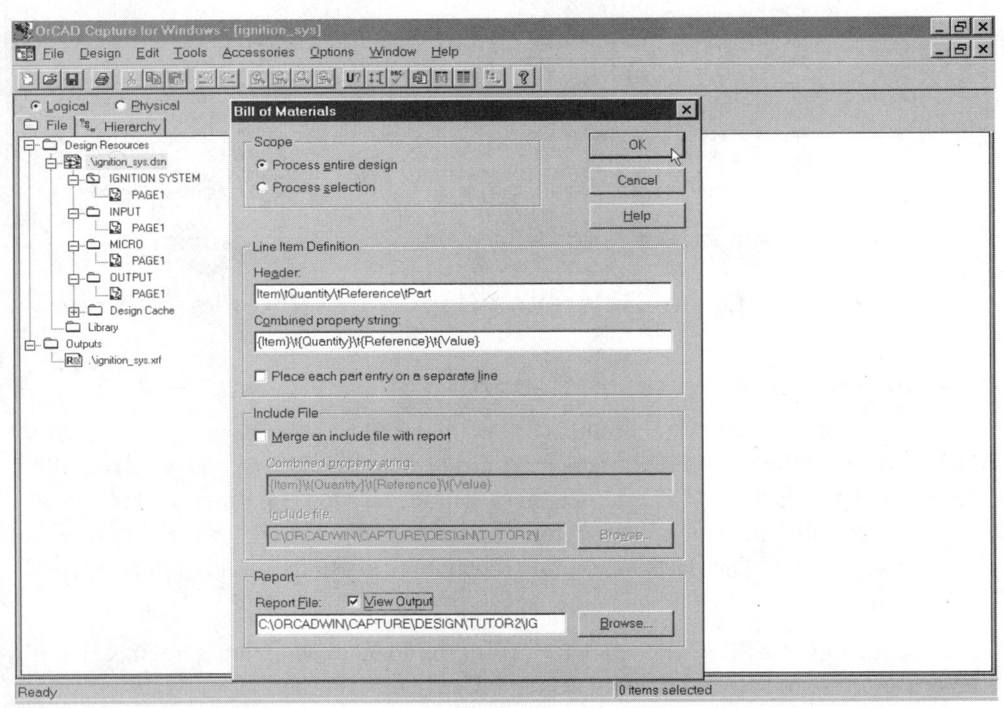

Figure 4-23 Dialog Box for the Bill of Materials Tool

Do not change any of the line-item definition fields. Make sure you select the View Output option and then click on OK to run the Bill of Materials tool. The tool requires only a few seconds to process a small design such as this exercise. When the tool has completed processing the report, the Windows WordPad is automatically launched and appears as shown in Figure 4-24.

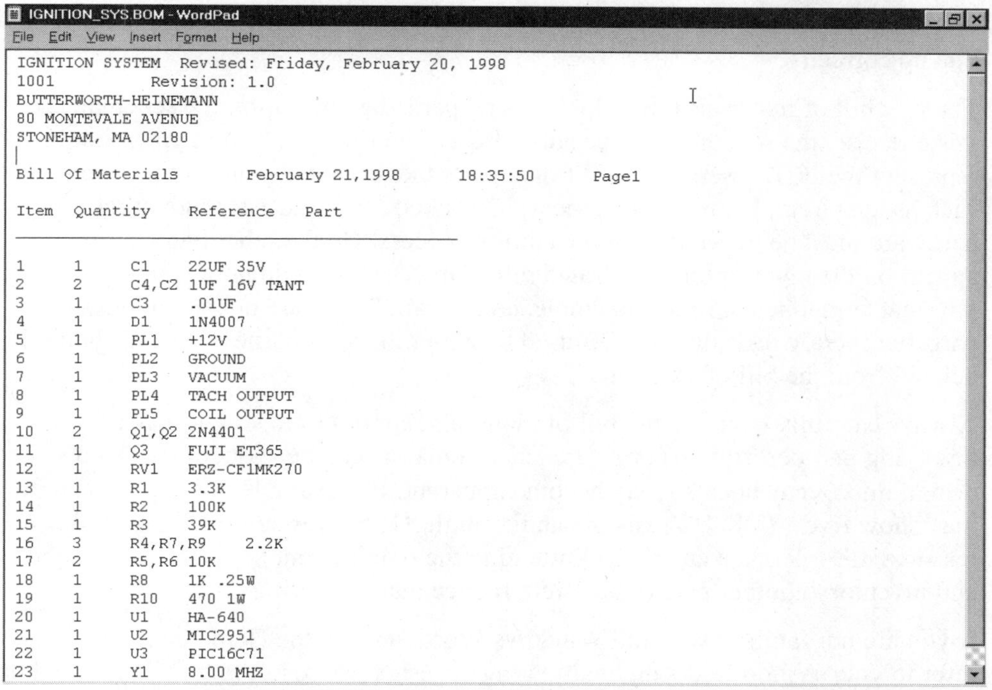

IGNITION_SYS.BOM - WordPad

File Edit View Insert Format Help

```
IGNITION SYSTEM  Revised: Friday, February 20, 1998
1001           Revision: 1.0
BUTTERWORTH-HEINEMANN
80 MONTEVALE AVENUE
STONEHAM, MA 02180

Bill Of Materials        February 21,1998      18:35:50       Page1

Item  Quantity    Reference   Part
______________________________________________

1      1      C1     22UF 35V
2      2      C4,C2  1UF 16V TANT
3      1      C3     .01UF
4      1      D1     1N4007
5      1      PL1    +12V
6      1      PL2    GROUND
7      1      PL3    VACUUM
8      1      PL4    TACH OUTPUT
9      1      PL5    COIL OUTPUT
10     2      Q1,Q2  2N4401
11     1      Q3     FUJI ET365
12     1      RV1    ERZ-CF1MK270
13     1      R1     3.3K
14     1      R2     100K
15     1      R3     39K
16     3      R4,R7,R9     2.2K
17     2      R5,R6  10K
18     1      R8     1K .25W
19     1      R10    470 1W
20     1      U1     HA-640
21     1      U2     MIC2951
22     1      U3     PIC16C71
23     1      Y1     8.00 MHZ
```

Figure 4-24 Bill of Materials Report in Windows WordPad

The body of the report is tab delimited, and the columns do not line up perfectly. If you want to print a neat copy, you must add tab characters. The identification header at the top of the report contains properties taken from the title block on the root schematic. Parts with the same description and reference designator prefix are grouped together (such as all 10K resistors or .1UF capacitors). Each line in the parts listing is given an item number and contains quantity, reference designators, and part value. The Bill of Materials tool alphabetically sorts the parts list by reference designator prefix. For example, capacitors with prefix C are followed by diodes with prefix D. Unfortunately, the tool does not effectively sort by part value.

You will have to expend considerable effort editing and cleaning up Capture bill of materials reports. Chapter 11 covers the use of a special sort utility that correctly sorts Capture-generated bills of materials by part value. The disk supplied with the book includes this sort utility. The bill of materials report file is in ASCII format. You can use most text editors, word-processing programs, and spreadsheet programs, including Microsoft Word and Excel, to edit the bill of materials report. This subject is also covered in detail in Chapter 11. If you use one of these programs, select a fixed pitch (nonproportional) font. Otherwise, columns will not line up correctly.

The raw bill of materials report includes all parts that appear on the schematic. In this exercise, the schematic represents a PCB. The final bill of materials should represent the PCB assembly. Additional parts that do not appear on the schematic, such as the circuit board, wire harness, IC sockets, and transistor mounting hardware must be added during the editing process. On the other hand, parts may appear on the raw Capture-generated bill of materials that do not actually exist as physical components. In this example, connectors PL1-5 are not actual physical parts but merely pads on the PCB used to solder the wire harness. PL1-5 should be deleted from the bill of materials.

Always carefully examine the bill of materials report. Often simple errors involving part descriptions or reference designators can be spotted. Potential design improvements also may become apparent. For example, the bill of materials may show five 1.00K 1% resistors and a single 1K 5% resistor. Making all the resistors 1% tolerance and thus eliminating the requirement to purchase, inspect, and inventory another part value might reduce manufacturing costs.

If you are not familiar with the Windows WordPad, use the Help command or refer to your Windows documentation (a suggested Microsoft reference is listed on page 41 in Chapter 2). Try printing the report file from WordPad. Use the Print command from the File menu in WordPad.

Using the Capture Text Editor

Capture provides a basic text editor that you can use to edit report files. You can also use the text editor to create or edit other types of ASCII text files, including VHDL and simulation models.

After you have run the parts cross reference and bill of materials reports, these files appear as outputs in Project Manager file view, as shown in Figure 4-25. Double click on the bill of materials report, ignition_sys.bom. This action launches the Capture text editor and loads the report file. The text editor window shown in Figure 4-26 appears.

The Capture text editor offers only limited capabilities. Font and tab settings are taken from the Capture preferences (see page 58 in Chapter 2). Standard file and printing commands are available from the tool bar or File menu. Cut and paste edit commands are available from the tool bar or Edit menu.

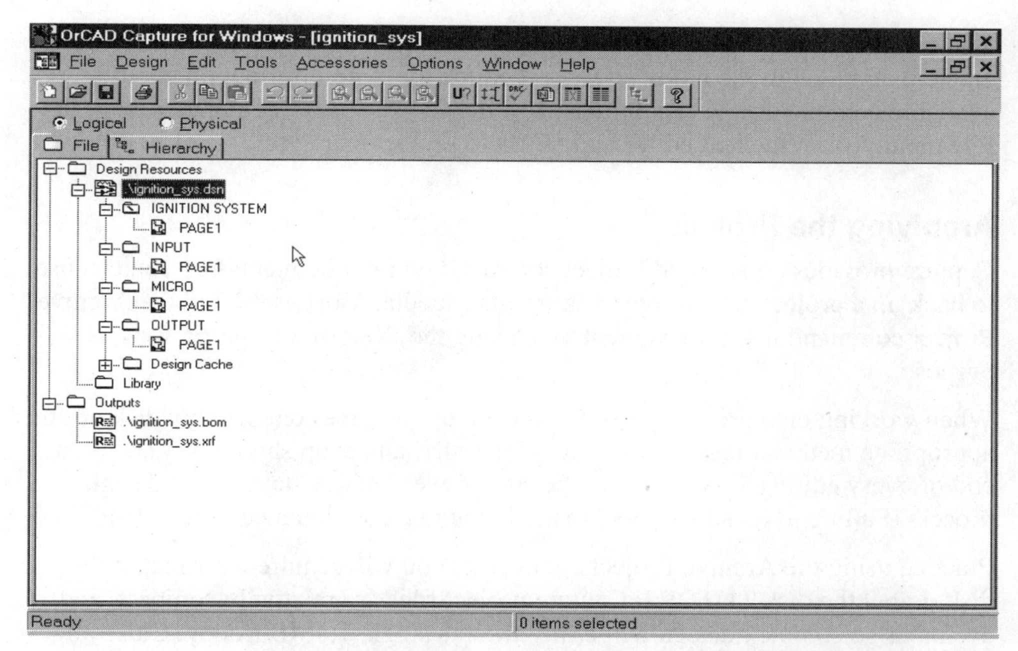

Figure 4-25 Project Manager File View with Report Outputs

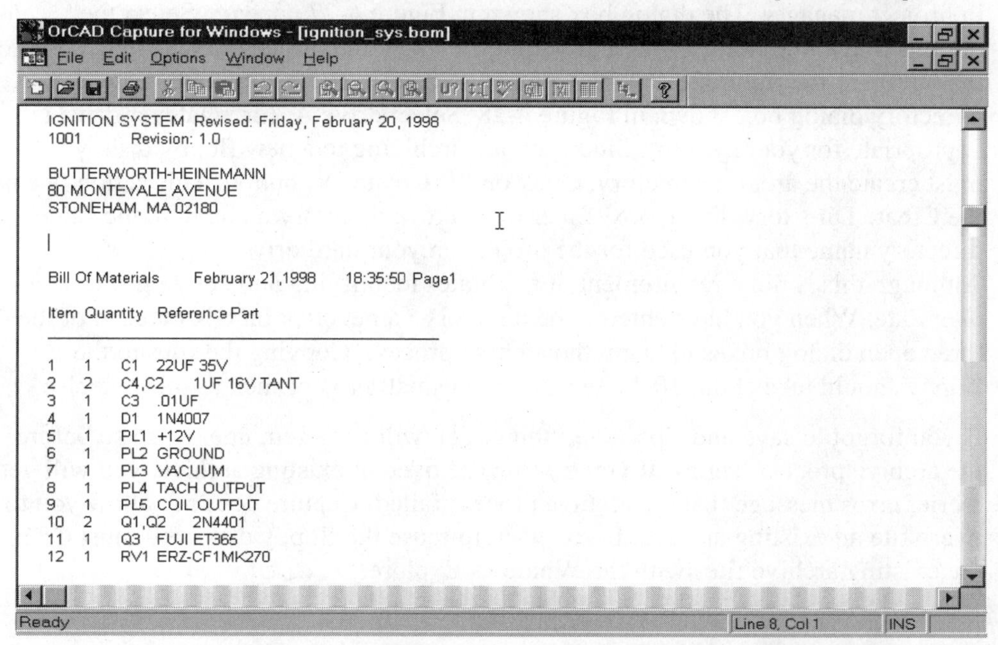

Figure 4-26 Bill of Materials File Loaded into the Text Editor

You can also access basic find and replace commands from the Edit menu. Click on Help and look in the index under text editor for more information about available commands and capabilities. Use the Close or Save commands from the File menu to exit the text editor and return to project manager.

Archiving the Project

Capture provides an Archive Project command on the File menu that you can use to back up a project onto floppy disk or other media. Most users find the Archive Project command more convenient than using the Windows Explorer for this purpose.

When working on a project, you should back up (archive) onto floppy disk or other appropriate media at regular intervals. Generally, a backup should be done at the end of every editing session and at the end of every major stage in the design process. Failure to do so will lead to the inevitable consequence of lost data.

Practice using the Archive Project command. You will require a formatted floppy disk. Label the disk TUTOR2. Capture project files are relatively compact, and even a large complex design will easily fit on a single 1.4MB floppy. Close all open schematics and files. Run the Archive Project command from the File menu in project manager. The dialog box shown in Figure 4-27 appears. Select the options to include library files and output files with the project archive. Click on the button to the right of the Archive Directory field. This action opens the Select Directory dialog box shown in Figure 4-28. Select drive A: (or whatever drive is appropriate for your system). Since you are archiving to a new floppy disk, you must create the archive directory. Click on the Create Dir button. This action opens the Create Directory dialog box. Enter the archive directory name. Use the same directory name that you used for the project on your hard drive, i.e., Tutor2. Although this is not a requirement, it facilitates identifying and copying files at a later date. When you have entered the directory name, click on OK on each of the three open dialog boxes to begin the archive process. Copying the files to the floppy should take about 10-15 seconds for a small project such as this exercise.

If you forgot to save and close everything, you will get a reminder to do so before the archive process begins. If you try to write over an existing archive, you will get a brief error message that the archive process failed. Capture does not allow you to overwrite an existing archive. If you want to reuse the floppy disk, you must delete the existing archive files with the Windows Explorer.

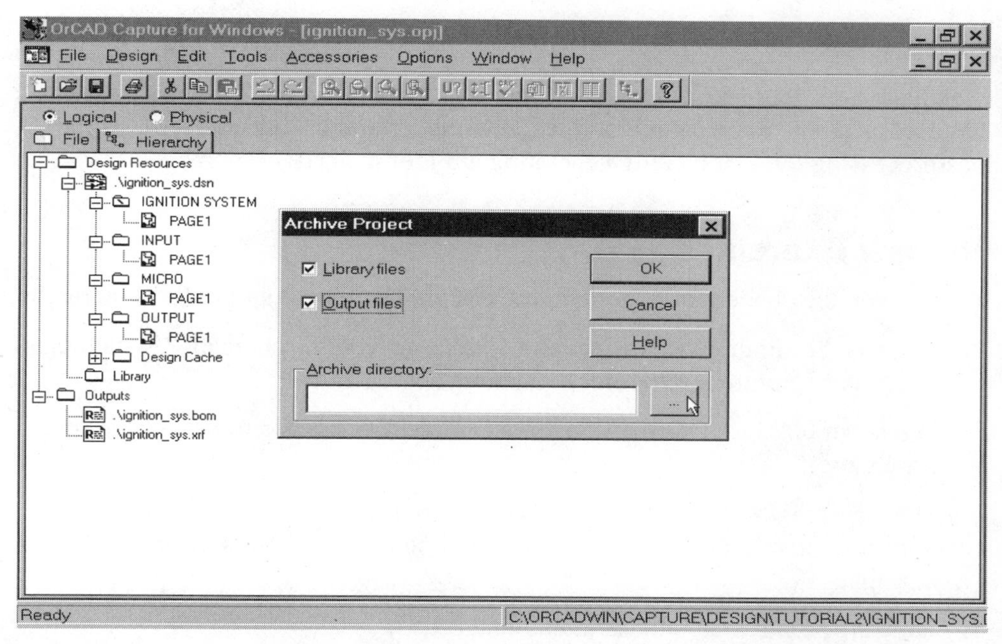

Figure 4-27 Dialog Box for the Archive Project Command

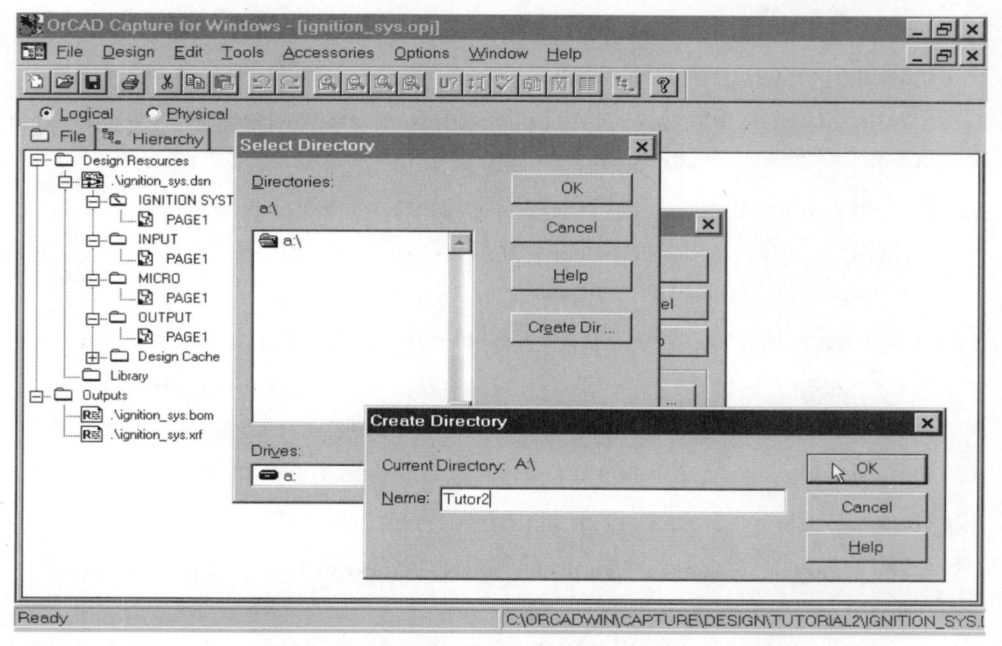

Figure 4-28 Entering the Archive Directory Path

Conclusion

You have now completed the hierarchical schematic exercise. At this point you have learned how to create hierarchical schematics and how to use some of the postprocessing tools, including generating a bill of materials report.

Review Exercises

1. Explain the concept of a hierarchical design. Describe a single-level hierarchy.
2. What is the significance of the terms *parent* and *child* schematic? What other term does Capture use to refer to a parent schematic?
3. What symbol does Project Manger use to identify the file folder for the root schematic?
4. How are hierarchical blocks used on the root schematic? What type of pins appear in hierarchical blocks? What corresponding symbol appears in the child schematics?
5. How are hierarchical blocks interconnected?
6. List and describe the various types of hierarchical pins. Describe three special types of output pins.
7. How do you navigate throughout a hierarchical design?
8. List and describe the various types of hierarchical ports commonly used. What are the suggested orientations?
9. Should you use hierarchical ports to represent connector pins?
10. Explain the concept of a net alias. What are the rules for placing net aliases? Can you copy and paste net aliases?
11. Which view in Project Manager gives an overview of the design structure?
12. How can you print an entire design? How can you select several pages for printing?
13. Describe two uses for the report generated by the Cross Reference Parts tool.
14. How can you generate a bill of materials report?
15. What Capture tools automatically load the Windows Notepad and WordPad for viewing output reports?
16. What tool does Capture provide for editing ASCII text files? How can you open a report file? How can you select the font properties?

17. Describe the suggested use of the Archive Project tool. What happens if you try to write over an existing Archive file?
18. What procedures does your organization recommend for file backup? If you are on a network, are certain areas of the network periodically backed up?

5

Postprocessing

At this point, you have learned most of the basic skills required to draw schematics using Capture. The exercise in this chapter continues to build on these basic skills and introduces additional postprocessing tools. The first session shows you how to create a complex schematic with address and data bus structures. The second session shows you how to use Capture's powerful postprocessing tools to check for electrical errors and to generate a netlist for input to a printed circuit board design system.

Your task will be to recreate the three-sheet hierarchical schematic shown in Figures 5-1 through 5-3. These figures represent the schematic for a PC XT (ISA bus) interface card. Make a copy of the figures and keep them handy for reference. Before starting, review the material in Chapter 1 on page 26 about bus structures.

First Session – Creating a Design with Bus Structures

The model for the exercise is a single-level hierarchy similar to the one that you completed in the previous chapter. The root schematic sheet shown in Figure 5-1 is a block diagram that shows interconnections between two child schematic sheets in the next level down.

Launch Capture from the Windows 95 desktop. The Capture session frame appears. Click on File and then New Design. Next, click on File and then Save. Create a new folder called Tutor3. Then save the design using the name PC_Interface.dsn.

The first task is to create the root schematic shown in Figure 5-1. Start by editing the title block. Use the title PC INTERFACE CARD. Next, draw the hierarchical blocks using the techniques that you learned in Chapter 4. Use the implementation names BUS DECODE and DIGITAL I/O. Place the hierarchical pins. All of the signals, except the data bus, are outputs on the BUS DECODE hierarchical block and inputs on the DIGITAL I/O block. Use the bidirectional signal type for the data bus. Use the name D[0..7]. Note the use of the square brackets and the two periods between the 0 and 7. This is one of the bus naming conventions allowed by Capture.

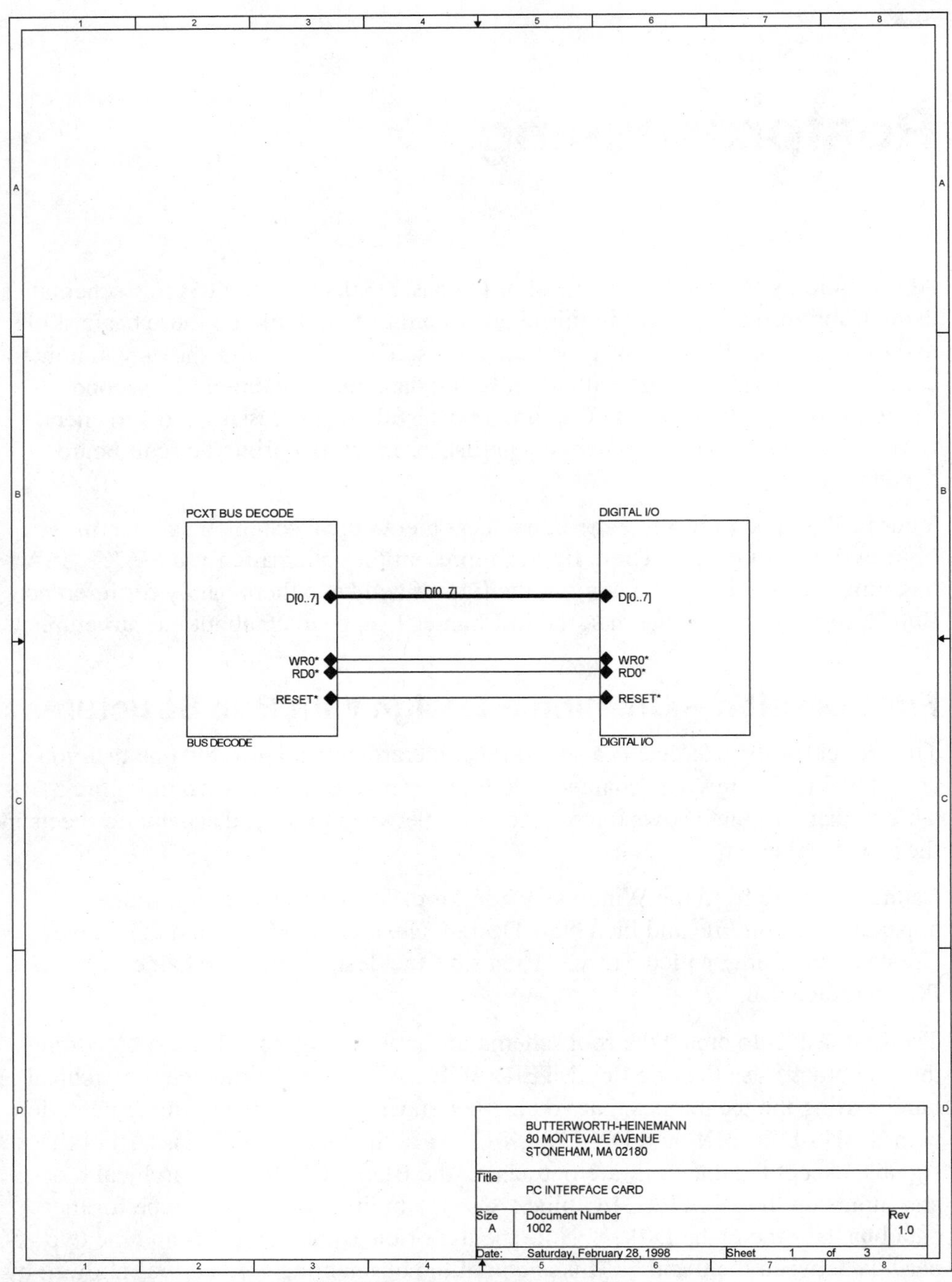

Figure 5-1 PC Interface Card (Sheet 1)

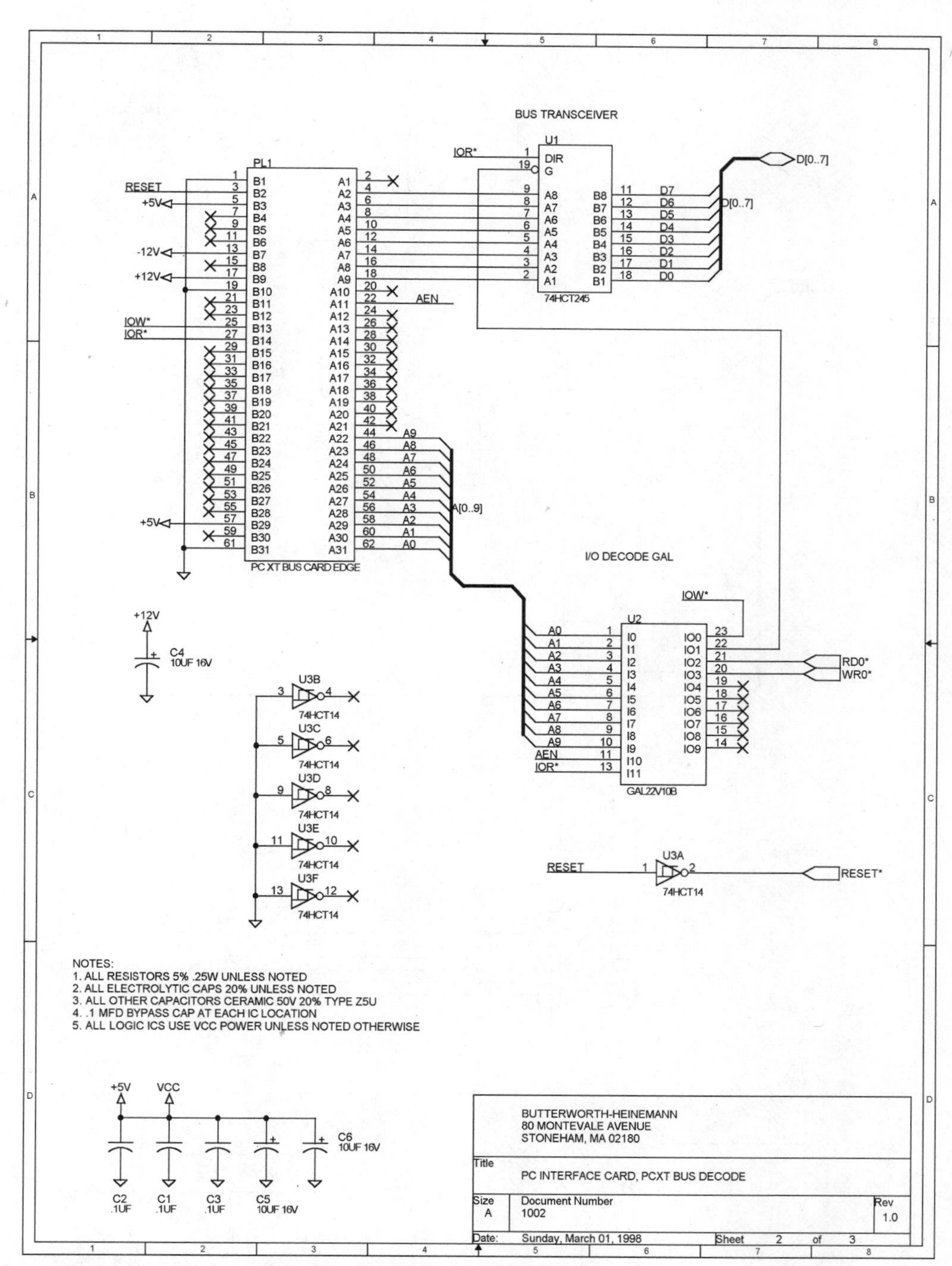

Figure 5-2 PC Interface Card (Sheet 2)

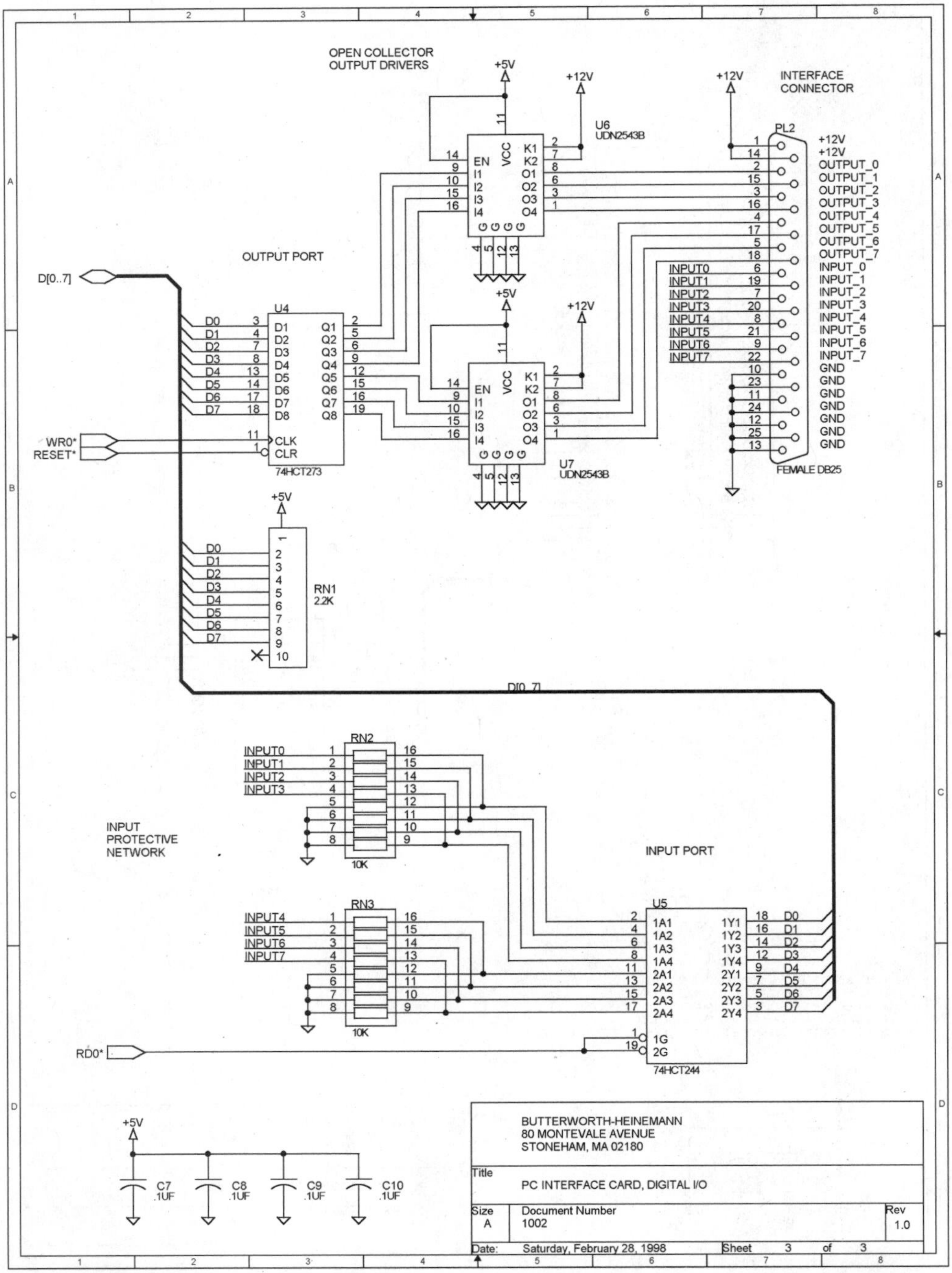
OPEN COLLECTOR
OUTPUT DRIVERS
+5V
+12V
INTERFACE
CONNECTOR
U6
UDN2543B
EN
VCC
K1
K2
I1
I2
I3
I4
O1
O2
O3
O4
PL2
+12V
+12V
OUTPUT_0
OUTPUT_1
OUTPUT_2
OUTPUT_3
OUTPUT_4
OUTPUT_5
OUTPUT_6
OUTPUT_7
INPUT_0
INPUT_1
INPUT_2
INPUT_3
INPUT_4
INPUT_5
INPUT_6
INPUT_7
GND
GND
GND
GND
GND
GND
GND
GND
FEMALE DB25
OUTPUT PORT
D[0..7]
U4
D0
D1
D2
D3
D4
D5
D6
D7
D8
Q1
Q2
Q3
Q4
Q5
Q6
Q7
Q8
WR0*
RESET*
CLK
CLR
74HCT273
U7
UDN2543B
INPUT0
INPUT1
INPUT2
INPUT3
INPUT4
INPUT5
INPUT6
INPUT7
RN1
2.2K
D[0..7]
RN2
10K
RN3
10K
INPUT
PROTECTIVE
NETWORK
INPUT PORT
U5
1A1
1A2
1A3
1A4
2A1
2A2
2A3
2A4
1Y1
1Y2
1Y3
1Y4
2Y1
2Y2
2Y3
2Y4
1G
2G
74HCT244
RD0*
C7
.1UF
C8
.1UF
C9
.1UF
C10
.1UF
BUTTERWORTH-HEINEMANN
80 MONTEVALE AVENUE
STONEHAM, MA 02180
Title
PC INTERFACE CARD, DIGITAL I/O
Size
A
Document Number
1002
Rev
1.0
Date: Saturday, February 28, 1998
Sheet 3 of 3

Figure 5-3 PC Interface Card (Sheet 3)

Draw the wires between the WR0*, RD0* and RESET* pins. Note that WR0* and RD0* end in a zero character followed by the star. Recall that the * character is one of the suggested standards for representing active low (inverted) signals.

The next step is to draw the data bus on sheet 1.

Bus Considerations

Bus objects are used to represent multiple signals, thereby eliminating clutter and improving readability. Good practice dictates always using bus objects to represent data and address signals that are routed between more than two ICs or that are routed between sheets.

Special rules apply to naming buses and bus signals, routing signals to and from buses, and routing buses between sheets:

- **Bus names**. Net aliases are used for naming buses. Every bus must have an associated net alias. The net alias name is of the form NAME[A..B] where NAME is any legal alphanumeric name, and A and B are integers that represent the wire numbers branching to and from the bus. Note that square brackets [ ] must be used. Two periods, a colon, or a dash can separate the numbers. Examples of valid names include: D[0..7], ADDR[16:31], and DATA[0-31]. A must be less than B. Negative numbers are not allowed. The name prefix should not end in a number. For example, do not use ADDR1 or DATA0 as this causes problems when generating a netlist.
- **Signal names**. Net aliases are also used to identify the signals branching from a bus. The signal net aliases are based on the bus net alias. For example, data bus D[0..7] has eight data signals D0, D1, through D7. A given signal can branch from a bus any number of times.
- **Bus entry**. Signals are normally routed to or from a bus via a special bus entry symbol, which appears as a 45-degree diagonal slash. Wires that cross a bus or end at a bus are not electrically connected to the bus unless you use a bus entry symbol or a junction symbol. Note that unlike OrCAD SDT, Capture allows you to use junction symbols in order to connect wires to a bus. However, this may lead to confusion. Suggested practice is to make all bus connections by means of bus entry symbols.
- **Bus signal net aliases on wire stubs**. If you draw a wire stub going to a pin and then place a net alias with a bus signal name on the wire, the wire and pin will be implicitly connected to the bus.

- **Bus routing**. Sections of the same bus can be joined together in a Y or can "cross." Capture will automatically place a junction at the intersection. Different buses that cross one another do not join, just as with wires. Bus are routed between sheets by module ports, again just as with wires, except the name of the module port must correspond exactly with the bus name.

- **Drawing conventions**. To help differentiate buses from wires, buses are traditionally drawn with 45-degree beveled corners. A 45-degree split is generally preferred at the point where a bus splits into two separate sections. You can draw a nonorthogonal bus segment by holding down the SHIFT key. Two styles of bus entry appearing as / or \ are available. Pick the entry style that flows with the direction of the bus. Use Figures 5-2 and 5-3 as examples of proper flow.

Completing the First Sheet

To complete the first sheet, draw the bus between the two hierarchical blocks as shown in Figure 5-1. Bus objects are drawn similarly to wires, except that you use the Bus tool from the tool palette or the Bus command from the Place menu. Use the Net Alias tool or command to place the bus name. Sheet 1 is now complete.

Starting the Second and Third Sheets

Use the skills you learned in the previous chapter to start the remaining two sheets. At this point you should be able to navigate the hierarchy to open the sheets; place parts, power, and ground symbols, no-connect symbols, wires, junctions, net aliases, and text; and edit the part descriptions and title block areas. Use the schematic names shown in Figure 5-1. Leave the reference designators alone for now, as you will be learning to use the Update Part References tool, which automatically numbers the reference designators in sequence. Also leave the buses, bus entries, and wires for bus signals for now, as drawing these objects will be covered in more detail later.

The following are some hints on new parts:

- **IBM PC XT bus card edge connector**: CONNECTOR IBM PC XT from the Custom library.

- **Female DB 25 pin connector**: CONNECTOR DB25 from the Device library. Vertically mirror.

- **74HCT series logic ICs**: these are in the TTL logic library Ttl.olb (since they have TTL family pinouts). Use the suffix number preceded by the star (*) wild card character to help locate parts. For example, *244 to find the 74HCT244. You will then get a list of the different "244" versions (such as 74LS244 or 74HCT44). Use the 74HCT14 part from the Custom library, which is a smaller outline part than that in the TTL library.

- **GAL22V10B generic array logic IC**: GAL22V10B from the Custom library.
- **UDN2543B power driver IC:** UDN2543B from the Custom library.
- **Resistor networks**: 10-pin SIP is RN10 and 16 pin DIP is RN16 DIP ISO, both from the Custom library.

Power Pins

You will recall from chapter 1 that Capture does not normally show ground and power pins for many logic ICs. Most TTL logic ICs have ground pins defined as GND and power pins defined as VCC. CMOS ICs use VSS and VDD. The 74HCT series ICs come from the TTL library (even though they are CMOS parts) and still use the original TTL ground and power pin names. The GAL22V10B also uses GND and VCC.

With today's trend toward multiple voltage designs, invisible ground and power pins can prove to be a real nuisance. If you need to check or verify invisible power pin names for a part, zoom in close, double click to edit part properties, and temporarily select the power pins visible option.

Invisible ground and power pins are automatically tied to ground and power planes with the same name. All the parts in this exercise that have invisible pins use GND and VCC. GND pins are no factor, because all the ground objects are also named GND and tied to the same plane. However the +5 volt power supply is a different situation. Visible +5 volt power objects on bypass capacitors and pull-up resistors are named +5V, as is the incoming +5 volt supply on the card edge connector. The VCC plane must be connected to this +5 volt plane. You can accomplish this by placing a VCC power symbol on the sheet and connecting it to a +5V power symbol. Refer to the lower left-hand corner of Figure 5-2. Note that you need only to do this once in the design, not on every sheet.

An alternative approach would have been to use power symbols named VCC throughout the entire schematic and then add a note indicating that the VCC supply is +5V. This was once common practice. In light of today's trend toward multiple voltage levels, you should always use the actual power supply voltage as the name for the power symbol to reduce the chance of misinterpretation.

Completing the Data Buses

Your last task in completing the second and third sheets is drawing the data bus structures. The term *bus structure* refers to the data bus and associated entries, wires, and net aliases. Let's use the data bus structure at the top right side of sheet two as an example. First, draw the bus and place the net alias that identifies the bus as shown in Figure 5-4. Since the bus entry for D0 will angle down, the bus ends one grid location above pin 18 on U1. In practice, you cannot always predetermine precise end points for a bus and may need to drag the end points after completing the bus structure.

The next step is to draw the first bus entry and wire. For reasons that will become apparent in a moment, always start with the lowest numbered element, D0 in this case. Use the Bus Entry tool or Place Bus Entry command. When you click on the tool, the bus entry initially appears in the wrong orientation for this particular situation. Press the right mouse button to bring up the shortcut menu shown in Figure 5-5 and click on the Rotate option. This will rotate the bus entry 90 degrees into the correct orientation for D0. Drag the bus entry to the bus and place it into position as shown in Figure 5-6. Then draw the wire from pin 18 to the bus entry.

You can use the Copy and Repeat commands to quickly copy the bus entries and wires for D1 through D7. Drag the mouse pointer over the first bus entry and wire to select these objects. Then hold down the CTRL key and drag a copy into position for D1 as shown in Figure 5-7. Make sure you keep the copy selected (highlighted with resize handles visible) as shown in the figure. Next, use the Repeat Copy command from the Edit menu. Each time you use Repeat Copy, a new bus entry and wire will appear one grid position above the previous copy. Capture remembers the offset you used for the first copy operation (D1) and repeats this offset for subsequent copies. Note that the Repeat command is context sensitive. If you previously copied an object or selection, the command appears as Repeat Copy on the Edit menu. Complete the remaining data lines through D7.

The final step is to add the net aliases for D0 through D7. Place the net alias for D0 as shown in Figure 5-8. As with most other commands, the Net Alias command remains modal. When placing successive net aliases ending with a numerical suffix, Capture automatically increments the suffix each time you place another net alias. When you start with D0, you can just keep clicking to place D1 through D7. This feature saves a considerable amount of time. When you are finished, the data bus structure should appear as shown in Figure 5-8.

Complete the second and third sheets by drawing the remaining bus structures using the time saving techniques just described.

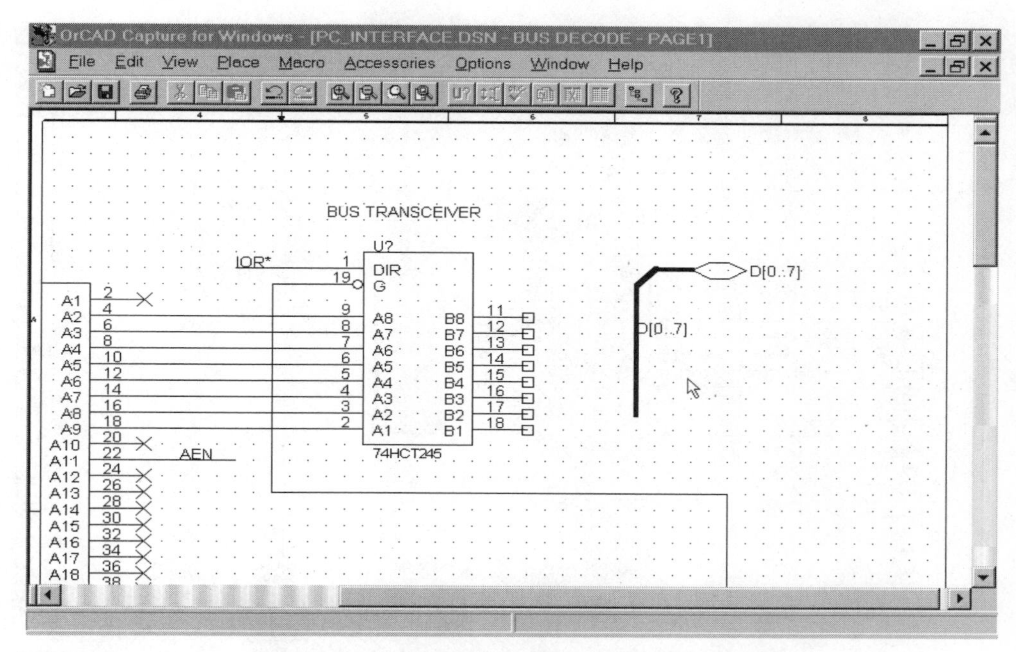

Figure 5-4 Starting the Data Bus on Sheet 2

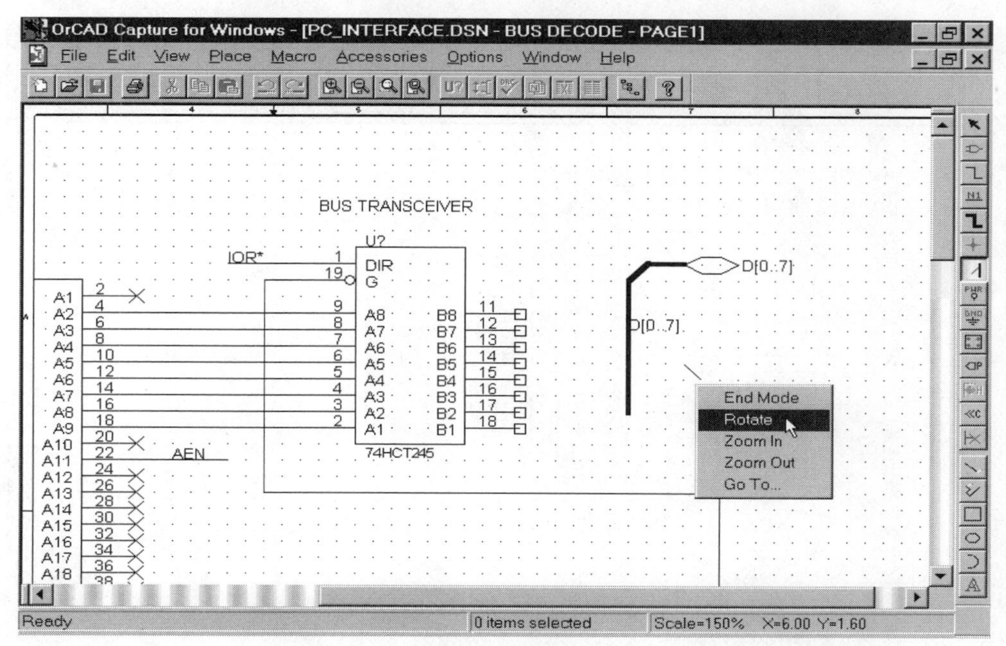

Figure 5-5 Rotating an Entry Object

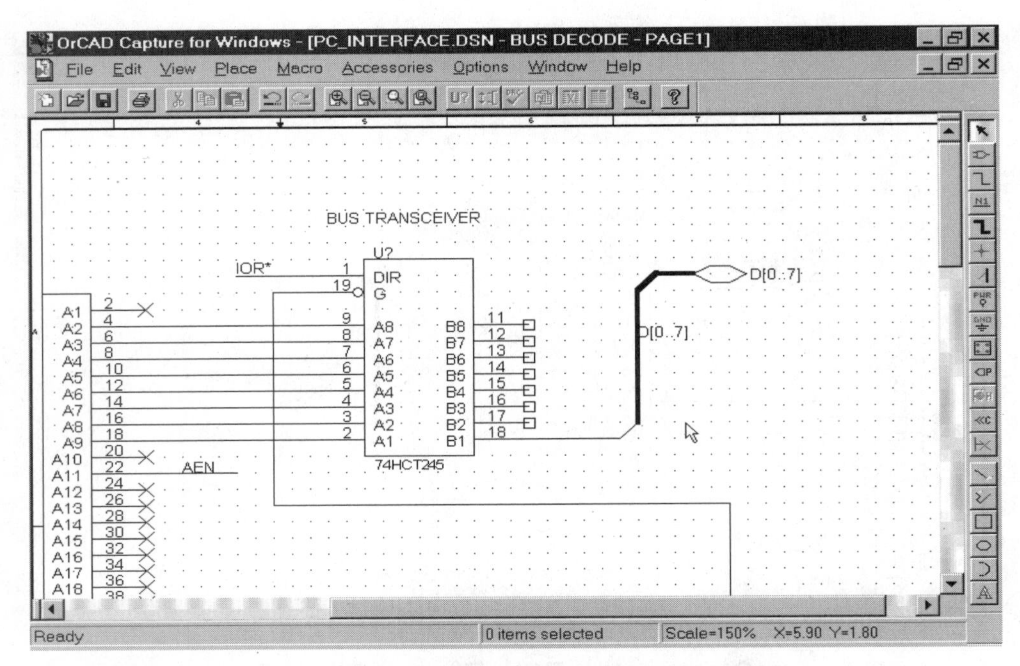

Figure 5-6 Completed First Bus Entry and Wire

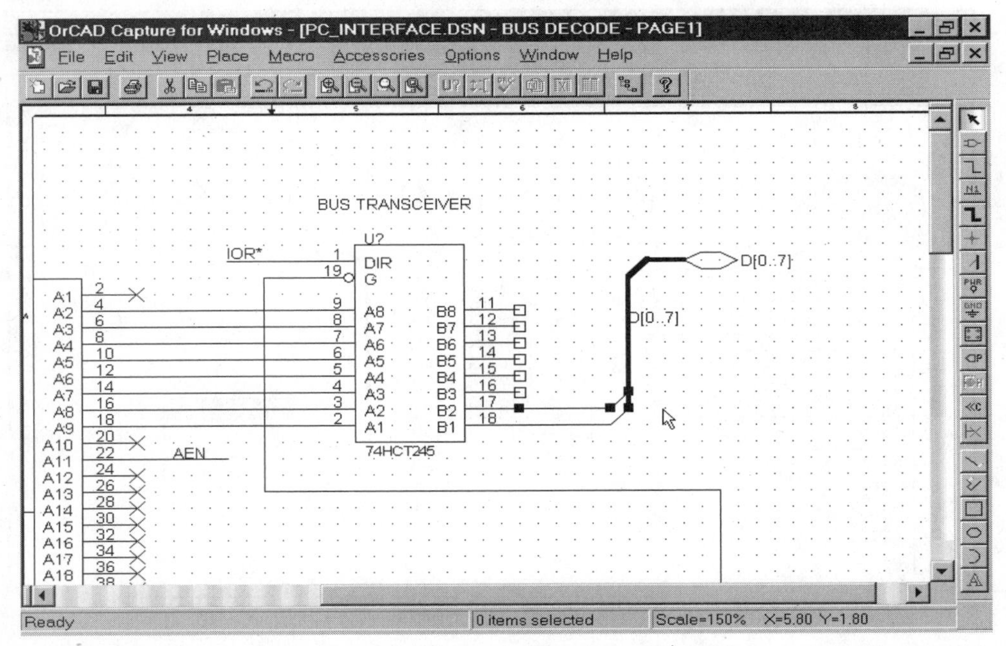

Figure 5-7 Setting Up to Use the Repeat Copy Command

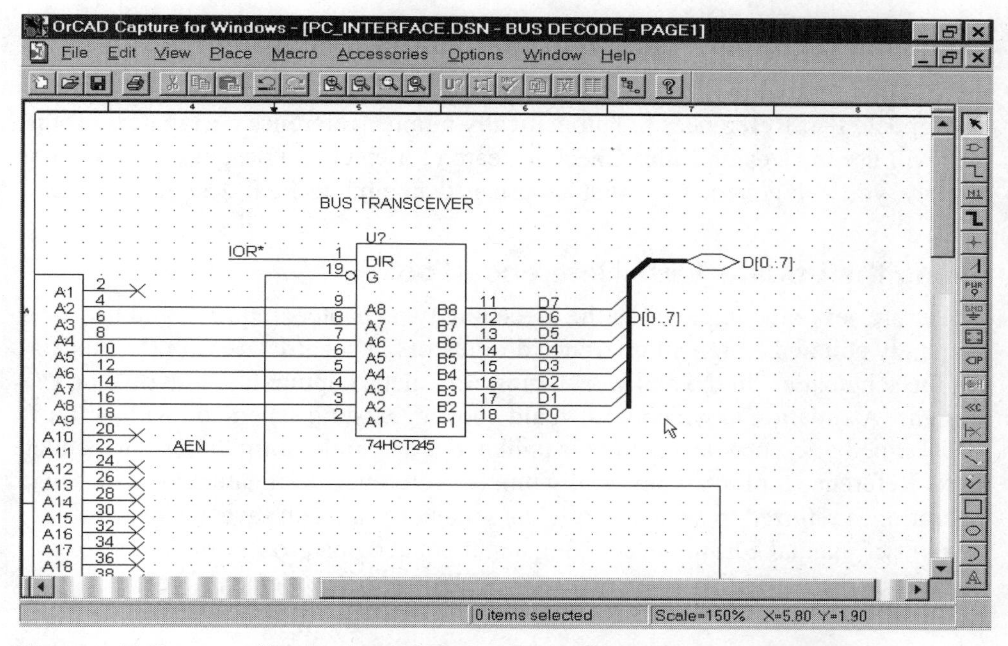

Figure 5-8 Completed Data Bus Structure on Sheet 2

Wrapping up the First Session

You have now finished most of the “manual labor” related to capturing the schematic for this exercise. Save and close the sheets and use Project Manager to print a hard copy. Carefully check your work and make any required corrections At this point all three sheets should appear exactly as in Figures 5-1 through 5-3 except for the reference designators, which will be automatically assigned in the next session.

Before proceeding, make an archive copy of the entire project to floppy disk.

Second Session – Postprocessing

The second session introduces several new postprocessing tools. You will learn to use Update Part References to automatically number reference designators. Then you will use the Design Rules Check to search for electrical design rule violations. Finally, you will generate a netlist for use as data input to PCB design software.

Using the Update Parts Reference Tool

The exercises you completed in the preceding two chapters required you to manually edit and assign all reference designators in accordance with the models for the schematics. In the real world, most schematic capture tasks involve new designs. A situation in which you would copy an existing schematic with preassigned reference designators is unlikely. You would normally use the Update Parts Reference tool to automatically number reference designator suffixes. This operation is referred to as *annotating the schematic* and can save a considerable amount of manual editing time. Using an automated postprocessing tool also eliminates the possibility of errors, such as skipped or duplicate reference designators.

The Update Parts Reference tool offers a high degree of flexibility. One extremely useful feature is the capability to perform an incremental annotation. In this case, only reference designators that still have an unassigned suffix (such as R?) are annotated. The tool automatically starts numbering from the last assigned reference designator. The need for incremental annotation often arises when an engineering change is made to an existing product and parts are added to the PCB.

Use the Update Parts Reference tool to annotate the design that you completed in the first session of this exercise. In Project Manager, click on the design. Note that the postprocessing tools near right side of the toolbar now become available. Click on the Update Parts Reference tool. The dialog box shown in Figure 5-9 appears. The following options are available:

- **Scope**. You can select whether the tool processes the entire design or just the selected schematic folders or pages. In most cases, you should avoid processing individual folders or pages as this can easily result in duplicate reference designators in an unselected part of the design.
- **Action**. Incremental reference update is the safest course of action. In many cases, you will preassign reference designators for certain parts such as connectors. Incremental update does not affect preassigned parts. Incremental update is also useful when performing engineering changes. The only caveat is that numbering start from the highest pre-assigned value. Gaps are not filled.

For example if the design contains R1, R2, and R4, incremental update will assign new resistors starting with R5. Unconditional update is useful for cleaning up a design with gaps in assignments or serious errors such as duplicate assignments. The last option, which resets parts references, can be used to clean up a schematic that requires preassignments. First reset everything back to a "?," make the required preassignments and then run an incremental update to complete the task.

- **Physical Packaging.** You can enter the properties that must match for parts to be grouped in a single package. In most cases, the default values give proper results. For more information, click on Help within the dialog box.
- **Reset reference numbers option**. Starts numbering at "1" on each page. This option has limited usefulness. Do not use this option if you plan to run a bill of materials listing or PCB design netlist for the entire design.
- **Renumber all pages**. This option also has limited usefulness, as most users prefer to manually assign pages in some logical order.

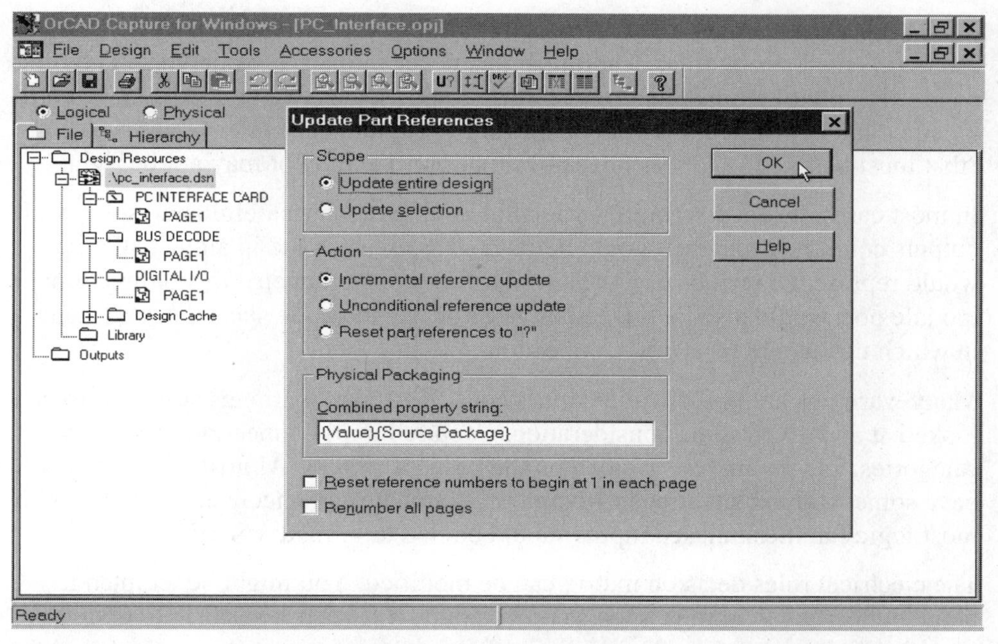

Figure 5-9 Dialog Box for the Update Parts Reference Tool

After you have selected the appropriate options as shown in Figure 5-9, click on OK to run the tool. A warning screen appears and asks if you want to continue. Click on OK again. When the tool has completed processing, another warning screen appears and asks if you want to save changes to your project. Click on OK. Your entire schematic is now annotated. Print out a hardcopy and examine it. The reference designators Capture assigned for your schematic may not exactly match those in Figures 5-2 and 5-3.

Creating a Bill of Materials

Use the techniques learned in Chapter 4 to run the Bill of Materials tool. Print out the bill of materials and carefully examine it. At this point, you can catch many simple errors such as wrong part descriptions or reference designator assignments. If you find any errors, go back and correct the design. Remember always to save and back up all your project before making major changes or going on to the next postprocessing step.

Electrical Design Rules Check Overview

The Design Rules Check (DRC) tool checks for possible violations of basic electrical connectivity rules. Capture uses a decision matrix to analyze all possible connection permutations. Incorrect connections, such as two outputs connected together can be flagged as errors or warnings. Capture defines errors as situations "that must be fixed" and warnings as "situations that may or may not be right."

In most cases, errors do require correction or at least very careful analysis. Two IC outputs connected together would be flagged as an error and in almost all cases would represent a serious design flaw. An IC output connected to a bidirectional module port would also be flagged as an error, yet there are special circumstances in which this would be a valid connection.

Many warnings are in fact valid connections. However, warnings still need to be looked at and given some consideration. Capture lumps connections into broad categories, and no analysis is done on the basis of signals. Almost all designs will have some warning situations. A common example is an unconnected input. With most logic families, unused inputs should be tied to ground or +5V.

The electrical rules decision matrix can be modified. You might be tempted to blank out certain warnings, such as open inputs. But what if a sensitive CMOS input is left floating? Experience shows that for the most part Capture's default decision matrix is well thought out and best left alone. An occasional nuisance error or warning is preferable to letting a serious mistake slip by.

Running Design Rules Check

Before running the tool, let's deliberately create an error in the design for the tool to catch. From Project Manager, double click on the PCXT Bus Decode schematic (sheet 2) to open it. Then add a wire connecting the top two unused 74HCT14 inverter outputs as shown in Figure 5-10. Save and close the schematic page to return to Project Manager.

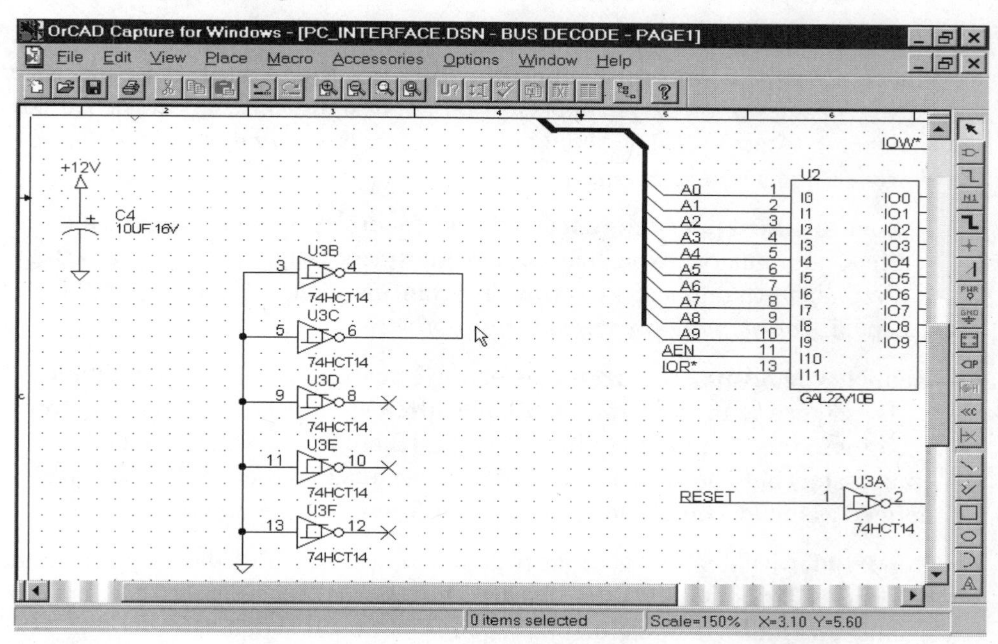

Figure 5-10 Creating a Deliberate Error in Sheet 2

In Project Manager, click on the design. Next click on the DRC tool. The dialog box shown in Figure 5-11 appears. The following options are available:

- **Scope**. You can select whether the tool processes the entire design or just the selected schematic folders or pages.
- **Action**. Check design rules is the default. Violations are flagged with DRC markers that appear on the schematic page. You can also run the DRC tool to delete existing DRC markers. If you select this action, the report options are dimmed out and unavailable.

The following additional report options are available:

- **Create DRC markers for warnings.** Note that errors are always flagged with markers. You also can use this option to flag warnings. Most users find that that the markers are not nearly as useful as the report generated by DRC.
- **Check hierarchical port connections**. Verifies one-to-one correspondence between the name and type of hierarchical pins on the parent schematic and hierarchical ports on the child schematics. *You should always select this option for hierarchical designs.*
- **Check off-page connector connections**. Verifies one-to one correspondenece between names of off-page connectors on flat designs (this subject is covered in detail in Chapter 7). *You should always select this option for any multipage designs using off-page connectors.*
- **Report identical part references**. Checks for errors involving duplicate reference designators. Note that Capture considers assignments such as U1 and U1A to identical duplicates. Normal assignments such as U1A and U1B are accepted. *You should always select this option.*
- **Report type mismatch parts**. This option catches errors in which multiple parts assigned to the same package have different properties, such as different PCB footprints. For example if U1A and U1B are both LM2903 (a dual comparator) but one has an 8-pin DIP and the other an 8-pin SOIC package defined as the PCB footprint. *You should always select this option.*
- **Report hierarchical ports and off-page connectors**. Lists these objects in the report file. This option is useful in some situations, such as cross checking against a written specification that requires particular signal names.
- **Check unconnected nets**. Checks for floating pins, including any nets without a drive signal. Also checks for identically named nets on different pages that are not properly interconnected with hierarchical ports or off-page connectors. *You should always select this option.*
- **Check SDT compatibility**. Checks for certain objects and properties that cannot be written to an older OrCAD SDT format file. Only useful if you plan to transfer designs to another system running SDT.
- **Report off-grid objects**. Lists all off-grid objects. This does not include part values and reference designators, which are frequently placed off-grid for neat alignment. Off-grid electrical objects can cause serious problems, such as unconnected nets. *You should always select this option.*
- **Report all net names**. Net names refers to net aliases. This option is useful in some situations, such as cross checking against a written specification that

requires particular signal names. If you do not assign specific net aliases, Capture will assign random names and the report information will be of little use.

- **Report File**. If you select the option to view the output, Capture launches the Windows Notepad with the report file. The report file path defaults to the current design directory, and the file name defaults to the design name with a .DRC extension. You can change these defaults if required. *You should select the View Output option.*

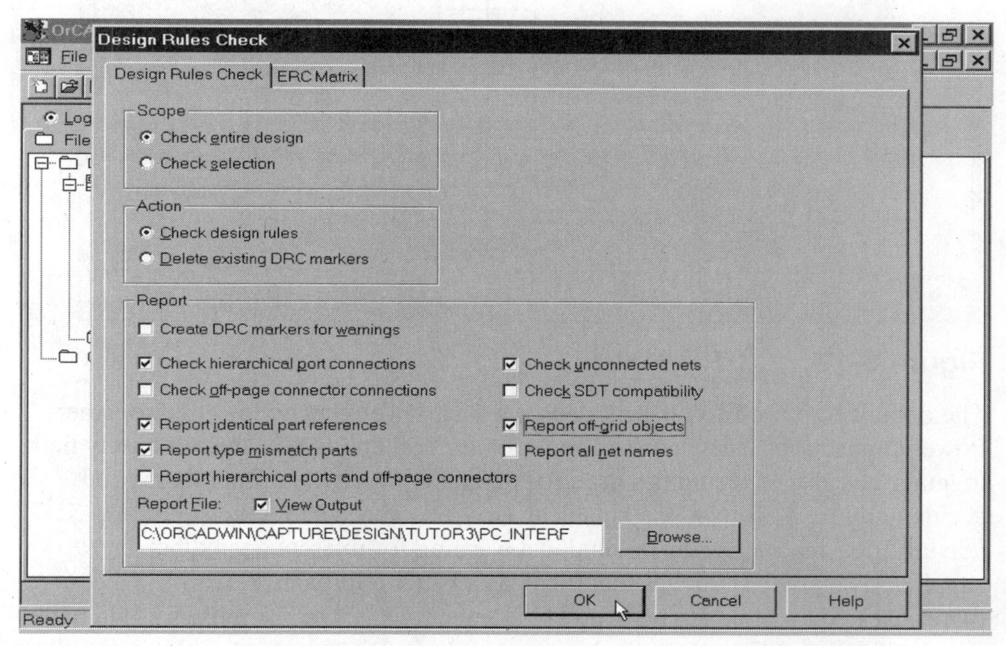

Figure 5-11 Dialog Box for Design Rules Check Tool

Select the options shown in Figure 5-11 before proceeding. Then click on the ERC Matrix (Electrical Rules Check) tab at the top of the dialog box. This brings up the ERC matrix shown in Figure 5-12. Empty blocks on the decision matrix represent valid connections. Blocks labeled W or E represent warnings and errors. If you click on a box, it cycles through empty, W, and E. Note that ERC matrix screen also provides a box near the bottom of the screen on which you can click to restore the defaults for the matrix.

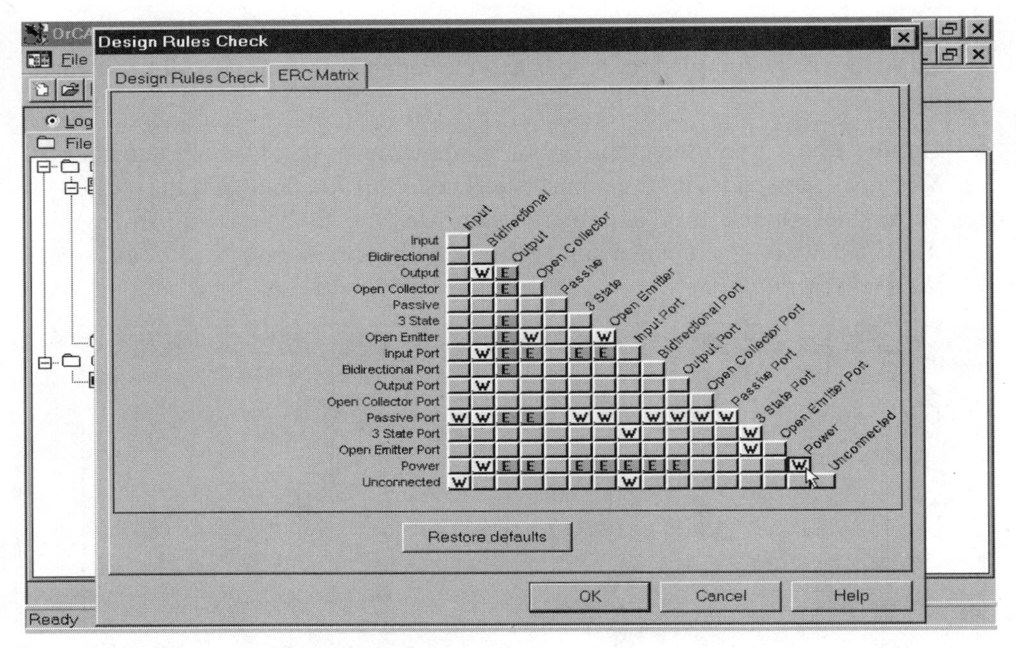

Figure 5-12 ERC Matrix

The default Capture ERC matrix does not flag power connections tied together. Power connections, such as VCC and +5V or VSS and GND, are commonly tied together because of the names used for invisible power pins. But what if you accidentally tie VSS to +5V? This is an easy mistake. The previous OrCAD SDT version ERC matrix defaults included a warning for interconnected power supplies. The author believes that this was a better approach. A warning message that causes you to double check power connections is a minor nuisance compared with a prototype PCB that goes up in smoke.

Click on the box at the intersection of the power row and column to set a warning for interconnected power supplies as shown in Figure 5-12. Then click on OK to return to the DRC dialog box.

Begin the DRC process by clicking on OK in the DRC dialog box. For a small design, such as the exercise in this chapter, the DRC tool only takes a few seconds to complete the check. When the tool has completed processing the report, the Windows Notepad is automatically launched and appears as shown in Figure 5-13.

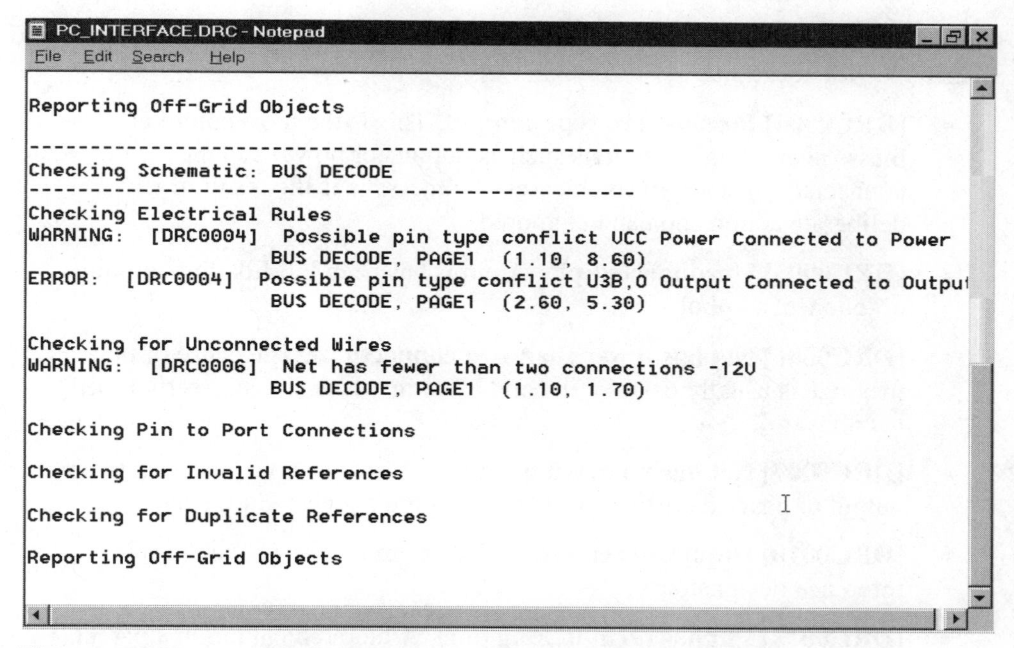

Figure 5-13 DRC Report in Windows Notepad

Check Electrical Rules flags any errors and warnings (if this option is selected) by placing a donut-shaped DRC marker at the affected location on the schematic. All errors and warnings and any additional information generated for selected reporting options are written to the DRC report file in the form of ASCII text.

Note that two warnings and one error message appear for the BUS DECODE schematic. The first warning has flagged the interconnected VCC and +5V power supplies (part of the line scrolls off the screen). You can ignore this warning, since you interconnected these supplies to provide power to invisible IC pins. The error message refers to the signal conflict deliberately created by tying the 74HCT14 inverter outputs together in Figure 5-10. This is a serious error which must be corrected. The second warning flags the −12V supply at position B7 of the PCXT bus edge connector. The −12V supply is present at the pin but not used. This warning can be ignored.

Use Notepad to print out a hardcopy of the DRC report. You should always do this if any error or warning messages occur. Cross out the messages as you dismiss them or correct the underlying mistakes. Then rerun the DRC tool.

The following are some of the more common message codes. Note that a given message can be either a warning or an error depending on the ERC matrix. As shown in Figure 5-13, the messages typically include signal, part, or pin names and schematic X,Y locations.

- **[DRC0004] Possible pin type conflict.** This is the most common message and flags situations such as outputs or power supplies connected together. In some cases, the message is the result of a deliberate action and can be ignored.
- **[DRC0005] Unconnected pin**. A pin is has been left floating without a no-connect symbol.
- **[DRC0006] Net has fewer than two connections.** The cause of this message is usually easy to find and often involves an incorrectly typed net alias.
- **[DRC0007] Net has no driving source**. A net has no connection to any output or passive type pin, or two inputs are connected together.
- **[DRC0010] Duplicate reference**. Two or more parts have the same reference designator.
- **[DRC0012] Pin has no matching port**. A hierarchical pin in a parent schematic has no corresponding hierarchical port in a child schematic. This message usually occurs with DRC0013.
- **[DRC0013] Port has no matching pin**. A hierarchical port in a child schematic has no corresponding hierarchical pin in the parent schematic. This message usually occurs with DRC0012.

After examining the DRC report, exit Notepad and open the BUS DECODE schematic. You should see a donut-shaped DRC marker on one of the outputs of the 74HCT14 inverters you tied together. If you double click on the DRC marker, a message window appears as shown in Figure 5-14.

Correct the deliberate error by deleting the wire between the two outputs of the 74HCT14. The next step is to remove the DRC marker. Markers can be deleted using the Delete command, just like any other object. The disadvantage of this approach is that a complex schematic may have many DRC markers related to trivial warnings that you are going to ignore. Finding and manually deleting all these markers is tedious and time consuming. A much easier approach is to rerun the DRC tool with the delete markers option selected. This will automatically remove all markers.

Go ahead and run the DRC tool again to remove the DRC marker on sheet 2.

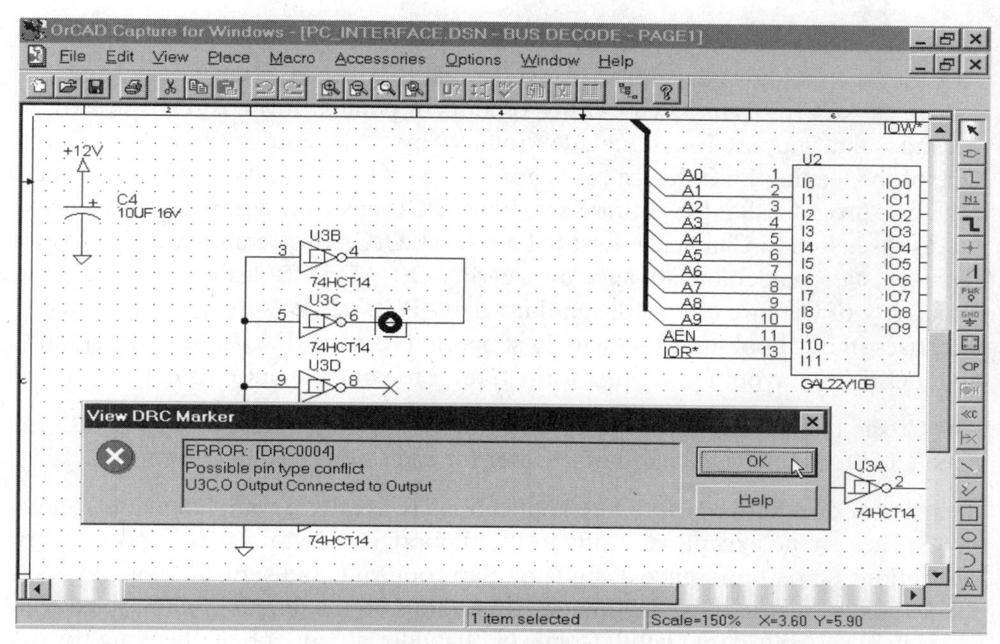

Figure 5-14 Viewing a DRC Marker Message

Netlist Overview

The final topic in this chapter is netlist generation. This is one of the major benefits of schematic capture with a program such as OrCAD. The primary use for a netlist is as input to a PCB design system. Netlists also can be used as input to circuit analysis programs such as SPICE or as input to PLD (programmable logic design) programs. The focus of this exercise is on generating a netlist for third-party PCB design software.

The term *netlist* is used for a file that contains data representing the parts and electrical interconnections found in a design. The electrical interconnections are referred to as *nets*. A net consists of part pins (referred to as *nodes*) that are connected together. Power and ground are also considered to be nets. Parts properties include the reference designator, part value, and PCB footprint. The PCB footprint refers to the part outline and pad pattern that appears on the PCB.

No industry standard exists for netlist data formats. Most PCB design software vendors use their own proprietary format, and may support one or two other popular formats. At last count, Capture supports more than forty different netlist formats. Many of these formats are obsolete, the original vendors having long since gone out of business or having reorganized. The most widely supported netlist format for low-cost Windows 95 based PCB design software is Tango. ACCEL Technologies developed Tango format for their PCB design package, known as ACCEL Tango PCB. Other vendors of popular low cost PCB design packages that support Tango include Protel 98 by Protel and Eagle by Cadsoft. One of the leading PCB design packages is PowerPCB by Pads Software. This is a high-end program that the author uses. PowerPCB has a proprietary netlist format that is discussed in Chapter 8. Last but not least, OrCAD offers Layout Plus, which represents a considerable advance over earlier OrCAD PCB design software. Since a majority of Capture users still run third-party PCB design software, the netlist exercises in this book focus on widely supported formats. For more information about OrCAD Layout Plus netlist format, refer to your Layout Plus documentation.

The Create Netlist tool is used to generate netlists from Capture design data. Create Netlist includes a unique formatter for each supported netlist format.

Transfer of data between software applications is usually a complex matter, and netlist files are no exception. Many of the supported netlist formats impose severe restrictions on part reference designator, part footprint, and signal (net alias) names. These restrictions may include limitations on the length of names and what characters are considered valid. Create Netlist does a good job of checking for invalid names and attempts to make corrections where possible.

In most cases, the user can expect to spend some time manually editing the netlist file generated by Capture (or any other schematic software package) before transferring the data. Time spent editing the netlist file is often a good tradeoff for even more time that would otherwise have to be spent setting things up in Capture or in the PCB design software. You must make sure that the library parts defined in Capture will match those defined in the PCB design software. Both footprints and pin names must match. Enhancements to Capture, such as a spreadsheet-style properties editing tool, greatly simply this task compared with earlier OrCAD SDT versions.

Pin Names and Pin Numbers

The subject of pin names and pin numbers requires additional discussion. Pin names and numbers are pin properties that are discussed in detail in Chapter 6, which covers the library part editor. All Capture parts have pin names, but some parts are defined without pin numbers. ICs are examples of parts that have both pin names and pin numbers. Most discrete parts, including capacitors, inductors, resistors, diodes, and transistors, are examples of parts that have no pin numbers. These parts have pin names, but the pin names are invisible.

Users sometimes become confused by the use of numeric pin names in Capture library parts. Capacitor, inductor, and resistor pins are named 1 and 2. Polarized electrolytic capacitors have pin number 1 assigned to the positive pin. Discrete semiconductor devices typically have alphanumeric pin names. Diode pins are named CATHODE and ANODE. Transistor pins are named EMITTER, BASE, and COLLECTOR. In all of these examples of discrete parts, the information is in the pin name field. The pin number field is not used.

Netlist formats may use either the pin name or pin number field to identify pins. If the netlist format uses pin numbers and the pin number field for a particular part is undefined, Capture will generally substitute the pin name. This is where the problems start. Not all netlist formats or PCB design software support alphanumeric pin numbers, such as would occur with diodes and transistors.

Parts such as small ICs have pin numbers defined by the manufacturer or industry standards. Pin numbers for such standardized parts are usually not an issue. Large PGA (pin grid array) ICs can pose problems. These parts have alphanumeric pin numbers such as A11, in which the alpha character corresponds to the vertical axis of the grid (A to Z from top to bottom) and the numeric character(s) corresponds to the horizontal axis starting from 1 at the left. If the PCB design software does not allow alphanumeric pin names, one possible workaround is to arbitrarily number the pins in rows and columns.

Discrete semiconductor devices are especially troublesome because of the variety of possible pin arrangements and case styles. For example, let's consider transistors. Although a single standard symbol is used to represent all NPN transistors on the schematic, many different case styles are available. Case styles range from SOT-23 surface mount devices to large industrial versions with screw terminals. The pin arrangement varies with case style. A given case style can even have several different pin arrangements. For example both 2N4401 and BF224 transistors come in TO-92 cases. The 2N4401 has pin arrangement EBC (emitter, base, and collector) wheras the BF224 has pin arrangement CEB from left to right, viewing the flat front surface of the parts.

Let's assume the design requires a 2N4401 and a BF224. There are three alternatives for dealing with the pin arrangement situation:

- Create two custom Capture library parts, one for each NPN transistor pin arrangement. Assign numeric pin names 1,2,3 in order from left to right. Create a single TO-92 part in the PCB design software, also using pin numbers 1,2,3 from left to right. Transfer the netlist without editing.

- Create a single Capture library part, representing a NPN transistor with emitter as pin 1, base as pin 2, and collector as pin 3. Create two TO-92 parts in the PCB design software, using pin arrangements 1,2,3 and 3,1,2. Transfer the netlist without editing.
- Use the existing Device library part for an NPN transistor with pin names EMITTER, BASE, and COLLECTOR. Create a single TO-92 part in the PCB design software using pin numbers 1,2,3 from left to right. Edit the netlist and replace the pin names with numbers 1,2,3 or 2,3,1 as required.

Which approach is best? Since the Capture part editor is very quick and easy to use, the first alternative might seem to be the logical choice. In the real world you will encounter many different types of transistors. MOSFETs and some ICs also come in TO-92 packages. The OrCAD libraries already have symbols defined for these parts. Likewise, most PCB design software packages come with a parts library that includes common case styles such as TO-92.

If you use either of the first two alternatives, you will still have to double-check the pin arrangements at some point during the design process to make sure a mistake did not occur. The author's experience suggests that staying with the standard libraries and using the third alternative is the most efficient approach.

Part Properties

The minimum part information that the netlist must contain before loading into a PCB design package includes reference designator, part value, and PCB footprint. Reference designators identify particular parts both on the schematic and on the PCB layout. Few problems occur transferring reference designator information via the netlist.

Capture reserves a special part property, the PCB footprint, to determine the part shape and pad layout for PCB design purposes. For example, a design might have two 10K 5% .25 watt resistors. These two parts have the same part value, but one might be a conventional axial lead package and the other surface mount. You would use the PCB footprint property to differentiate the two package styles.

In most cases, the PCB footprint property is entered during the schematic capture process and automatically extracted when the netlist is generated. However, this is not always practical. The electronics engineer doing the schematic may not know all the correct PCB footprint names if another department or a subcontractor does the PCB design. The design may also require new parts for which PCB footprint names have not yet been assigned.

In this case, an alternative approach is to use a dummy value or the part value as a placeholder for the PCB footprint field in the netlist. The PCB design entity then assumes responsibility for editing the netlist and inserting the appropriate PCB footprints.

Entering PCB Footprint Properties

You can double click on individual parts and edit part properties, including the PCB footprint. However, this approach is time consuming and prone to errors and omissions. Using the Browse tool is a more convenient means of editing properties. This tool provides a convenient spreadsheet-type interface. You can quickly edit the properties of all parts in the design. You can also check and edit net names and the properties of hierarchical ports and off-page connectors.

In Project Manager, click on the design. Then click on Edit, Browse, and Parts. The parts browse window appears. Label buttons appear at the top of each column. If you click on a button, parts are sorted and redisplayed in alphanumeric order based on that column. To select all parts for editing properties, hold down the left mouse button and drag a selection box across the window as shown in Figure 5-15. Then click on Edit and Properties.

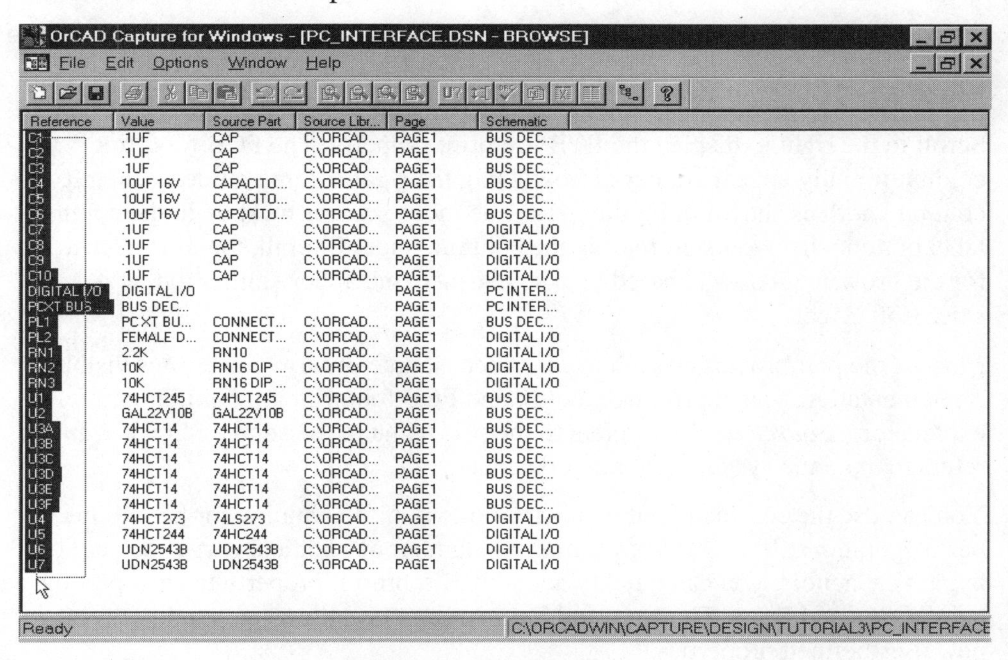

Reference	Value	Source Part	Source Libr...	Page	Schematic
C1	.1UF	CAP	C:\ORCAD...	PAGE1	BUS DEC...
C2	.1UF	CAP	C:\ORCAD...	PAGE1	BUS DEC...
C3	.1UF	CAP	C:\ORCAD...	PAGE1	BUS DEC...
C4	10UF 16V	CAPACITO...	C:\ORCAD...	PAGE1	BUS DEC...
C5	10UF 16V	CAPACITO...	C:\ORCAD...	PAGE1	BUS DEC...
C6	10UF 16V	CAPACITO...	C:\ORCAD...	PAGE1	BUS DEC...
C7	.1UF	CAP	C:\ORCAD...	PAGE1	DIGITAL I/O
C8	.1UF	CAP	C:\ORCAD...	PAGE1	DIGITAL I/O
C9	.1UF	CAP	C:\ORCAD...	PAGE1	DIGITAL I/O
C10	.1UF	CAP	C:\ORCAD...	PAGE1	DIGITAL I/O
DIGITAL I/O	DIGITAL I/O			PAGE1	PC INTER...
PCXT BUS ...	BUS DEC...			PAGE1	PC INTER...
PL1	PC XT BU...	CONNECT...	C:\ORCAD...	PAGE1	BUS DEC...
PL2	FEMALE D...	CONNECT...	C:\ORCAD...	PAGE1	DIGITAL I/O
RN1	2.2K	RN10	C:\ORCAD...	PAGE1	DIGITAL I/O
RN2	10K	RN16 DIP ...	C:\ORCAD...	PAGE1	DIGITAL I/O
RN3	10K	RN16 DIP ...	C:\ORCAD...	PAGE1	DIGITAL I/O
U1	74HCT245	74HCT245	C:\ORCAD...	PAGE1	BUS DEC...
U2	GAL22V10B	GAL22V10B	C:\ORCAD...	PAGE1	BUS DEC...
U3A	74HCT14	74HCT14	C:\ORCAD...	PAGE1	BUS DEC...
U3B	74HCT14	74HCT14	C:\ORCAD...	PAGE1	BUS DEC...
U3C	74HCT14	74HCT14	C:\ORCAD...	PAGE1	BUS DEC...
U3D	74HCT14	74HCT14	C:\ORCAD...	PAGE1	BUS DEC...
U3E	74HCT14	74HCT14	C:\ORCAD...	PAGE1	BUS DEC...
U3F	74HCT14	74HCT14	C:\ORCAD...	PAGE1	BUS DEC...
U4	74HCT273	74LS273	C:\ORCAD...	PAGE1	DIGITAL I/O
U5	74HCT244	74HC244	C:\ORCAD...	PAGE1	DIGITAL I/O
U6	UDN2543B	UDN2543B	C:\ORCAD...	PAGE1	DIGITAL I/O
U7	UDN2543B	UDN2543B	C:\ORCAD...	PAGE1	DIGITAL I/O

Figure 5-15 Parts Browse Window

The parts properties edit window appears as shown in Figure 5-16. You can use the scroll bars to view additional parts or properties.

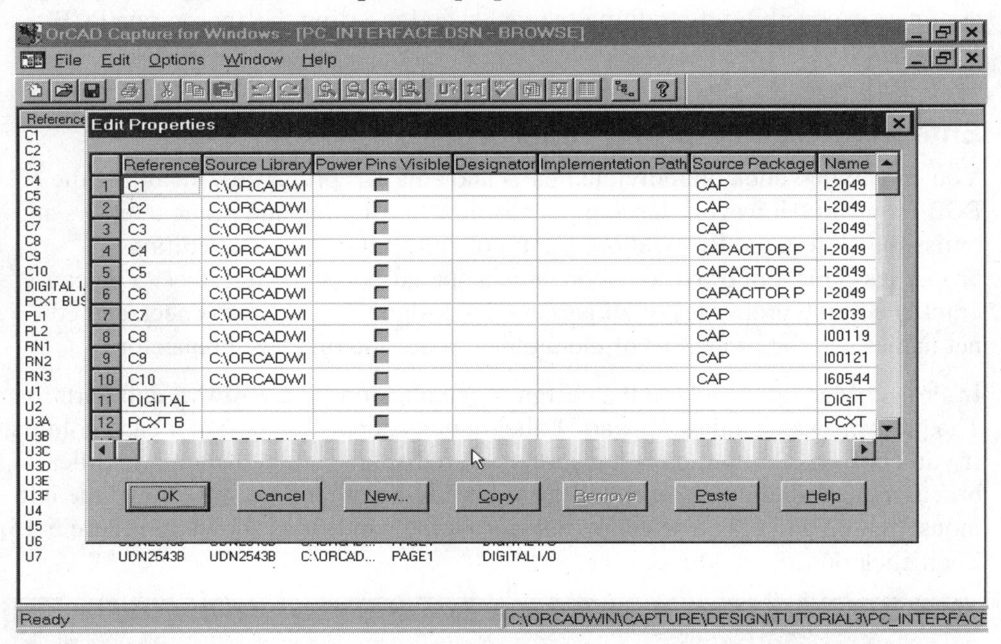

Figure 5-16 Initial View of Parts Properties Edit Window

Scroll to the right to display the PCB footprint property. The PCB footprint column initially appears rather narrow. Drag the right column guide to enlarge the column width as shown in Figure 5-17. The mouse cursor must be in the column label button when you start to drag the column guide. Overall, the user interface for the browse window is based on a spreadsheet metaphor similar to that of Microsoft Excel.

Most of the part properties such as reference, source library, power pins visible, implementation, value, primitive, color, and PCB footprint are familiar and self-explanatory. *Source package* refers to the part name in the source library. *Name* refers to a unique system-generated ID code.

You can use the command buttons at the bottom of the window for basic copy and paste operations. If your design contains a number of identical parts, such as .1 UF bypass capacitors, you can quickly add a PCB footprint property to each part by using Copy and Paste. The New and Remove buttons allow you to add or remove new user-defined properties.

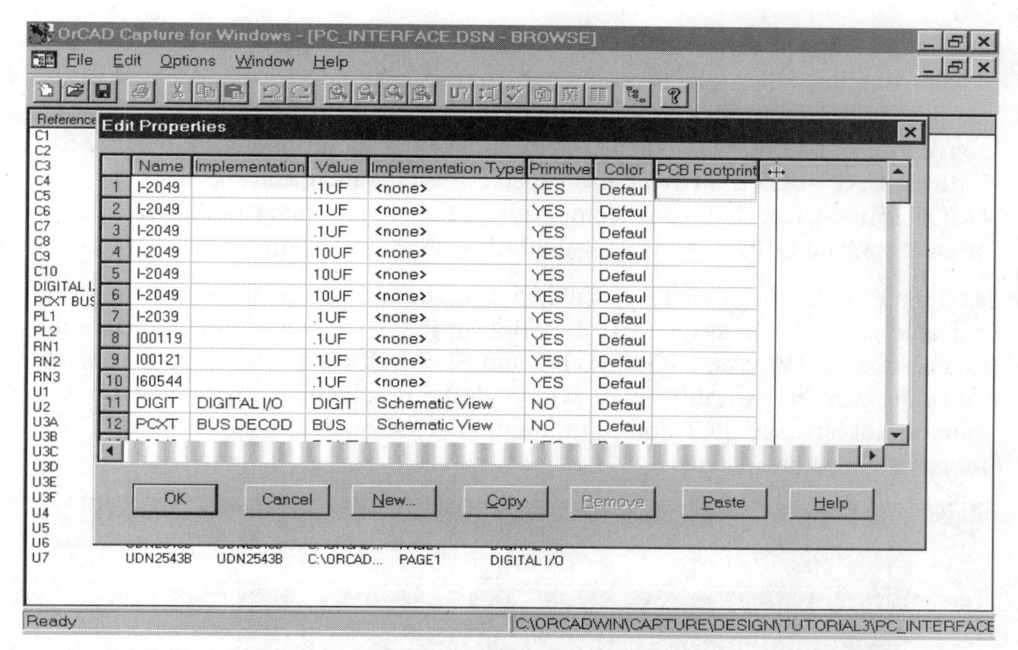

Figure 5-17 Dragging the PCB Footprint Column Guide

You are now ready to start adding the PCB footprint values to each part. Table 5-1 lists the PCB footprint for each part:

Table 5-1 PCB Footprint Cross Reference

Reference Designator	Part Value	PCB Footprint
C1,C2,C3,C7,C8,C9,C10	.1 UF	CAP\40LS
C4,C5,C6	10 UF 16V	ECAP\10LS25A
PL1	PC XT BUS	CON\IBMPCXT
PL2	FEMALE DB25	DCON25\FRA
RN1	2.2K	RN\10P
RN2,RN3	10K	DIP16
U3	74HCT14	DIP14
U5	74HCT244	DIP20
U1	74HCT245	DIP20
U4	74HCT273	DIP20
U2	GAL22V10B	DIP24\300
U6,U7	UDN2543B	DIP16

The PCB footprint values in Table 5-1 are for illustrative purposes only and are taken from the author's PCB design parts library. No industry standards exist for naming PCB footprints, and your organization undoubtedly has its own naming conventions and library. You can interpret the PCB footprint names in Table 5-1 as follows: CAP\40LS is an axial lead capacitor with .4 inch lead spacing, DCON25\FRA is a D-sub connector with 25 female pins and right-angle orientation, and DIP24\300 is a 24-pin DIP with .3 inch width.

Add the PCB footprint values in Table 5-1 to your design. You can cut and paste repeated values, such as the PCB footprint for the .1 UF capacitors. Note that the two hierarchical blocks, DIGITAL I/O and BUS DECODE, also appear as parts. Since these are nonprimitive parts with underlying schematic implementations, they cannot have any PCB footprint. Figure 5-18 shows the window after PCB footprints have been added.

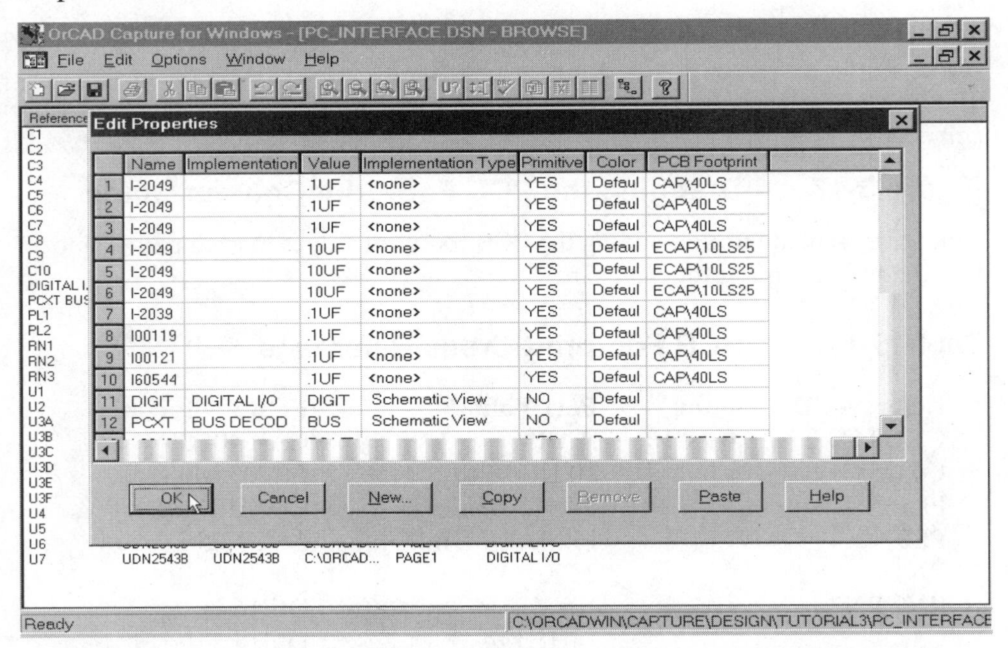

Figure 5-18 Edit Window with Completed PCB Footprints

When you are finished, click on OK to continue. You will return to the browse window (Figure 5-15). You can then click on Project Manager. The next step will be to create the netlist.

Note that if you double click on a part in the browse window, you will automatically open the schematic in which the part appears. The selected part is highlighted on the schematic.

How OrCAD Creates a Netlist

The exercise in this chapter focuses on creating a "flat" netlist. Most PCB design systems require this type of netlist. A flat netlist contains unique names for all parts and signals (nets) in the design. Any structural information about the design hierarchy or reuse of nonprimitive objects is removed. This means that all instances of hierarchical blocks are exploded. The entire design is resolved into one giant sheet with all the connectivity information required for a PCB.

The Create Netlist tool incrementally processes design information to generate the final netlist. The tool generates some intermediate files that remain in the project directory. These are binary files with .INS, .RES, and .PIP extensions. The .INS file contains parts information The .RES file contains connectivity information. The .PIP file contains any pipe commands present in the schematic. Pipe commands are special instructions placed on the schematic in the form of text. Pipe commands appear when schematics are generated for input to electrical simulation packages such as SPICE. You do not need to concern yourself about these intermediate files and you can safely delete them from your project directory.

In some cases, multiple names may be associated with a given electrical signal as it courses its way through the schematic structure. However, the netlist requires one unique name for each signal net. Capture prioritizes names and assigns the highest priority name to the net. Names are ranked in priority from highest to lowest as follows: named nets, hierarchical port names, off-page connectors, power object names, net aliases, and system-generated names.

Running Create Netlist

In Project Manager, click on the design. Next, click on the Create Netlist tool. The dialog box shown in Figure 5-19 appears. You can select a netlist format by clicking on a tab at the top of the screen. Certain formats including EDIF, SPICE, and Layout (used for OrCad PCB design software) have their own tabs. Additional formats, including the Tango format used in this exercise, appear on the "Other" tab. Click on the Other tab. Use the scroll box to select Tango format.

The part value and PCB footprint boxes allow you to select the combination of properties that will be output in these netlist fields. In most cases, you can use the default properties. If you have defined user properties, you can combine these with the part value or PCB footprint.

The Options box allows you to select netlist specific options. Not all netlist formats have options. Note that no options appear for Tango format.

If you select the option to view the output, Capture launches the Windows Notepad with the netlist file. The netlist file path defaults to the current design directory, and the file name defaults to the design name with a .NET extension. You can change these defaults if required. Note that some netlist formats also generate a second file with additional information.

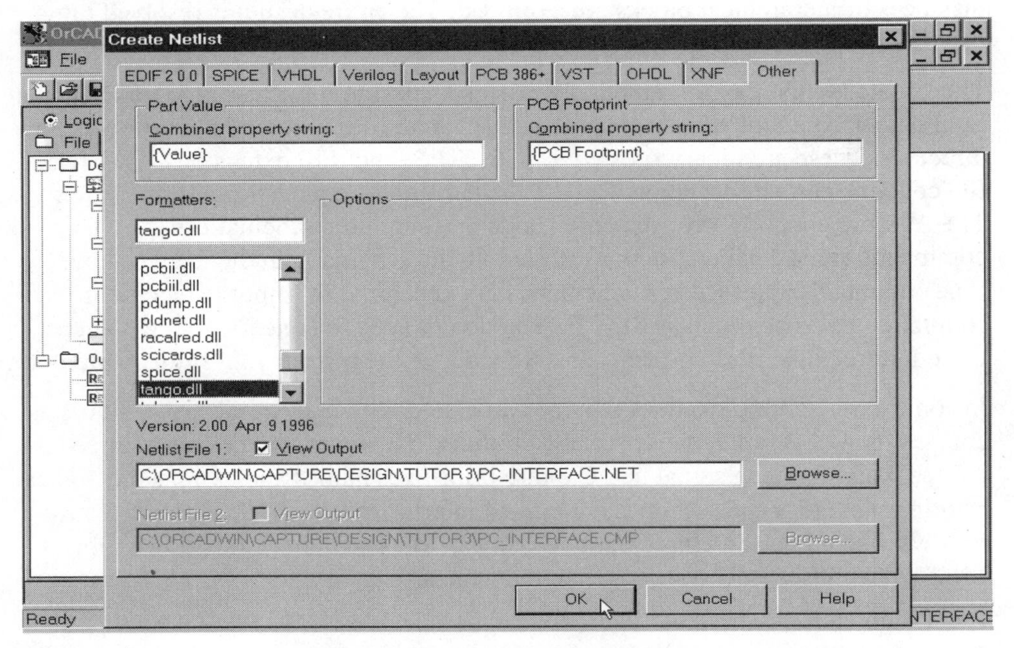

Figure 5-19 Dialog Box for the Create Netlist Tool

You should select the View Output option. Then click on OK to continue. For a small design, such as the exercise in this chapter, the Create Netlist tool requires only a few seconds to run.

During the netlist process, you will get a message "Netlist formatter reported warnings – check Session Log." This is a routine warning that you will examine later.

When the tool has completed processing the netlist, the Windows Notepad is automatically launched and appears as shown in Figure 5-20.

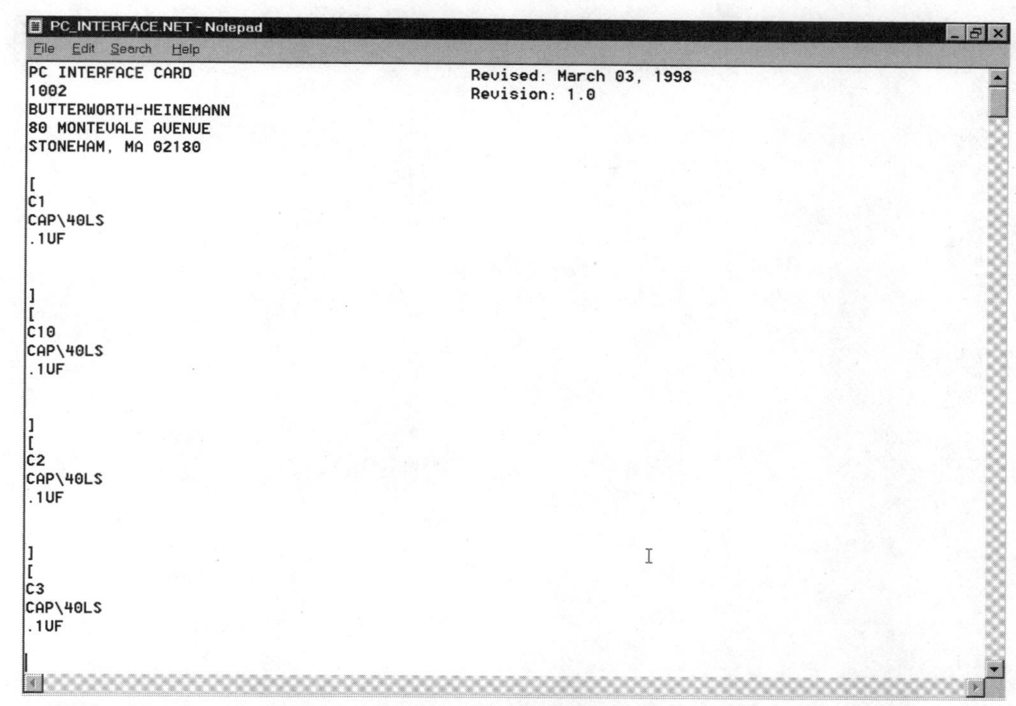

```
PC INTERFACE CARD                              Revised: March 03, 1998
1002                                           Revision: 1.0
BUTTERWORTH-HEINEMANN
80 MONTEVALE AVENUE
STONEHAM, MA 02180

[
C1
CAP\40LS
.1UF


]
[
C10
CAP\40LS
.1UF


]
[
C2
CAP\40LS
.1UF


]
[
C3
CAP\40LS
.1UF
```

Figure 5-20 Parts Section of Tango Netlist

Overview of Tango Netlist Format

Some understanding of Tango netlist format will be very helpful if the requirement arises to check or edit a netlist file. Other netlist formats are similar but use different syntax.

The Tango netlist is an ASCII text file that starts with an identification header similar to the header in a bill of materials file. The identification header contains properties taken from the title block of the root schematic, including the design name, revision code and date, drawing number, and company name and address.

Immediately following the identification header is the parts section. Each part data block is enclosed in square brackets ([and]). Part data blocks include lines for the reference designator, PCB footprint, and part value. A few blank lines appear for reserved Tango properties not supported by OrCAD Capture.

The signal net section follows the parts section. Figure 5-21 shows a portion of the signal section in our sample netlist. Note that the order of parts and signals and system-generated net names may be slightly different in your design.

PC_INTERFACE.NET - Notepad

File Edit Search Help

```
(
D6
U4,17
RN1,8
U5,5
U1,12
)
(
D7
U4,18
RN1,9
U5,3
U1,11
)
(
INPUT6
PL2,9
RN3,3
)
(
N00155
U6,15
U4,6
)
(
+12V
U6,2
U7,2
PL2,1
U6,7
PL2,14
U7,7
C4,1
```

Figure 5-21 Signal Net Section of Tango Netlist

Each signal net data block is enclosed in parentheses. Signal net data blocks begin with the signal name. Additional lines list each node in the net. The node description consists of the part reference designator and the pin number. For example U4,17 refers to pin 17 of U4. Note that N00155 (middle of Figure 5-21) is a system-generated signal net name. Such system-generated names will begin with "N" (for net) followed by a unique numeric suffix.

Tango netlist format has the following limitations:

- Names are truncated to 16 characters.
- System-generated net names are truncated to five digits following the "N" prefix.
- All characters should be uppercase.
- Reserved ASCII characters include: ( ) [ ] – (dash) , (comma).

As you can see, Tango format is straightforward and easy to interpret.

Go ahead and print out the netlist from Notepad. You should closely examine the netlist for obvious errors. On several occasions the author has spotted an error on the netlist that slipped by on the schematic, such as missing signals or connections. You should also keep a copy of the netlist handy when starting on a PCB design.

Viewing the Session Log

Recall that a warning message appeared during the netlist process. Exit Notepad and return to Project Manager. Details of warning and error messages and processing results from tools appear in the session log. View the session log by clicking on Window and then Session Log. The session log appears as shown in Figure 5-22.

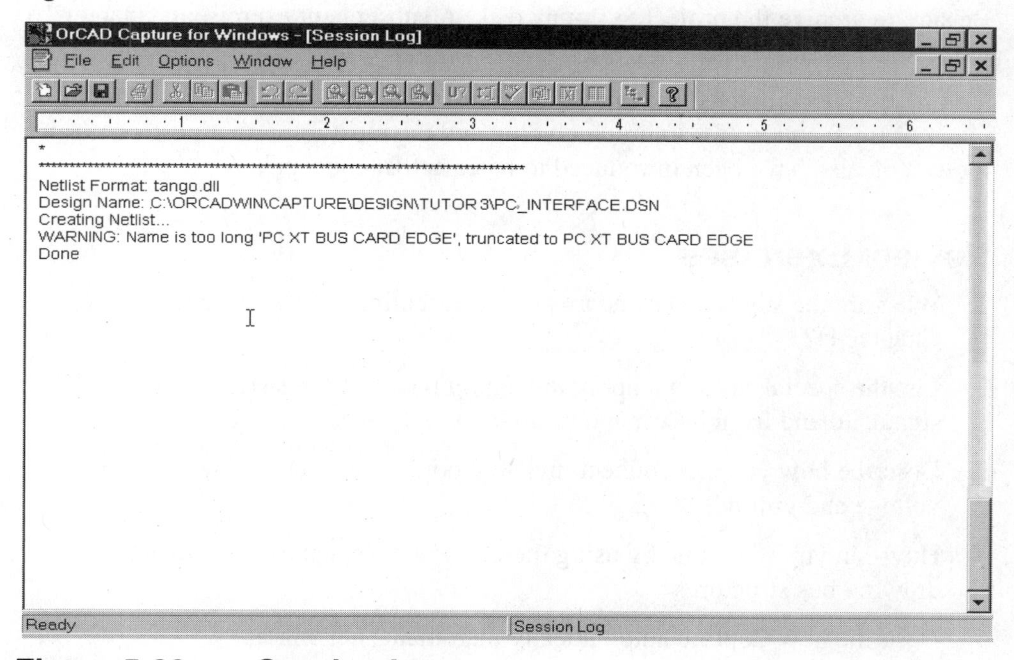

Figure 5-22 Session Log

In this case, the warning message is relatively benign. Create Netlist has truncated the name of the card edge connector to meet the 16 character limitation for Tango format.

Your session log may appear different. The session log is cleared every time you start Capture. You can also use the Clear Session log command from the Edit menu. The session log shown in Figure 5-22 was cleared before the Create Netlist tool was run.

Context-sensitive help is available for session log messages by placing the cursor on the message line and then pressing F1. You can also search the session log for particular information by using the Find command from the Edit menu.

The session log may prove useful when requesting technical support from OrCAD. You can print the session log by using the Print command. You can also save the session log. The file name defaults to SESSION.TXT.

Conclusion

Be sure to archive the project to floppy disk. A subsequent exercise in Chapter 10 will use the same design.

You have now completed the third exercise. At this point you have learned to use many of the features of Capture, including and the most important postprocessing tools. You also have been introduced to basic netlist concepts.

Review Exercises

1. What are the suggested standards for representing active low signals (see Chapter 1)?
2. List the special rules that apply to naming buses and bus signals, routing signals to and from buses, and routing buses between sheets.
3. Describe how you can connect invisible power pins to the power supply voltage and ground.
4. How can you save time by using the Copy and Repeat commands when drawing bus structures?
5. Describe how Capture automatically increments net aliases
6. Describe the function of the Update Parts Reference tool. Explain the difference between unconditional and incremental updates. If you are adding a few parts to a schematic, which type of update would you use?
7. Describe the function of the DRC (Design Rules Check) tool. What options should you typically select?
8. Experiment with creating deliberate errors such as those listed on page 168 and then running DRC.
9. What are the pros and cons of modifying the ERC matrix to warn of interconnected power supplies as suggested in the tutorial exercise?

10. Describe the function of the Create Netlist tool. What two basic types of information appear in a typical netlist?
11. Experiment with creating a netlist in PADS-PCB format (select padspcb.dll). Examine the output and compare it with the Tango format.
12. What two approaches can be used in Capture to enter PCB footprint properties?
13. What intermediate netlist files can you safely delete?
14. What ASCII characters are reserved in Tango format?

6

Part Editor

So far you have accessed existing library parts to complete the previous exercises. Although Capture comes with an extensive parts library, sooner or later you will have to create new parts. One reason is that the electronics industry constantly introduces new components, such as complex ICs. Also, many discrete components such as switches and transformers appear in an almost endless variety. The part editor is used to edit existing parts and create new parts. More often than not, you can create a new part by simply editing an existing one.

The library editor tutorial is divided into three sessions. In the first session, you will learn the basics by making a minor edit to an existing part in the Device library and then saving the part to your Custom library. The second session shows you how to create a new IC. In the third session, you will create a device with multiple parts.

Many of the skills you have learned in the previous exercises are directly applicable to the part editor as it has a similar user interface and shares many common commands. You will find that the ease of creating new parts is one of the great strengths of Capture. This chapter concludes with some tips on library management.

Overview of Library Parts

Capture stores all graphic data for library parts in vector form. Older versions of OrCad SDT stored graphic data for library parts in both raster (OrCAD used the term *bitmap*) and vector form. Note that you cannot convert and load older OrCAD bitmap libraries into Capture. Raster data consists of patterns of dots called *pixels*. Vector data consists of entities representing straight lines, arcs, circles, and text characters. Vector data is highly compact and easy to edit. Windows graphics drivers convert the vector data to raster form for display. This process is entirely transparent to the user.

First Session – Editing a Part

Your task will be to create the transformer with center tapped primary and secondary windings shown in Figure 6-1.

The Capture installation process described in Chapter 2 included copying the Custom library on the disk supplied on this book into your library directory. If you have not already done so, use the Windows Explorer to copy the Custom.olb file from the disk onto your C:\Orcadwin\Capture\Library directory.

Launch Capture from the Windows 95 desktop. The Capture session frame appears. Click on Open and then Library. Select the Device library. Project Manager will display a list of library contents as shown in Figure 6-2. The library icon resembles two sheets with parts symbols near the top left of the screen.

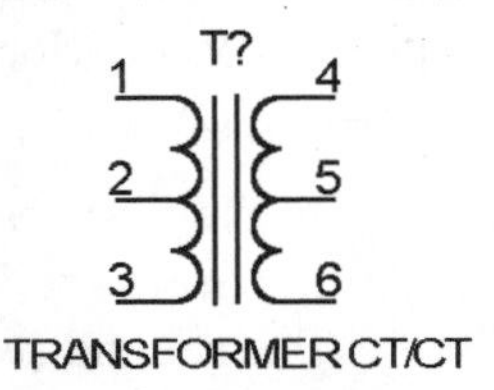

Figure 6-1 Transformer with Center Tapped Windings

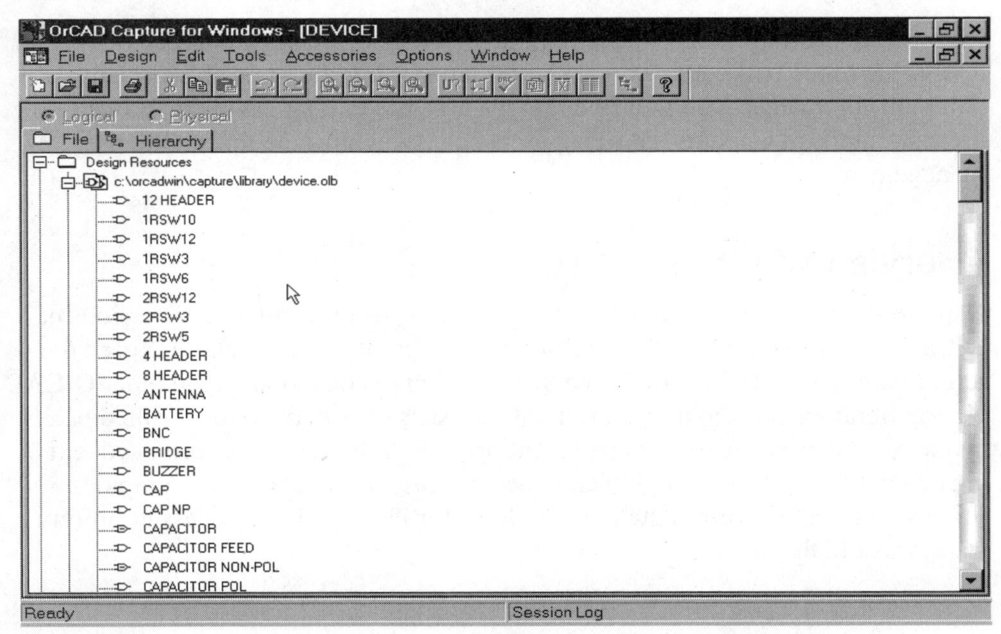

Figure 6-2 Setting Edit Library Configuration Options

Scroll down the list to TRANSFORMER CT. You will copy and paste this part to the Custom library and then edit it to create the new transformer. Double click on TRANSFORMER CT. This launches the part editor and opens the part, similar to the action that would occur if you double clicked on a schematic page. Note that the part editor also uses the same preference settings as the schematic editor (see Chapter 2 for details).

The part appears as shown in Figure 6-3. Note that TRANSFORMER CT has a center tapped secondary winding but no center tapped primary. It also lacks pin numbers.

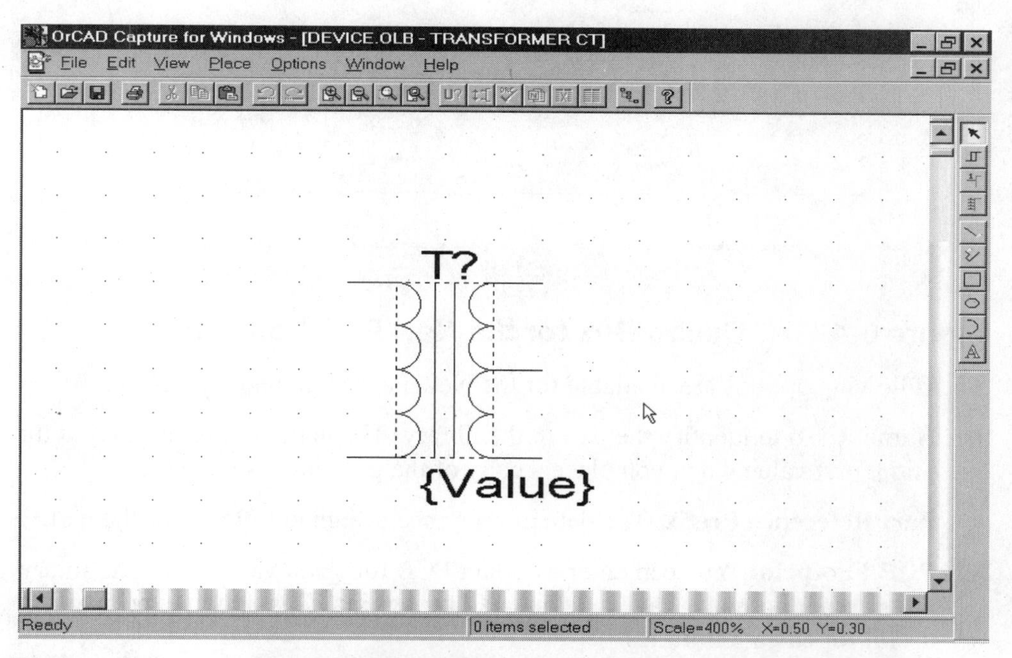

Figure 6-3 TRANSFORMER CT from the Device Library

Copy the part to the clipboard. You can easily do this by clicking on Edit and Select All. The Select All command selects the entire screen contents. You could also drag the mouse pointer across the part to select it. Then use the Copy tool.

Click on the project manager and then click on File and Close Project to close the Device library. Your selected part remains on the Windows clipboard.

Next, open the Custom library. In Project Manager, click on the library icon to select the entire library. Then click on Design and New Part. The New Part command allows you to add a new part to the library. The dialog box shown in Figure 6-4 appears.

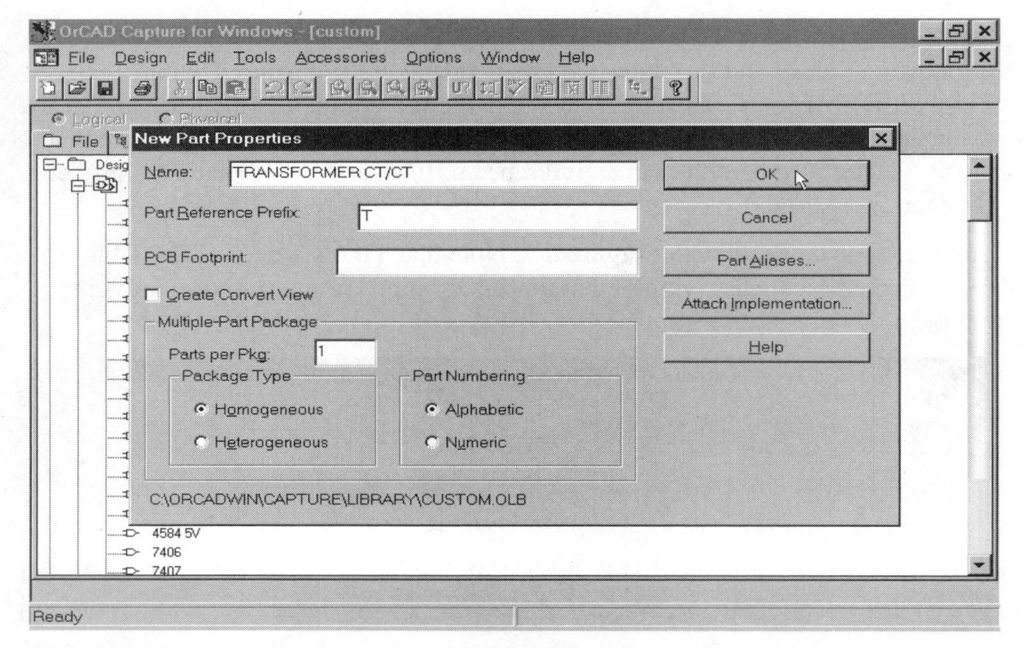

Figure 6-4 Dialog Box for the New Part Command

The following options are available for the New Part command dialog box:

- **Name**. Used to identify the part in the library. The name is also appears as the initial part value when you place a copy of the part into a schematic.
- **Part Reference Prefix**. The default reference designator prefix for the part.
- **PCB Footprint**. You can enter a default PCB footprint value. Capture library parts do not normally have a default PCB footprint. This option is primarily useful for custom libraries.
- **Create Convert View**. Checking this box allows you to create an alternative convert view, such as the DeMorgan equivalent of a basic logic gate.
- **Parts per Pkg**. Used to specify the number of parts per package. Useful for logic ICs with multiple gates.
- **Package Type**. Applies to multiple parts per package devices only. Homogeneous devices have identical parts, such as the gates in a 74HCT14 hex inverter. Heterogeneous devices may have nonidentical parts. In the third session of this exercise, you will create a heterogeneous device: a relay with separate parts for the coil, normally open, and normally closed contacts.

- **Part Numbering**. Applies to multiple parts per package devices only. Specifies the identification scheme for the parts in the package. Generally accepted practice is to use alphabetic part numbering, i.e., U1A, U1B, U1C. You can also select numeric part numbering, i.e.. U1-1, U1-2, U1-3. Numeric part numbering may cause problems with netlists for some PCB design systems.
- **Part Aliases**. This option displays a dialog box that allows you to add or delete additional names for the part. The aliases appear in the library list. All graphics and properties except for the name remain the same. This option is useful for digital logic, where a function such as a hex inverter may have multiple names such as 74HCT14, 74LHC14, and 74LS14.
- **Attach Implementation**. You can use this option to attach a schematic or a text file representing data such as PLD source code.

Use TRANSFORMER CT/CT as the name for your new part. Use T as the part reference prefix. Note that the other options default to the correct values for a single part per package device. Leave these options set as shown in Figure 6-4. Click on OK to launch the part editor and create the new part. The new part will initially appear with a dotted outline as shown in Figure 6-5.

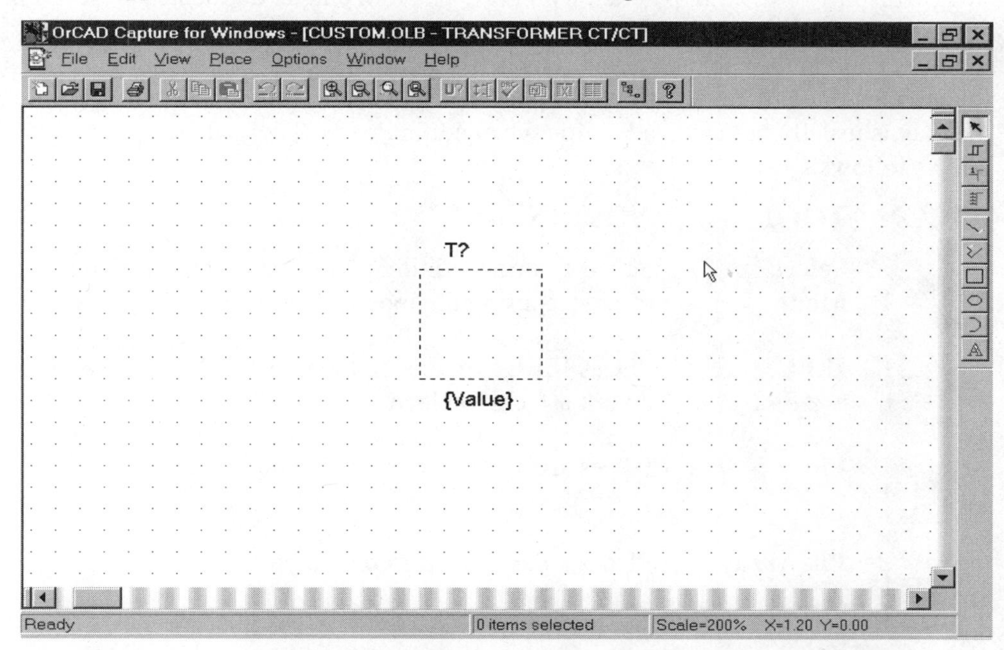

Figure 6-5 Part Editor with Initial View of the New Part

Introduction to the Part Editor Tools

The user interface for the part editor is very similar to that of the schematic editor. The familiar Capture toolbar appears at the top of the screen. Refer to page 73 in Chapter 3 for details about the toolbar. As with the schematic editor, icons for postprocessing tools remain dimmed (grayed out), since these tools can only be run from the project manager window.

The icons appearing at the right side of the screen comprise the part editor tool palette. This tool palette includes tools for placing IEEE symbols, pins, and drafting objects. The tools for drafting objects (line, polyline, rectangle, ellipse, arc, and text) function the same as in the schematic editor.

As in the schematic editor, the main toolbar and part editor tool palette can be docked and resized. Pressing the right mouse button brings up a context-sensitive shortcut menu, and the ESC key cancels most commands.

The Part Editor Tool Palette

The part editor tool palette is normally displayed when the Capture part editor is active. You can click on View at the top menu bar and then click on Tool Palette to toggle display of the tool palette on and off. The tool palette duplicates most of the commands on the top menu bar and associated pulldown menus. Most users find that clicking on the tool palette buttons is the quickest means of launching commands. Freeing up additional display area by turning off the tool palette display is usually not a consideration when editing parts. Tool palette commands are as follows.

ICON	TOOL	ACTION
	Selection Mode	Sets the normal editing mode and allows selection of objects using the mouse pointer.
	IEEE Symbol	Places IEEE symbols on the part. IEEE symbols do not affect pin electrical properties.
	Pin	Places signal or power pins on the part.
	Pin Array	Places an array of pins on the part.

	Line	Draws lines. Note that lines are not electrical objects.
	Polyline	Draws polylines (multisegment shapes and polygons). You can hold down the SHIFT key to force the lines to be orthogonal. Note that polylines are not electrical objects.
	Rectangle	Draws rectangles and squares. You can hold down the SHIFT key to force drawing a square. Note that these shapes are not electrical objects.
	Ellipse	Draws ellipses and circles. You can hold down the SHIFT key to force drawing a circle. Note that these shapes are not electrical objects.
	Arc	Draws arcs. Note that arcs are not electrical objects.
	Text	Places text using a dialog box. Note that text does not have electrical properties and should not normally be used in place of pin numbers or pin names.

All of these tools correspond directly to identical commands available on the Place menu. The term *tool* is used if the command is launched from an icon. An additional command, Place Picture, is available only from the Place menu. As with most Windows programs, you can also use keyboard shortcuts for menu commands. For example, you can use the key sequence ALT, P, P to launch the Pin tool.

Tool palette commands are modal. This means that the command remains in effect until cancelled. For example, once you click on the pin tool, you can place multiple pins. After a command is launched, you can press the right mouse button to bring up a shortcut menu to edit properties or end the current command. You can also use the ESC key to end or cancel commands.

You may also find that the hotkey combination (CTRL+T) for toggling snap to grid is very useful when precisely positioning graphics. However, you must never place pins off grid.

Editing the Transformer

The first step is to paste the original part, TRANSFORMER CT, that you copied onto the clipboard from the Device library. Use the Paste tool. The transformer appears on the screen.

Editing Pin Properties

The new transformer shown in Figure 6-1 requires pin numbers, so start by editing the pin properties. Start with pin 1. Double click on the pin to bring up the dialog box shown in Figure 6-6.

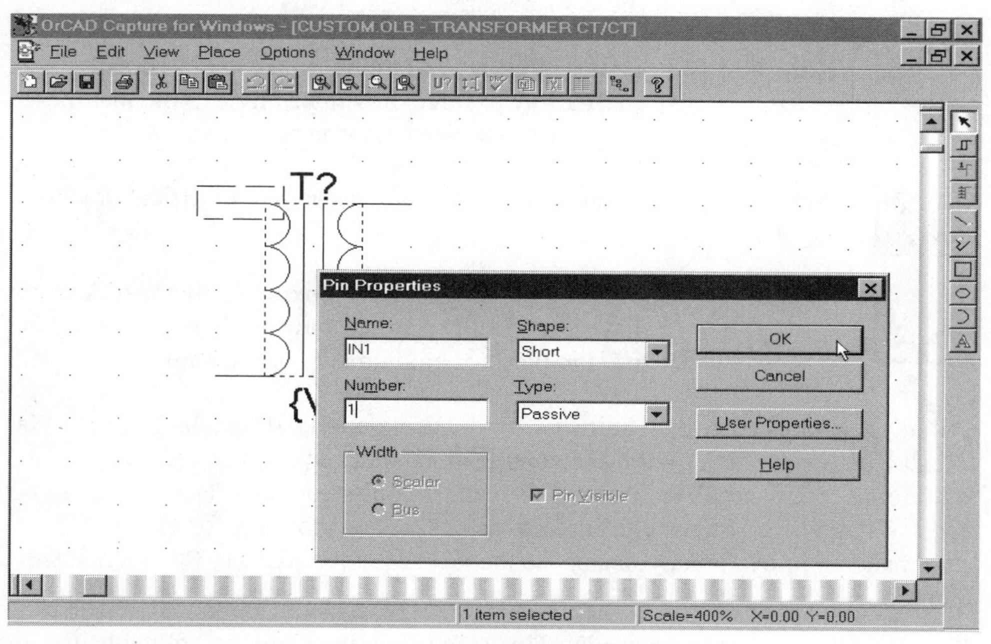

Figure 6-6 Dialog Box for Pin Properties

The dialog box allows you to set several pin properties:

- **Name**. Capture requires that all pins have a name. If you do not enter a name, a system-generated name is assigned to the pin. Invisible pin names are used for most passive components and discrete semiconductors, such as diodes and transistors. Part properties, set via the Options menu, determine pin name visibility. This subject is discussed later. General schematic drafting practice is to use an overbar to indicate negation or active low signal pins. To create an overbar above the pin name, type a backslash (\) character after every character in the pin name, i.e., type C\L\K\ for $\overline{\text{CLK}}$.

- **Number**. Assignment of a pin number is optional. In general, if a pin does not have a number, the Create Netlist tool will use the pin name. Many passive components such as capacitors and resistors and discrete semiconductors such as diodes and transistors do not use pin numbers. Pin number visibility is determined by means of the part properties, set via the Options menu.
- **Width**. Normal pins that connect to a wire are scalar. Bus pins represent a special case. Bus pins can only connect to a bus and must follow bus naming conventions for the pin name. The use of bus pins is not recommended, as problems may occur when generating a netlist.
- **Shape**. Available pin shapes are shown in Figure 6-7. General pin shape usage is as follows:

  Line. The standard pin shape. Recommended for most parts with numbered pins such as ICs. Three grid units length.

  Short. Use to save space on parts with complex graphics, such as the transformer in this exercise. One grid unit length.

  Zero length. Use for invisible power pins.

  Dot. Use for inverted or active low signals. Never use a dot pin and a name with an overbar together, as the double negation cancels out and leads to confusion.

  Clock. Use for positive edge triggered signals on logic ICs, such as counters and flip-flops.

  Dot Clock. Use for negative edge triggered signals on logic ICs. Never use a dot clock pin and a name with an overbar together, as the double negation cancels out and leads to confusion.

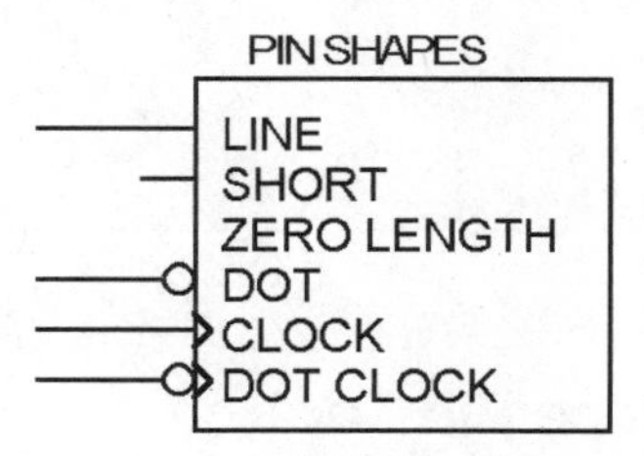

Figure 6-7 Pin Shapes

- **Type.** You can select the electrical properties of the signal associated with the pin. This information is used for electrical design rules checking. See pages 119-120 in Chapter 4 for a listing of the available pin types.

- **Pin Visible.** This checkbox is available only for power pins. You can click on the box to make power pins invisible. Use the zero length shape for invisible power pins. Other types of pins are always visible.
- **User Properties option**. Clicking on this option brings up a detailed list of properties. You can also add user-defined properties.

Enter "1" in the pin number field. You do not need to change any other fields or options. Then click on OK. The pin number will appear above the line for the pin. Add pin numbers for the remaining four pins. Use the numbers shown in Figure 6-1.

Placing a New Pin

The next step in editing the transformer is to add the primary center tap pin. Click on the Place Pin tool. The dialog box shown in Figure 6-8 appears.

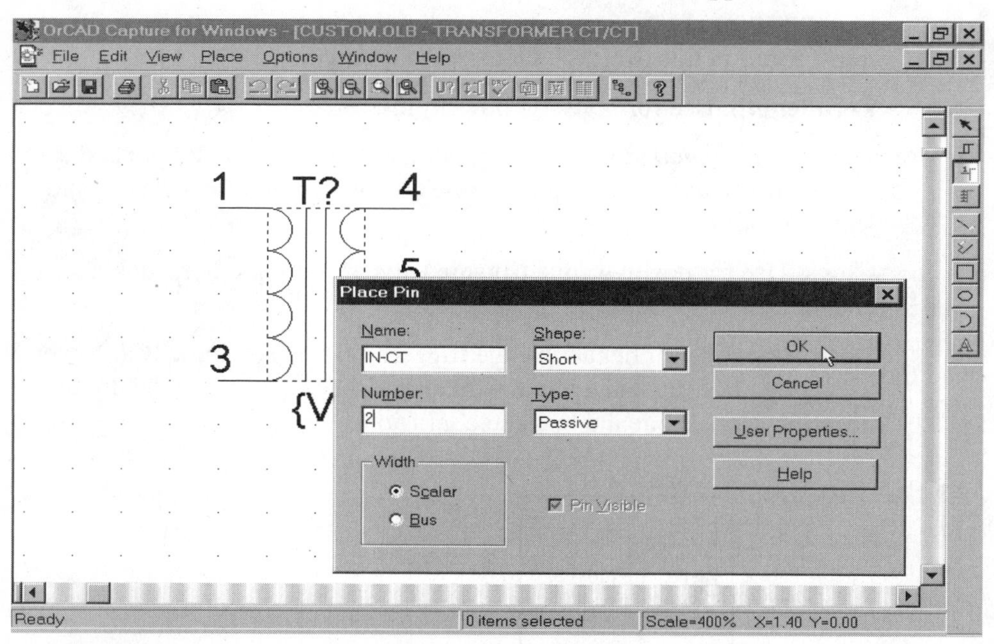

Figure 6-8 Dialog Box for Place Pin Tool

The Place Pin dialog box is identical to that used for editing pin properties. The only difference is that you are now entering properties for a new pin. Enter the values shown and click on OK. You can now drag the pin into position. The pin follows the part border (dotted line) as do hierarchical pins on hierarchical blocks. The finished transformer part now appears as shown in Figure 6-9.

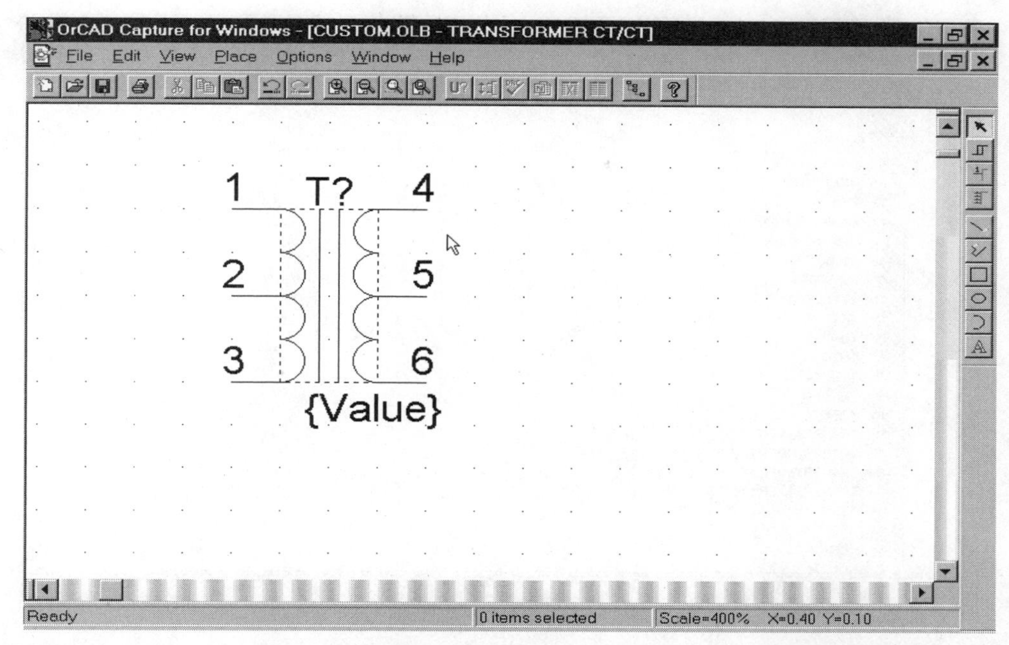

Figure 6-9 Finished Transformer with Primary Center Tap

Saving the New Part

Save your new part to the library. Click on the Save tool. Then click on Project Manager. You can then click on the Save tool again to save the library. Note that saving a new part is a two-step process. You must first save the part to the library and then save the library itself. When you are finished, the new part (TRANSFORMER CT/CT) appears on the library list as shown in Figure 6-10.

Archiving the Custom Library

You can use the Archive Project command to save an open library file. In a practical sense, the results of using Archive Project for a library file are no different from using the Windows Explorer to copy the library file to a floppy or network drive. Since making changes to OrCAD-supplied libraries is not recommended, you only need to archive any custom libraries you have created. Recommended practice is always to archive a custom library after making any changes.

At this point, go ahead and archive your Custom library. This completes the first session of the part editor exercise.

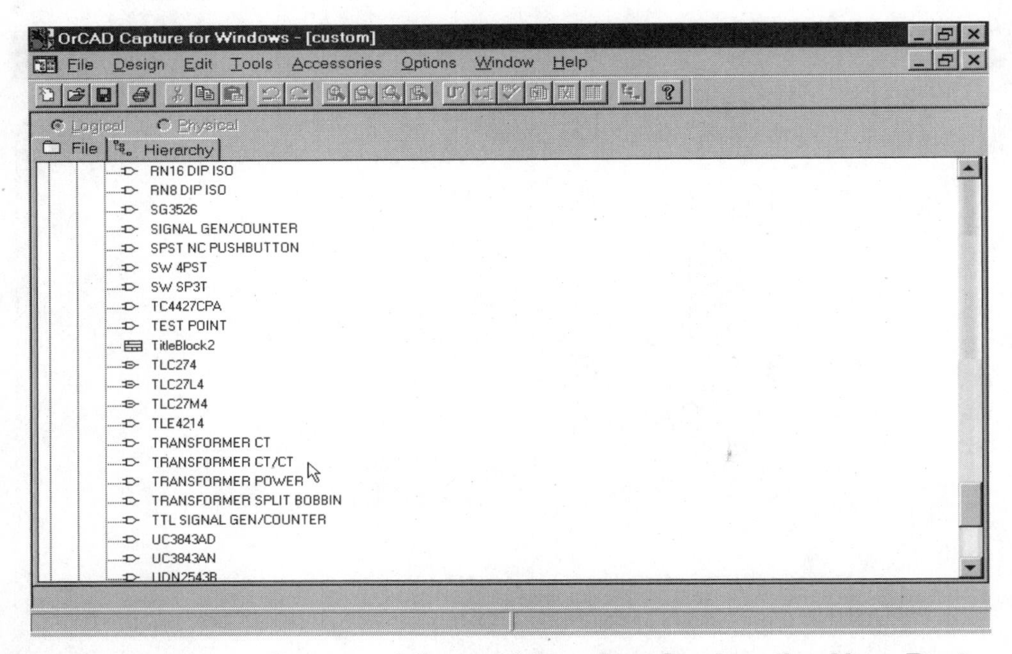

Figure 6-10 Custom Library List after Saving the New Part

Second Session – Creating a New IC Part

The second session involves creating the Maxim MAX877 voltage regulator IC shown in Figure 6-11. The MAX877 uses a novel switching regulator technology to generate a constant 5.0 volt supply from a 1.2 to 6.0 volt input and is ideal for battery-powered systems using four alkaline or Nicad battery cells. The MAX877 is not in the OrCAD-supplied libraries and will be required for the exercise in the next chapter.

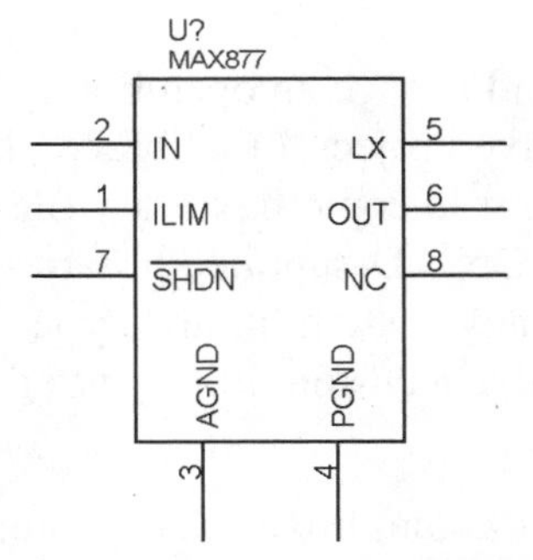

Figure 6-11 Maxim MAX877 Voltage Regulator IC

Signal names and pin arrangements for the MAX877 IC are taken from the Maxim data book. Recommended practice for analog and power ICs is to avoid invisible power and ground pins, thus all pins on the MAX877 are visible.

To create the MAX877, start by opening the Custom library. Click on the library in Project Manager. Then click on Design and New Part. Enter the name MAX877 in the dialog box. You can accept all the other default values in the dialog box. Click on OK to continue.

The new part initially appears as a small rectangular border. Drag the border so that the dimensions of the part are X=.8 and Y=1.1, as shown in Figure 6-12. You may experience some difficulty clicking on the border and dragging the resize handles. Note that the overall part dimensions are displayed in the lower right corner of the display.

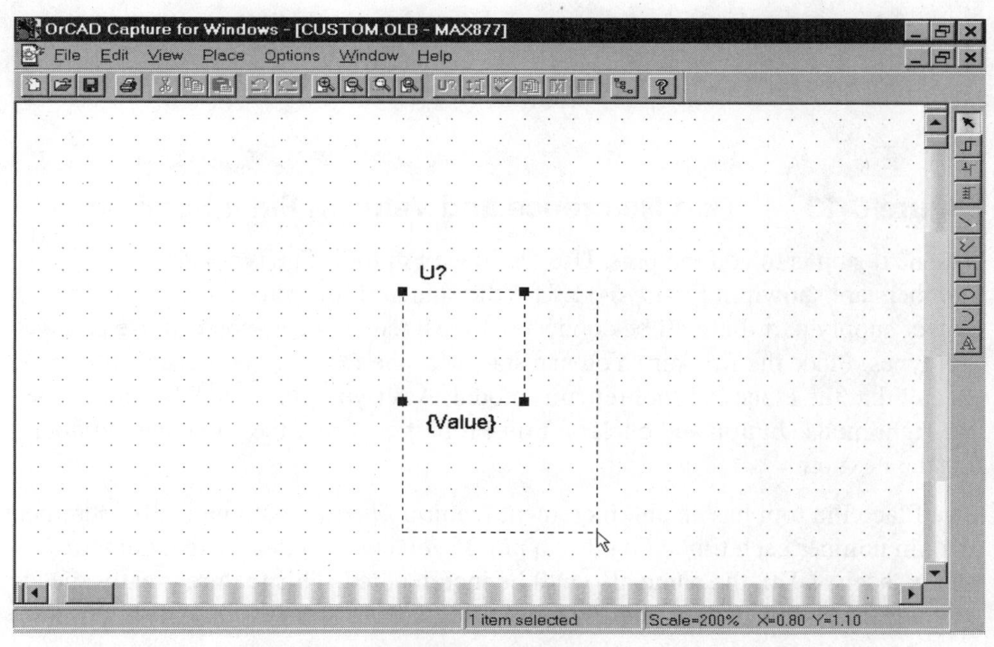

Figure 6-12 Dragging the Part Border

Since the part has ground pins on the bottom, you should reposition the reference designator and part value at the top of the part as shown in Figure 6-13. Just click on the objects and drag them into position. When you later place the part in a design, the reference designator and part value will appear in these positions.

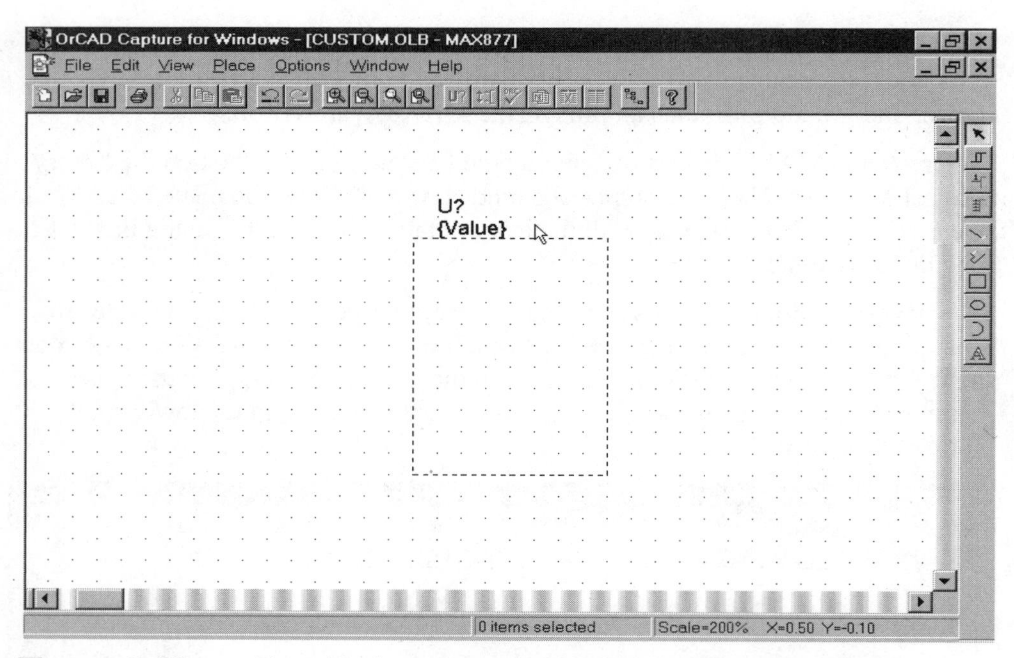

Figure 6-13 Part Reference and Value in Final Position

The next step is to add the pins. Use the Place Pin tool. The pin names and numbers are shown in Figure 6-11. Use the standard line pin shape. Since this is a power supply part that will be connected to discrete devices, you can use passive pin types. Place the first pin. You can start with the pin near the upper left corner. Recall that the Place Pin tool remains modal. After you place the first pin, press the right mouse button and click on Edit Properties. Enter the name and number for the next pin.

The Place Pin tool has an autoincrement feature. The tool automatically increments the pin number each time you place a pin. If you enter a name with a numeric suffix, such as D0, the name also will be incremented. This feature greatly speeds placing successive data or address pins on logic ICs.

After you have placed all the pins, the part should appear as shown in Figure 6-14. The next step is to draw the body outline as shown in Figure 6-15. You can use the Place Rectangle tool to draw the body outline for most ICs. Click on the tool and then drag the mouse pointer from the top left corner to the bottom right corner of the part border. Note that if you try to drag the rectangle past the part border, the border expands. If this happens, you will have to first resize the rectangle and then the border.

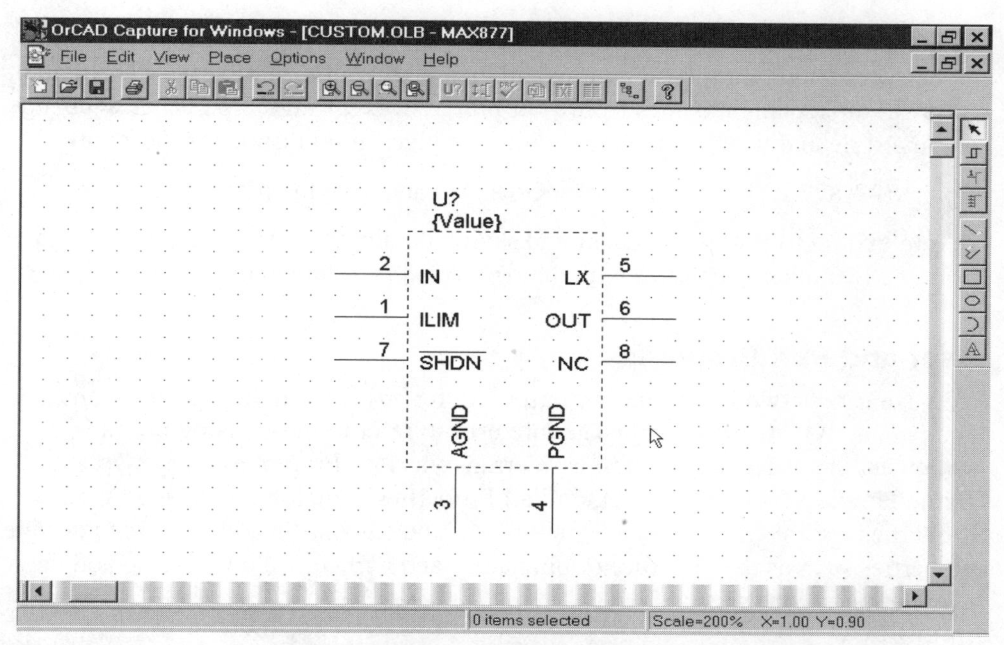

Figure 6-14 Part with Completed Pin Placement

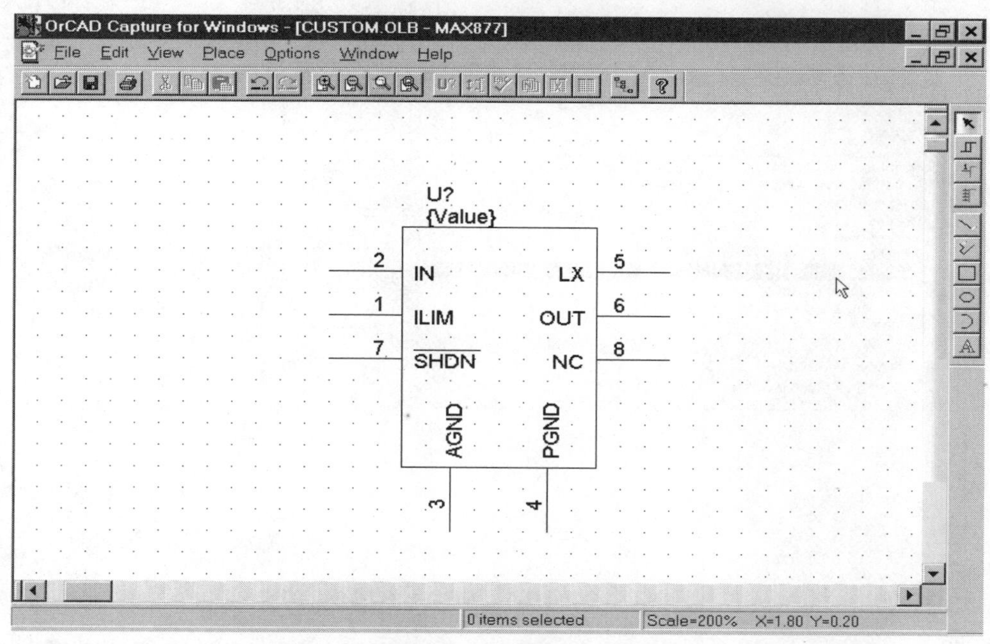

Figure 6-15 Part with Completed Body Outline

Graphic objects such as lines and rectangles have properties. You can select the object, press the right mouse button, and then click on Edit Properties. The default color for the rectangle comes from your preferences settings. You can also change the line style and width. The default thin line gives good results for most parts.

Your MAX877 IC is now complete. Go ahead and save the part.

Before exiting to Project Manager and saving the library, let's take a moment to examine two other properties with dialog boxes that relate to the part editor.

User and Part Properties

You can access two important properties dialog boxes from the Options menu. Unfortunately, the terminology Capture uses to refer to these dialog boxes is somewhat inconsistent. The first is referred to as Part Properties on the Options menu. However, when you click on Part Properties, a dialog box labeled User Properties appears as shown in Figure 6-16. You can use this dialog box to change properties such as the visibility of pin names and numbers. You can also add new user-defined properties.

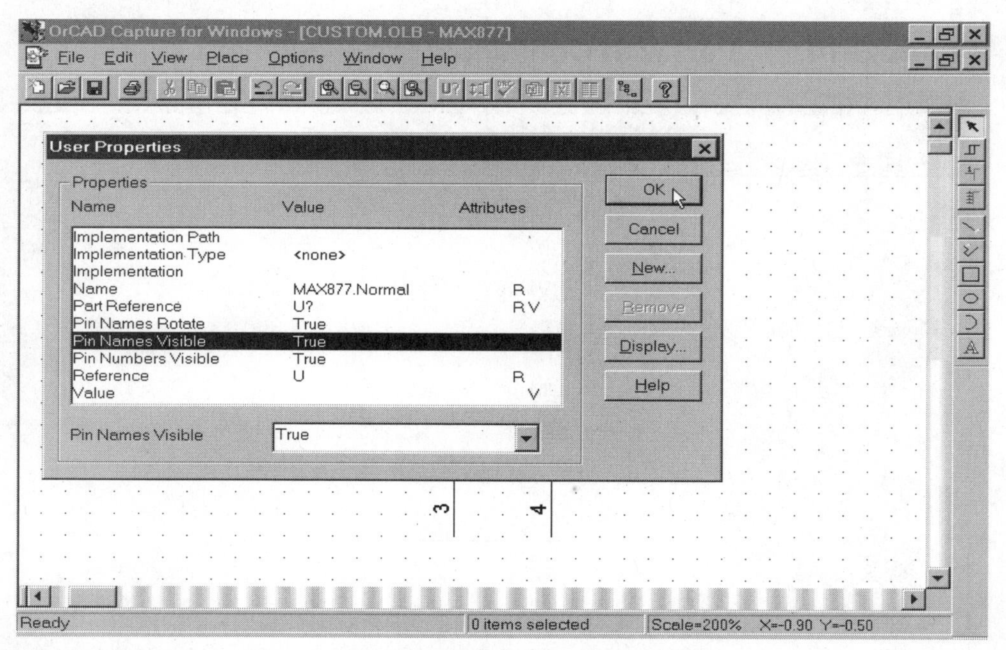

Figure 6-16 Dialog Box for User Properties

The second properties dialog box on the Options menu is referred to as Package Properties. Clicking on Package Properties brings up a dialog box labeled Edit Part

Properties as shown in Figure 6-17. This dialog box is similar to the one used for creating a new part. You can change the part name or reference designator. You can also add aliases (identical parts with different names).

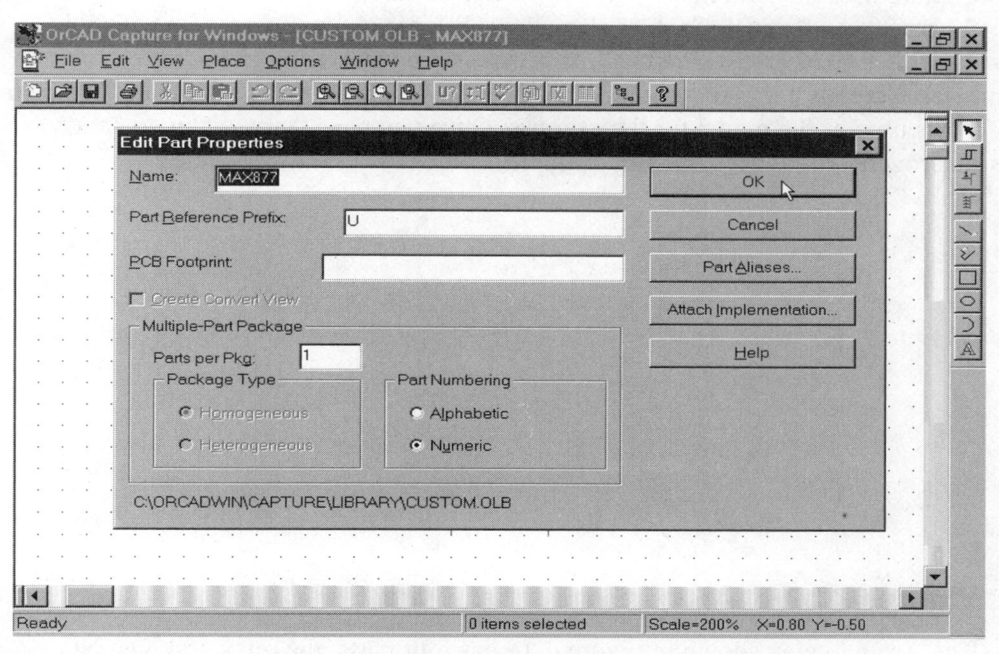

Figure 6-17 Dialog Box for Edit Part Properties

Third Session – Creating a Device with Multiple Parts

Recall from page 188 that devices can have multiple parts per package. The most common example is a logic IC such as a 74HCT hex inverter that has six gates. Capture uses the term *homogeneous* to refer to multiple-part devices in which all the parts are identical. Capture also allows heterogeneous devices, in which the individual parts are not identical. In this session, you will learn to create a heterogeneous part. An Aromat JH2A1B series industrial relay, shown in Figure 6-18, is used as the model for the exercise. Once you have learned how to create a heterogeneous device, creating a simpler homogeneous device is easy.

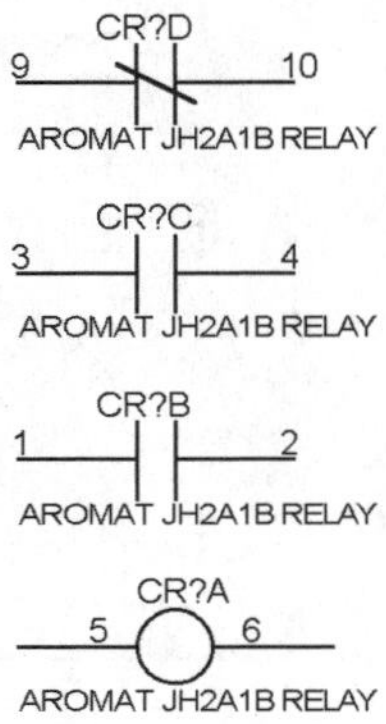

CR?D
9 10
AROMAT JH2A1B RELAY
CR?C
3 4
AROMAT JH2A1B RELAY
CR?B
1 2
AROMAT JH2A1B RELAY
CR?A
5 6
AROMAT JH2A1B RELAY

Figure 6-18 Aromat Industrial Relay

The Aromat relay shown in Figure 6-18 has four parts: the relay coil, two NO (normally open) contacts and an NC (normally closed) contact. On a side note, the term *form A* is used for NO contacts, *form B* for NC contacts and *form C* for changeover contacts (combined NO and NC). Hence the manufacturer's designation *2A1B* for this particular relay. The relay is drawn with symbols preferred for industrial controls. Part A is the coil, parts B and C are the NO contacts, and part D is the NC contact.

Relays in the Device library, such as RELAY 4PDT, are single-part packages drawn with conventional-appearing coil windings and switch contacts. These types of relay symbols are suitable for most electronic schematics. In the industrial controls field, relay logic is often drawn in a ladder style diagram. In this case, contacts for a given relay may appear in a different area or even a different sheet than the relay coil. You can accomplish this only with a heterogeneous package.

To create the Aromat relay, open the Custom library. Click on the library in Project Manager. Then click on Design and New Part. Enter the name AROMAT

JH2A1B RELAY in the dialog box. Use CR as the reference designator prefix. Enter 4 parts per package, and select a heterogeneous package. Then click on OK to continue.

The first part (part A) appears as a small rectangular border. Draw the coil circle as shown in Figure 6-19. Use the Place Ellipse tool. Recall that you can constrain the tool to draw a circle if you hold down the SHIFT key. Drag the mouse cursor from the top left corner of the border to the lower right corner (down and across two grid locations) as shown.

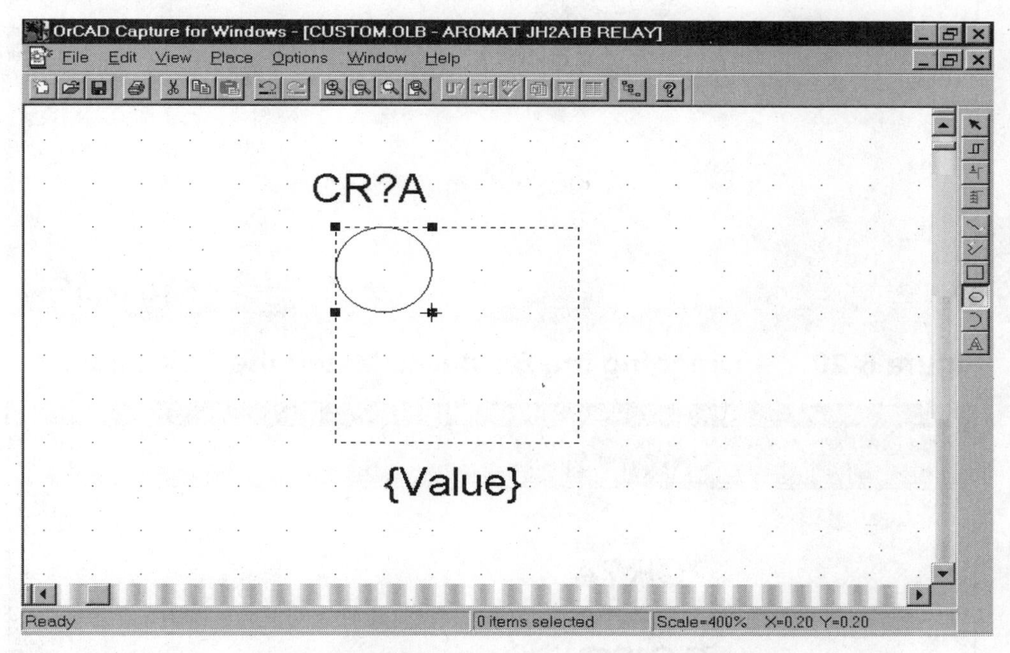

Figure 6-19 Drawing the Coil Circle for Part A

Next, drag the border so that it just encloses the coil circle. As with the MAX877 IC that you previously created, you may experience some difficulty clicking on the border and dragging the resize handles. The part should now appear as shown in Figure 6-20.

Place the coil pins as shown in Figure 6-21. Use the standard line pin shape and passive pin type. Use the following pin names and numbers: COIL1 for pin 5 and COIL2 for pin 6. Note that by coincidence, autoincrement works very nicely here.

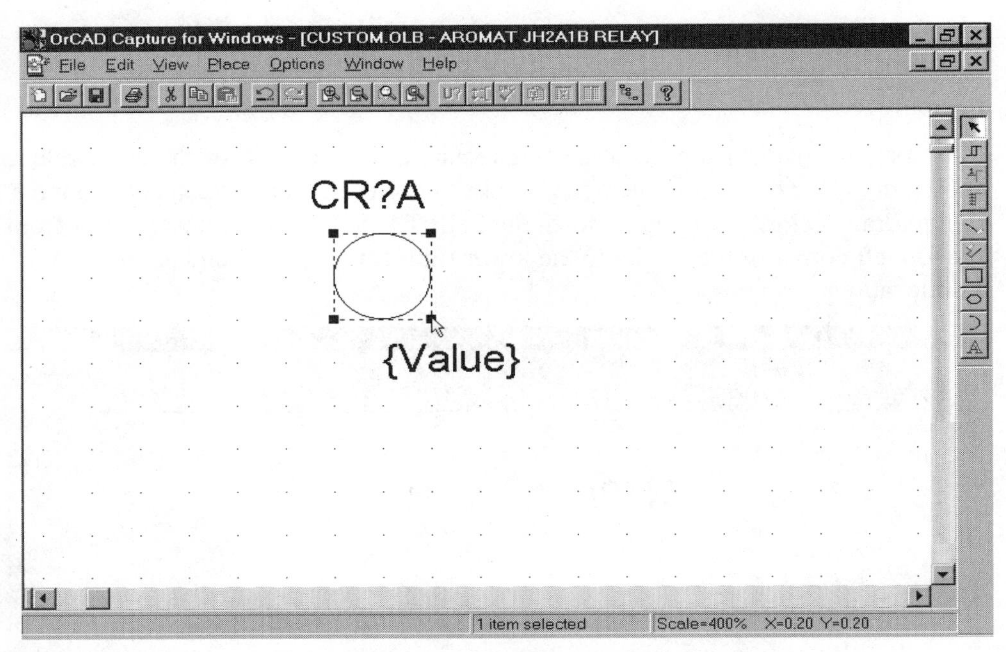

Figure 6-20 Dragging the Border to Match the Coil Circle

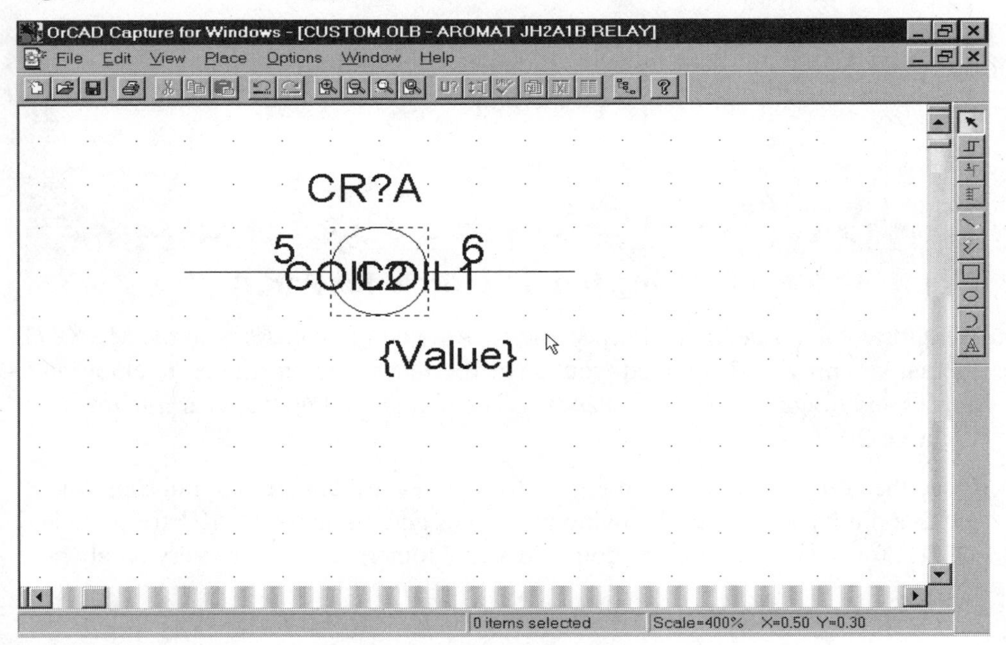

Figure 6-21 Part A with Coil Pins

The resulting part view now appears somewhat messy. The last step is to reposition the part value and reference designator and make the pin names invisible. Use the Part Properties command from the Options menu to make the pin names invisible. Your completed part A should now appear as shown in Figure 6-22.

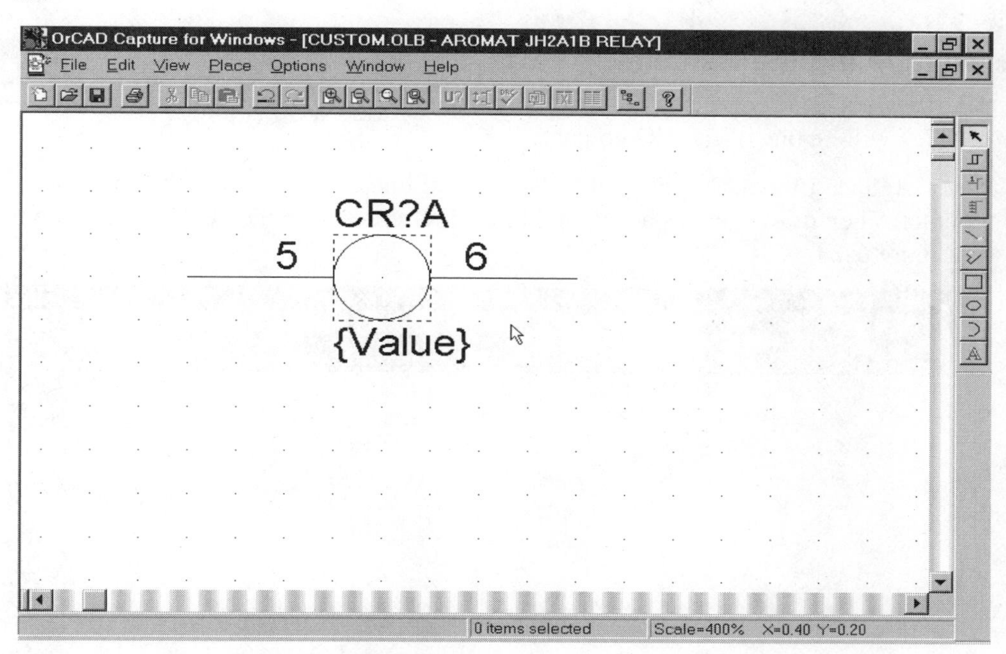

Figure 6-22 Completed Part A

Navigating between Parts in the Package

After you have completed one part in a multiple-part package, you can navigate to the remaining parts. The View menu provides several commands for this purpose:

- **Normal**. If you are creating or editing a device with a convert view, you can select which view to work in. Normal selects the normal view.
- **Convert**. This selection is available only for devices with a convert view.
- **Part**. Selects part view, in which a given part is displayed on the screen.
- **Package**. Selects package view. All parts contained in the device package are displayed. For heterogeneous parts, you can edit any object on any part. For homogeneous parts, you can edit only the pin names and pin numbers.

- **Next Part**. Available for multiple-part devices only. Displays the next part, i.e., part B if you have been working on part A.
- **Previous Part**. Available for multiple-part devices only. Displays the previous part, i.e., part A if you have been working on part B.

Completing the Remaining Parts

Use the Next Part command from the View menu to display the initial view of part B. This will be one of the NO contacts.

Use the Place Line tool to draw the two vertical lines that represent the NO contact. Then draw the two horizontal lines that represent pin extensions as shown in Figure 6-23.

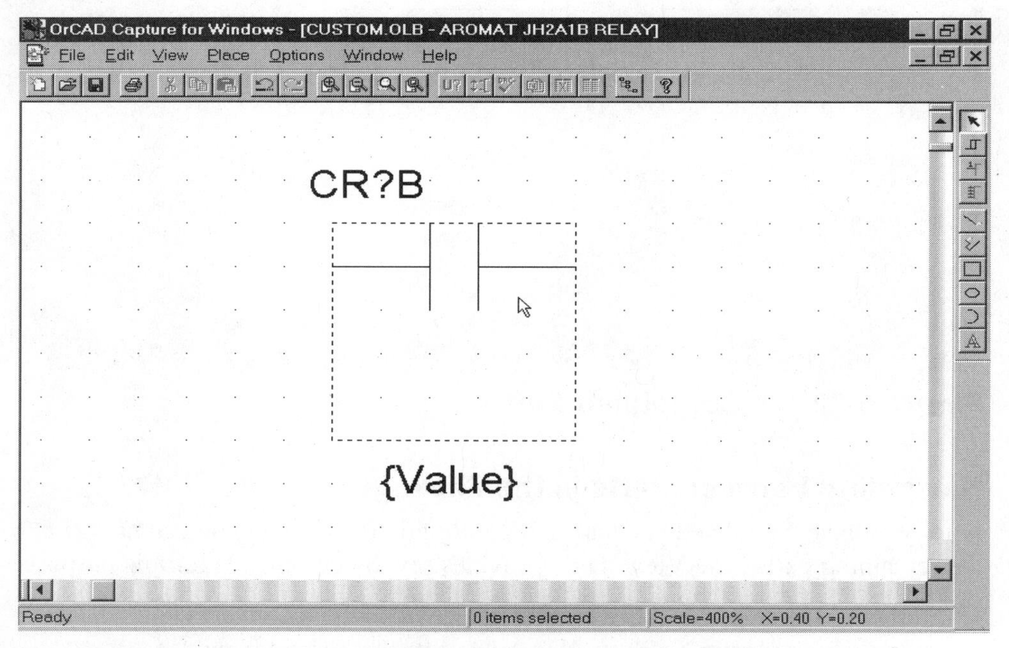

Figure 6-23 Drawing the NO Contacts for Part B

Next, drag the border so that it just encloses the NO contact as shown in Figure 6-24. Then place the contact pins. Use the short pin shape and passive pin type. Use the following pin names and numbers: NO1 for pin 1 and NO2 for pin 2. Again, autoincrement works very nicely for placing the second pin. Reposition the part value and reference designator and make the pin names invisible. Your completed part B should appear as shown in Figure 6-25.

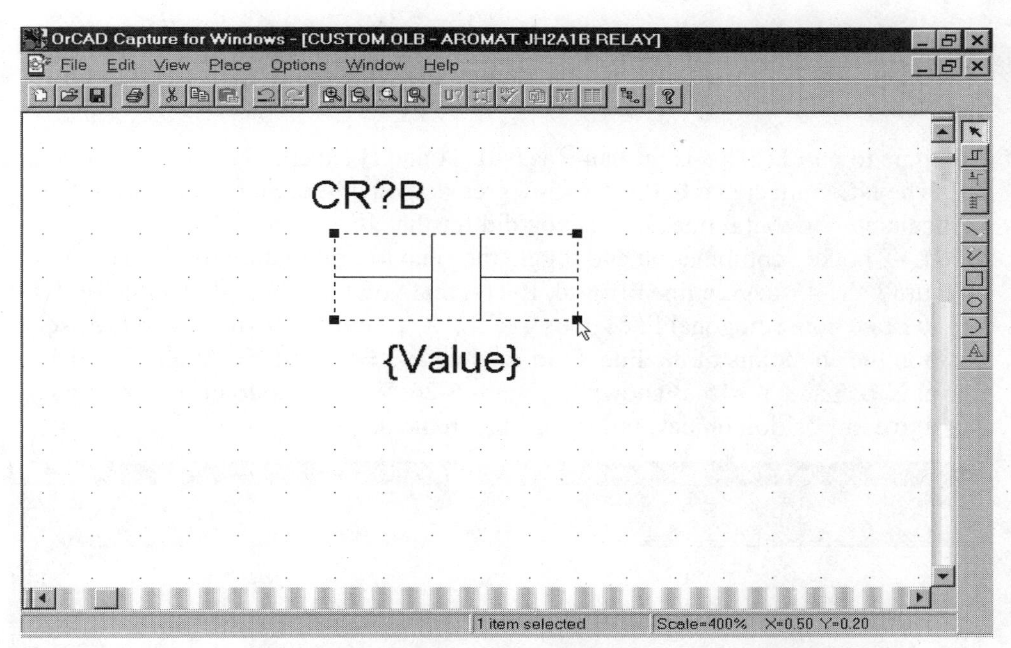

Figure 6-24 Dragging the Border to Match the Contact Outline

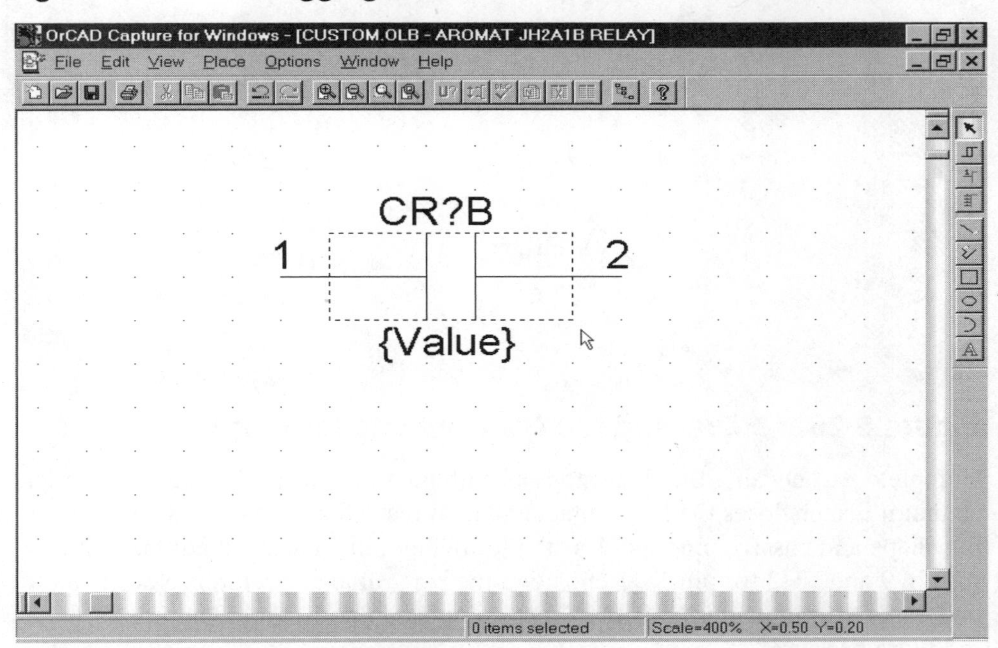

Figure 6-25 Completed Part B

After completing part B, follow the same steps to complete part C, the second set of NO contacts. Use the following pin names and numbers for part C: NO3 for 3 and NO4 for pin 4.

Continue to part D. This is the same as parts A and B except that the contacts are NC. The NC contact symbol has a 45-degree diagonal line. Start by drawing the vertical and horizontal lines just as you did for the NO contacts. The use the CTRL+T hotkey combination and toggle the snap to grid feature to off. You can now draw the 45-degree line off grid. Recall that you must hold down the SHIFT key to draw nonorthogonal lines. Observe the X,Y coordinate display to precisely position the endpoints of the line. Start the line at X=.15 and Y=.05 and end the line at X=.35 and Y=.15 as shown in Figure 6-26. When snap to grid is off, you can move or position objects in 1/10 grid increments.

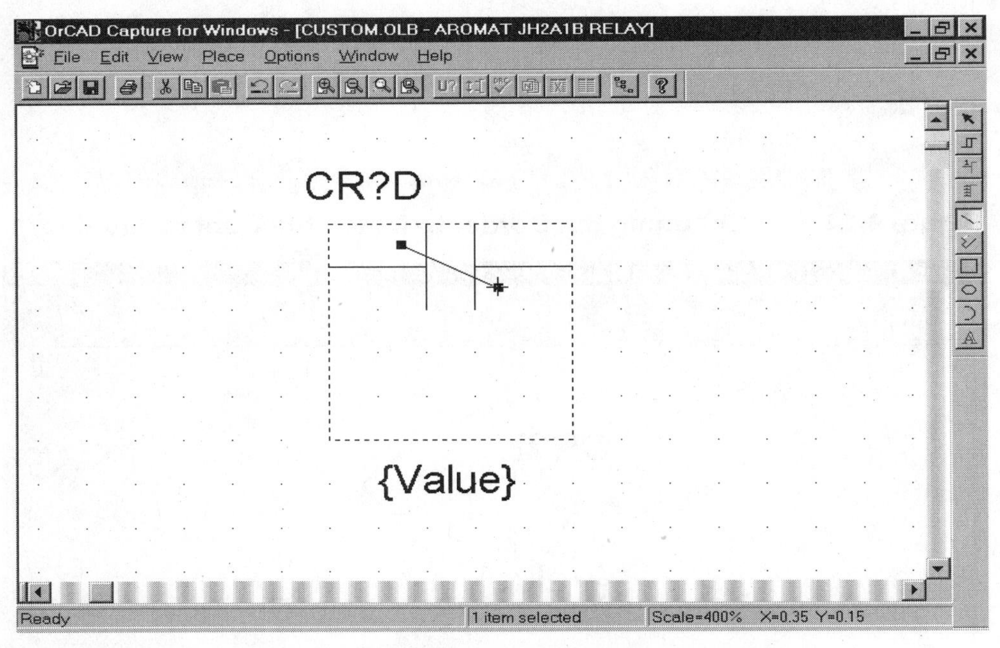

Figure 6-26 Drawing the NC Contacts for Part D

Complete part D using the same steps as with the previous parts. Drag the border so that it just encloses the NC contact and then place the contact pins. Use the short pin shape and passive pin type. Use the following pin names and numbers: NC1 for pin 9 and NC2 for pin 10. Reposition the part value and reference designator and make the pin names invisible. Your completed part D should appear as shown in Figure 6-27.

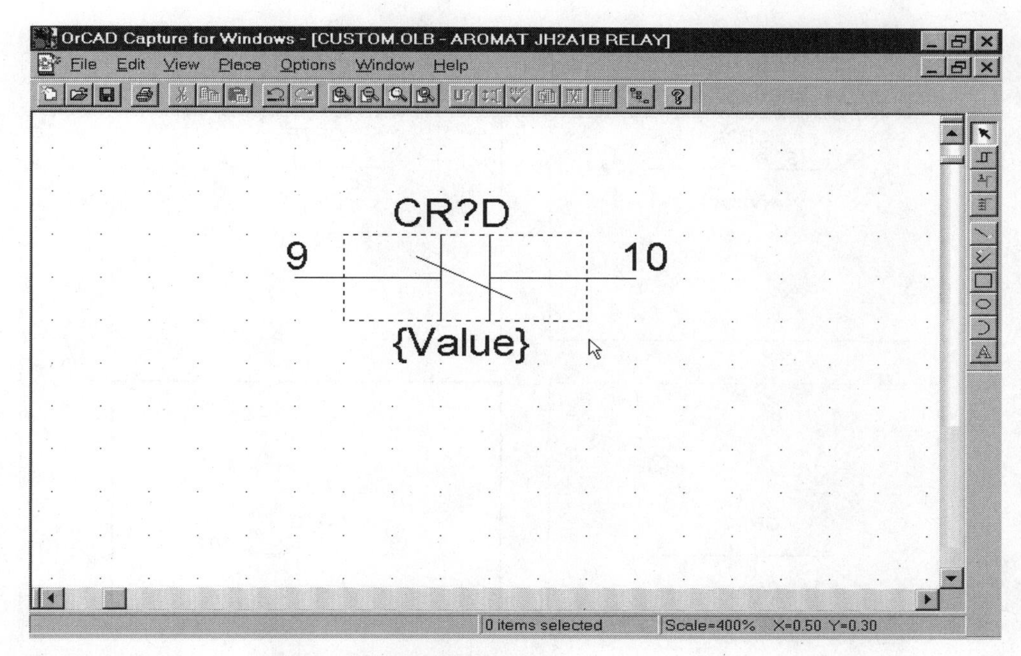

Figure 6-27 Completed Part D

You have now finished your first heterogeneous component. Use the package view feature from the View menu. The individual parts should appear as shown in Figure 6-28. Remember to save the changes to your Custom library.

On a final note about heterogeneous parts, try opening a new design and placing the various parts of the relay. Note that unlike homogeneous parts with more than one part per package, you cannot change the part after you have placed it. For example, if you place part B from the package, you cannot change it to part A.

You have now completed the tutorial exercise and learned how to use the part editor. This chapter concludes with some tips on creating parts and library management. Much of the focus of this chapter has been on the custom library. Creating new parts in the custom library is an integral step in the design process. Your custom library also represents a considerable investment in time and money. The following tips are intended to summarize the main points discussed in this chapter and to provide some guidelines for saving time and maximizing your return on investment.

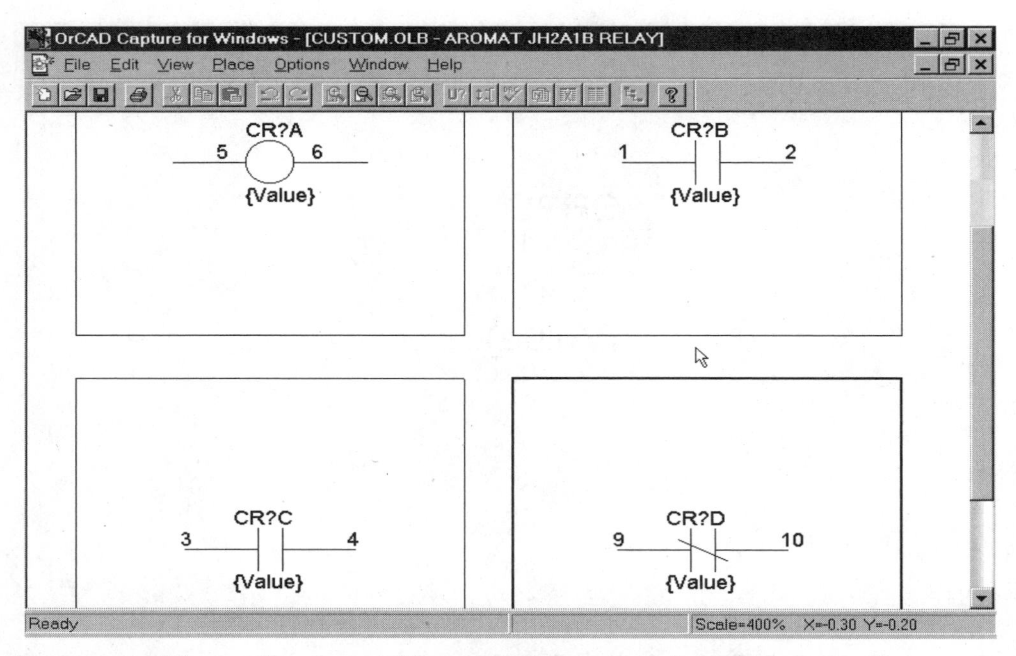

Figure 6-28 Package View of the Completed Relay

Tips on Creating Parts and Library Management

When creating a new part, spending some time up front to plan out the part usually pays off. Use the manufacturer's data sheet as a starting point for the body size and pin arrangement. Most component manufacturers now have Internet web sites that offer a fast and convenient means of downloading the latest part information.

Some data sheets show the pins arranged in the same order as on the device package. This is usually not the optimum arrangement from a circuit standpoint. The layout of a part should follow similar signal flow guidelines as used for schematics. Signal flow guidelines along with some suggestions on pin names are as follows:

- Signals should flow from left to right. This means you should place input pins on the left side and output pins on the right side.
- Bussed signals, such as multiple data, address, or device select signal lines, should be grouped together and ordered according to the signal, that is D0, D1, D3, and so on, regardless of the pin numbers. Signal order should always be from top to bottom.

- Power supply connections should be made at the top and bottom, with positive voltages at the top and negative voltages and ground at the bottom.
- Avoid invisible power pins except for simple TTL and CMOS logic devices used in single-supply designs. Never create analog or power devices with invisible power pins. Today's trend toward multiple voltage levels in logic circuitry renders invisible power pins more of a nuisance than a feature. Do not hesitate to edit existing logic parts to make power pins visible, if this will improve the clarity of your schematic. Copy and paste the logic parts to your custom library and then change their power pins to a passive type. This makes the pins visible and allows you to connect them directly to a power supply voltage.
- Use pin names as shown on the manufacturer's data sheet, unless an overwhelming reason exists to use a different name. Other technical documents, such as test and repair procedures, often reference the pin names used on schematics. When these pin names do not match part data sheets, confusion and errors can result.
- Pins with inverted (active low) signals can be represented either by a dot shape on the pin or by an overbar above the pin name. Do not use both representations, as they cancel each other out. Note that you cannot use a backslash at the end of a pin name, as it will result in an overbar above the last character in the name.
- When in doubt, use passive pins. Passive pins are general purpose and never cause electrical rules violations.

Never modify OrCAD-supplied libraries. Recall the caution in Chapter 1 about editing OrCAD-supplied libraries. If you edit an OrCAD-supplied library, all the parts you modified or created in that library will be lost when you load a subsequent software update that includes new library files. If you need to modify a part in one of the supplied libraries, first copy and paste the part to your custom library.

Avoid creating multiple custom libraries. They will become difficult to manage. Even a busy design department is not likely to wind up with more than a few hundred parts in the custom library since the OrCAD-supplied libraries are so extensive. Avoid creating a new part if the only change is to the part name and a library part already exists with the same pinout. This frequently occurs with logic devices as new logic families evolve. Many of the new 3V and 3.3V logic devices maintain the same pinout as standard TTL. Always save an updated archive copy of your custom library whenever you make a change or add a new part.

The Place IEEE symbol tool was mentioned briefly. Some companies use IEEE symbols to denote logic functions in accordance with ANSI/IEEE standards. IEEE symbology originated in the early 1980s and did not evolve to meet the needs of today's complex VLSI devices. Of all major vendors, only Texas Instruments has data books that still show IEEE symbols. Unless dictated by company policy, avoid using IEEE symbols.

Library files from older OrCAD versions such as SDT 386+ must be converted before use in Capture. This topic is discussed in Chapter 11.

Creating Special Symbols

You can use the part editor to create special symbols such as power and ground symbols, off-page connectors, hierarchical ports, and title blocks. Store these in your Custom library; do not edit the OrCAD supplied Capsym library. Use the New Symbol command from the Design menu. Then select the appropriate symbol type. You can use the power symbol selection for both power and ground symbols. No difference exists between power and ground symbols other than their name and intended usage. The name of the symbol determines its net association. Note that you can edit the symbol name when you place it. Power and ground symbols both use zero length invisible power type pins. Off-page connectors and hierarchical ports use zero length invisible pins. You can select the appropriate signal type depending on the intended usage.

Information Sources

Manufacturers' data sheets are the best source of information about part packages and pin arrangements. The electronics industry is rapidly transitioning away from printed data books to CD-ROM and online media. Many companies now publish several editions of their data on CD-ROM on a yearly basis. Manufacturers' and distributors' Internet web sites are the best source for immediate access to information about new parts. Several publishers specialize in compiling electronic part data. Two excellent sources are the *IC Master* published by Hearst Business Communications (phone: 516-227-1300; URL: www.icmaster.com) and the *D.A.T.A. Digests* published by Information Handling Services (phone: 800-447-4666). Both are published on a yearly basis. As the name implies the *IC Master* specializes in ICs. Individual *D.A.T.A. Digests* are available for various types of discrete semiconductors and ICs. Both the *IC Master* and *D.A.T.A. Digests* are available in printed or CD-ROM media. The *IC Master* CD-ROM includes a comprehensive directory of manufacturers' web sites. The *IC Master* and individual *D.A.T.A. Digests* are relatively expensive, each costing about two hundred dollars. Another source of information is the *EEM* (Electronic Engineers Master Catalog) also published by Hearst Business Communications (URL: www.eemonline.com) on a yearly basis. The *EEM* covers a wide range of electronic and electromechanical parts. Most engineers can qualify for a free copy of the *EEM*. An *EEM* CD-ROM is available for a nominal fee.

An additional resource worth mentioning is the *EITD* (Electronic Industry Telephone Directory) published by Chilton (phone: 610-964-4000). Most readers of *ECN* (Electronic Component News) qualify for a free copy of the *EITD*. The *EITD* lists some web sites, but a quick call to a manufacturer's home office is often the easiest way to get the web site URL or to request data books.

Review Exercises

1. Why should you avoid making changes to OrCAD-supplied libraries?
2. What is meant by the term *Convert View*?
3. Describe the two-part numbering schemes Capture uses for multiple-part packages. Which scheme is preferred? Why?
4. Describe the special tools unique to the part editor.
5. List the four basic pin properties.
6. What convention is used for names of inverted (active low) signal pins?
7. List and draw out the six available pin shapes. Describe the use of each pin shape.
8. What type of pins can be invisible?
9. What two steps are required to save a new part in your Custom library?
10. How do you make pin names and numbers invisible?
11. Describe the difference between homogeneous and heterogeneous multiple-part packages and give an example of each.
12. How do you navigate between the different parts in a multiple-part package?
13. Describe the suggested guidelines for bussed signal pins.

14. What is the difference between power and ground symbols? How is the net association determined?
15. If you have access to the Internet, visit the Maxim web site (www.maxim-ic.com) and download the data sheet for the MAX232A. This is an industry standard RS-232 interface IC. Create a custom library part for the MAX232A.

7

Advanced Features

The exercise in this chapter is the last of a series designed to get you off to a fast start using Capture. In this chapter, you will explore some of Capture's advanced features. The exercise is divided into three sessions. In the first session, you will draw a hierarchical schematic for a memory expansion board that requires isolated power in one section. The hierarchical structure is preferred for large and complex designs. Capture also provides for conventional multiple-sheet schematics, which are referred to as a *flat* design structure. In the second session, you will redraw the schematic using a flat design structure. The third session introduces advanced bill of materials and netlist techniques. The third session also introduces the gate and pin swap tool.

First Session – Using a Hierarchical Structure for a Small Design

In previous exercises, hierarchical schematics were drawn in a formal fashion, with the first sheet representing the hierarchical structure. Only hierarchical blocks (representing major circuit blocks drawn on subsequent sheets) and the interconnections between these blocks appeared on the first sheet.

For small designs, such as the memory expansion board used as the model for this session, a less formal approach improves readability. Use Figures 7-1 through 7-3 as the model for the first session. The main circuit block appears on the first sheet (Figure 7-1). Additional circuit blocks are shown as hierarchical blocks and then detailed on subsequent sheets (Figures 7-2 and 7-3). With the more formal approach, an additional sheet would have been required just to show the hierarchical structure.

As a general rule, try to keep the number of sheets to a minimum. For designs requiring more than four sheets, reserve the first sheet for hierarchical blocks showing the design structure. For less complex designs, place the main circuit block on the first sheet along with hierarchical blocks for the remaining sheets.

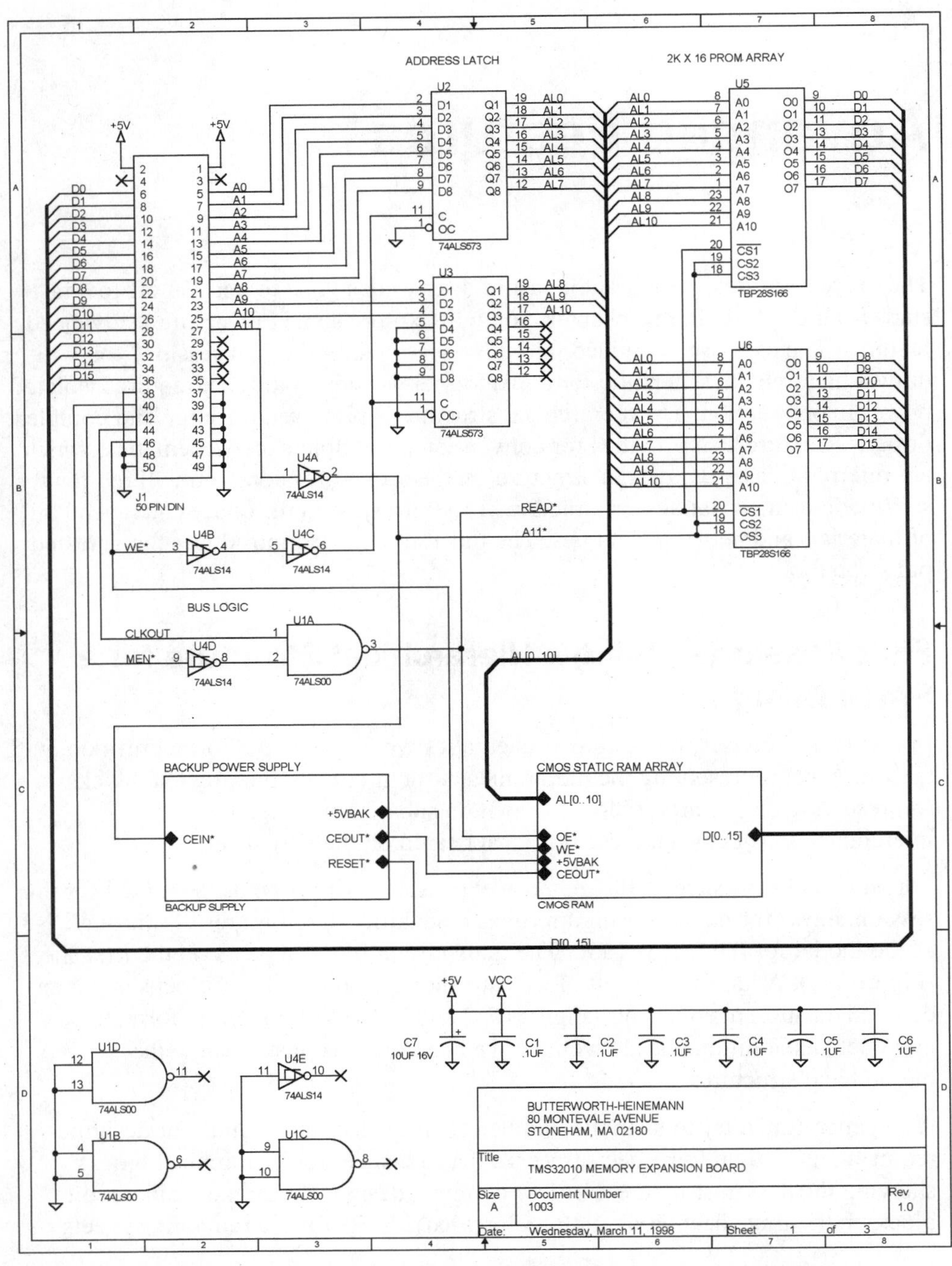

Figure 7-1 Memory Board (Sheet 1 – Hierarchical Design)

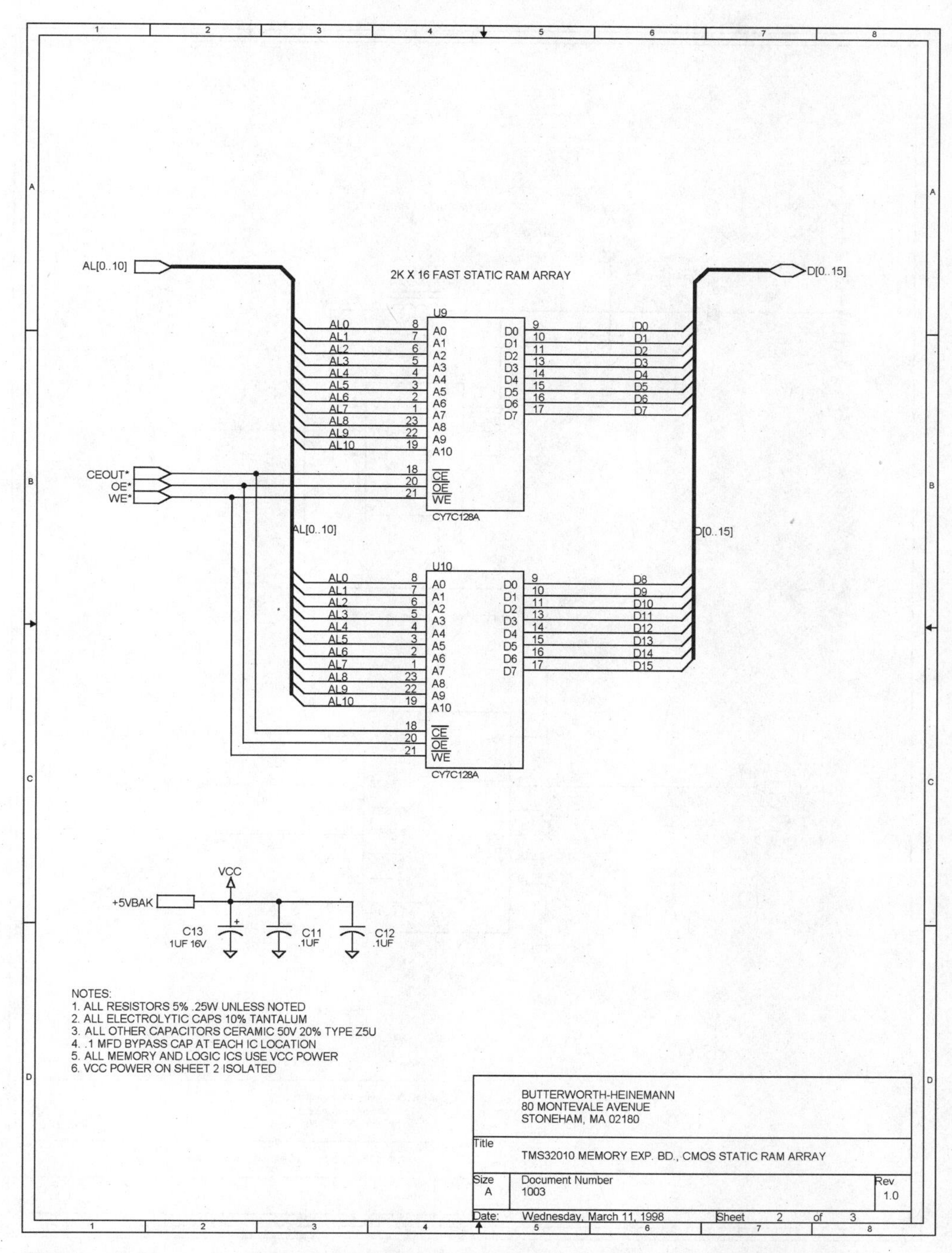

Figure 7-2 Memory Board (Sheet 2 – Hierarchical Design)

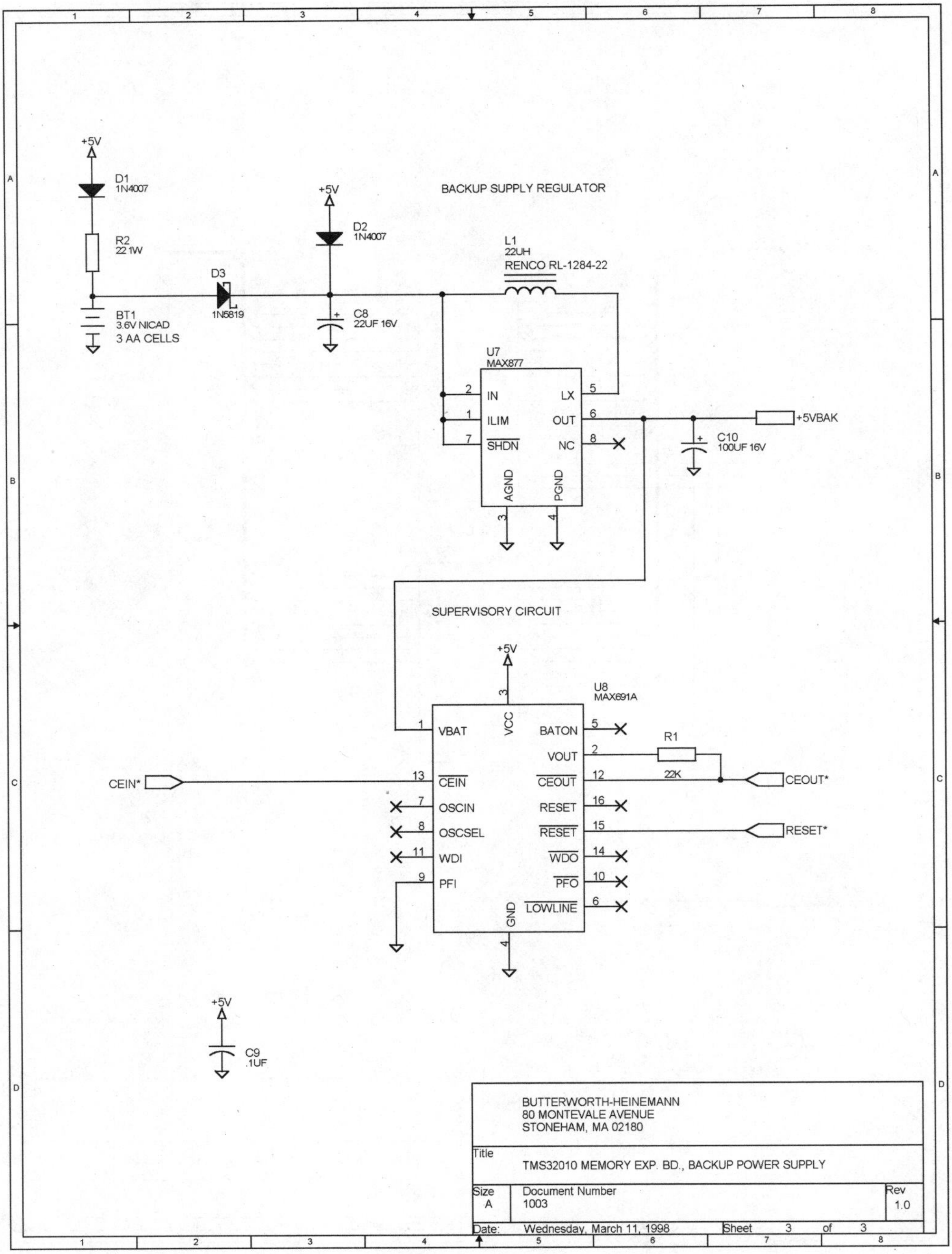

BACKUP SUPPLY REGULATOR

+5V
D1
1N4007
R2
22 1W
D3
1N5819
BT1
3.6V NICAD
3 AA CELLS
+5V
D2
1N4007
C8
22UF 16V
L1
22UH
RENCO RL-1284-22
U7
MAX877
IN
ILIM
SHDN
LX
OUT
NC
AGND
PGND
+5VBAK
C10
100UF 16V

SUPERVISORY CIRCUIT

+5V
U8
MAX691A
VBAT
VCC
BATON
VOUT
CEIN
CEOUT
OSCIN
RESET
OSCSEL
RESET
WDI
WDO
PFI
PFO
GND
LOWLINE
R1
22K
CEIN*
CEOUT*
RESET*
+5V
C9
.1UF

BUTTERWORTH-HEINEMANN
80 MONTEVALE AVENUE
STONEHAM, MA 02180

Title
TMS32010 MEMORY EXP. BD., BACKUP POWER SUPPLY

Size A
Document Number 1003
Rev 1.0

Date: Wednesday, March 11, 1998
Sheet 3 of 3

Figure 7-3 Memory Board (Sheet 3 – Hierarchical Design)

Drafting the Schematic

Start the exercise by creating a new folder called Tutor4. Then create a new design called TMS_Memory using the techniques you learned in the previous chapters.

By now you have mastered the basic skills required to complete the schematic shown in Figures 7-1 to 7-3. Since this exercise introduces some new techniques, just follow along for now. The following are hints on parts to help you get started:

- **50 Pin Din connector**: CON50A from the Device library.
- **74ALS series logic ICs**: these are in the TTL library. The 74ALS14 is shown with a smaller symbol (for convenience and fit) than the part in the TTL library. Use the 74HCT14 from the Custom library and edit the description.
- **TBP28S166 TTL PROM**: 28S166 from the Memory library.
- **CY7C128A CMOS Static RAM**: 7C128 from the Memory library.
- **3.6V Nicad battery**: BATTERY from the Device library.
- **1N5819 Schottky diode**: DIODE SCHOTTKY from the Device library.
- **22UH Inductor**: INDUCTOR IRON from the Device library.
- **MAX877 and MAX691A ICs**: these are in the Custom library. The MAX877 IC was created as part of the exercise in the previous chapter.

Creative Use of Copy and Paste

Previous exercises introduced the copy and paste tools. Since digital schematics tend to contain repeated circuit elements, creative use of copy and paste operations can save a considerable amount of drafting effort. This is especially true of address and data bus structures and memory arrays. Plan ahead, even to the extent of first drawing a very rough sketch by hand, before starting on the schematic. If you have some idea of what the finished schematic will look like, you can intelligently copy and paste repeated circuit elements.

When you select multiple objects for copy and paste or move operations, you can also use the Group command from the Edit menu to associate the objects. When you then select one of the objects in the group, the entire group is selected. The Ungroup command dissociates grouped objects.

Let's look at how you can save time on sheet 1. Place the 50-pin DIN connector and the 74ALS573 address latch along with signal wires and buses as shown in Figure 7-4. Before copying, set your schematic page editor Area Select preference to Fully Enclosed (see page 56 in Chapter 2 for details).

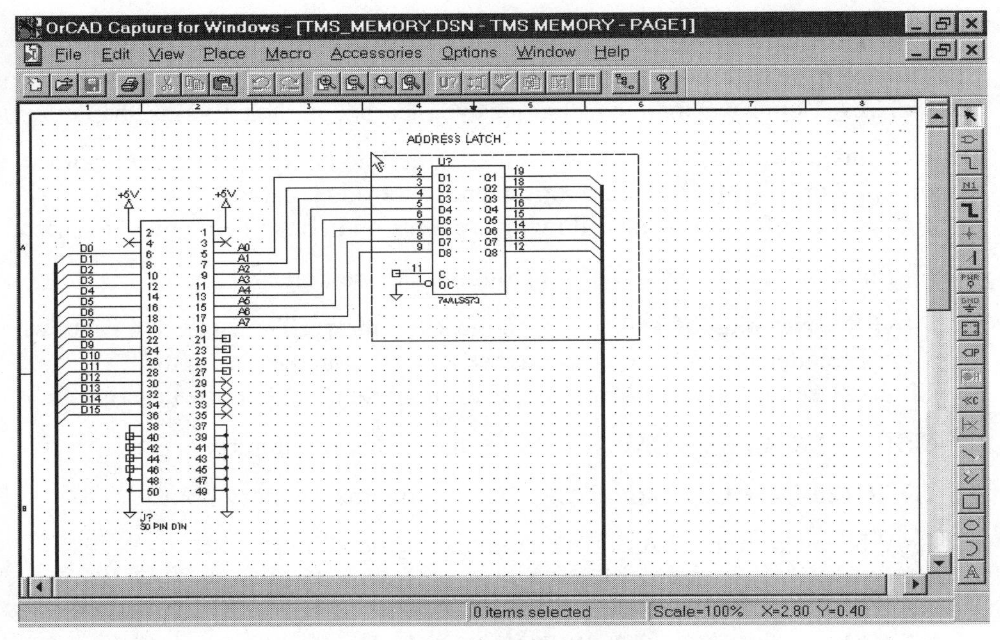

Figure 7-4 Selecting the First Address Latch

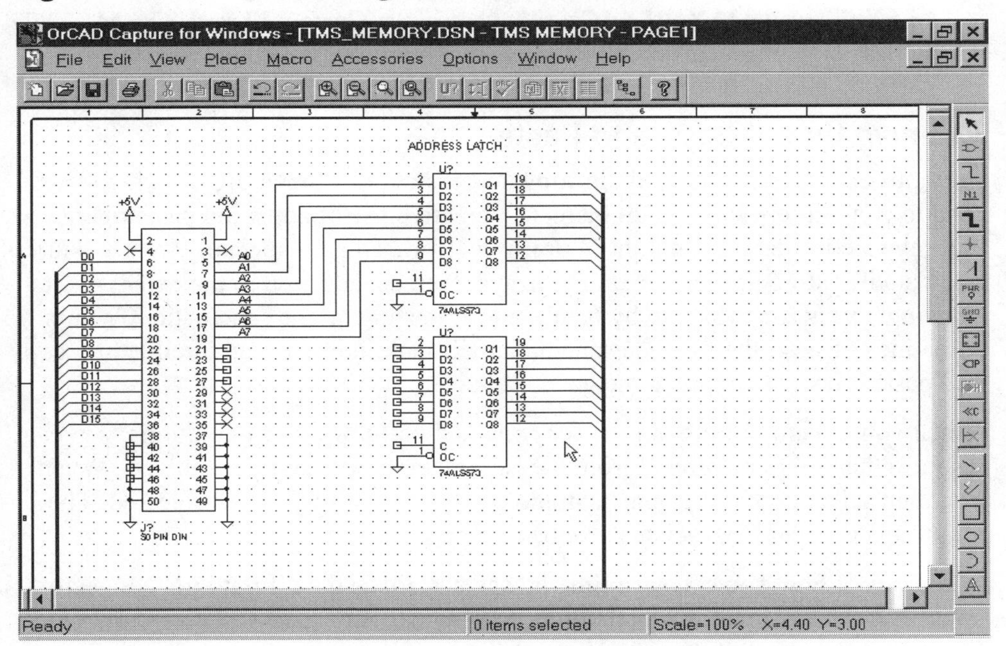

Figure 7-5 Pasting a Copy of the Address Latch

This means that only objects fully enclosed in the selection box will be copied. Select the 74ALS573 address latch and bus structure as shown in Figure 7-5 and click on Copy. Then click on Paste and position the objects as shown in Figure 7-5. The next step is to delete the wires and bus entries to Q4 through Q8 of the second address latch, since these are not used. Change the area-select preference to intersecting. Then select and delete the objects as shown in Figure 7-6.

Complete the address latch circuitry by placing the appropriate wires, no connects, and net aliases. When placing the net aliases, start with A8 and AL0 so that you can make best use of the autoincrement feature. The completed address latch circuitry should appear as shown in Figure 7-7.

Next, place the first TBP28S166 PROM and associated bus structure as shown in Figure 7-8. Change the area-select preference back to fully enclosed. Then select and copy the objects shown in the figure. Note that the address bus entries are not enclosed in the selection area, since bus entries with a different orientation are required for the second PROM. Paste a copy of the objects as shown in Figure 7-9.

Complete the PROM array by placing bus entries and net aliases. When placing the net aliases, start with D0 so that you can make use of the autoincrement feature. The completed PROM array should appear as shown in Figure 7-10.

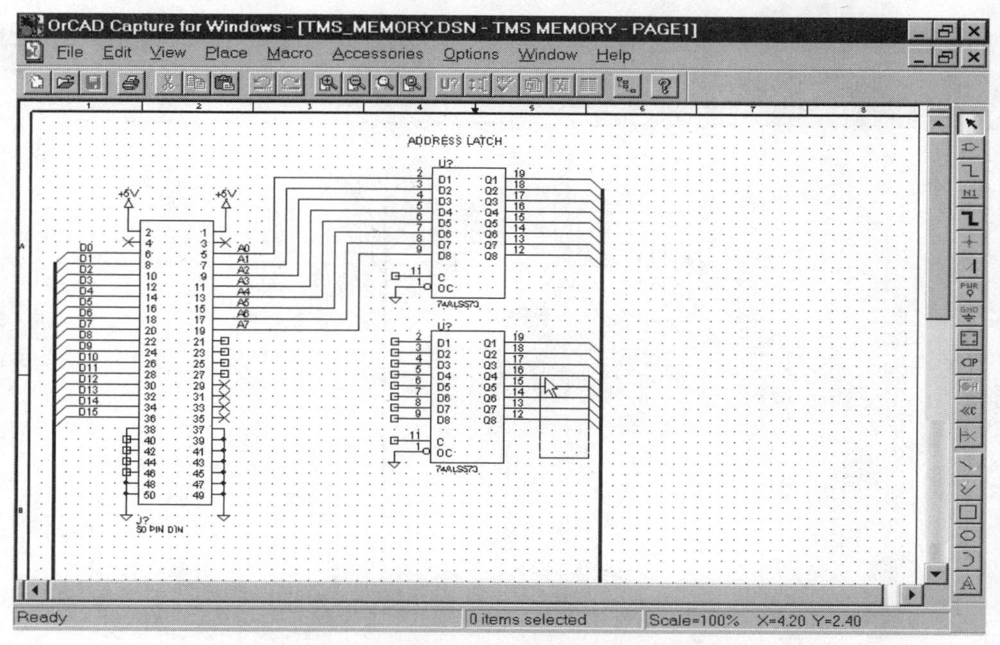

Figure 7-6 Deleting a Part of the Bus Structure

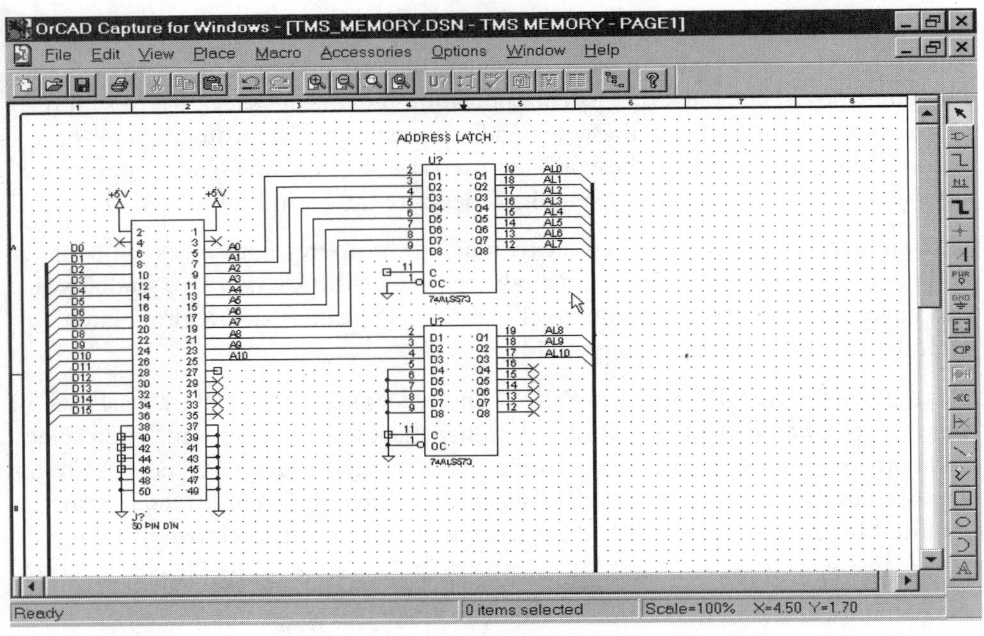

Figure 7-7 Completed Address Latch Circuitry

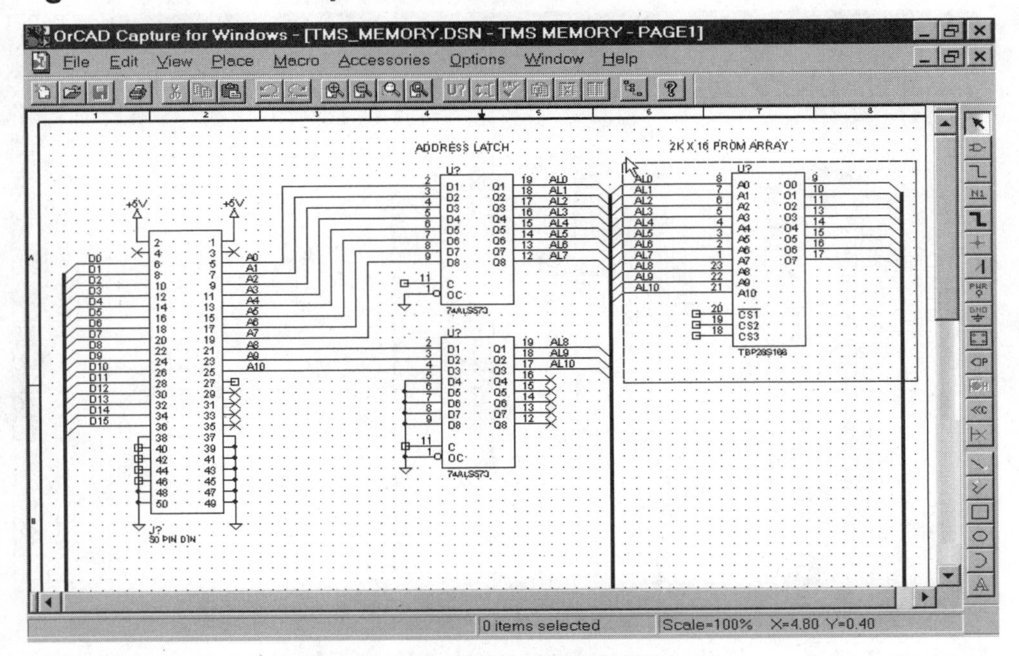

Figure 7-8 Selecting the First PROM

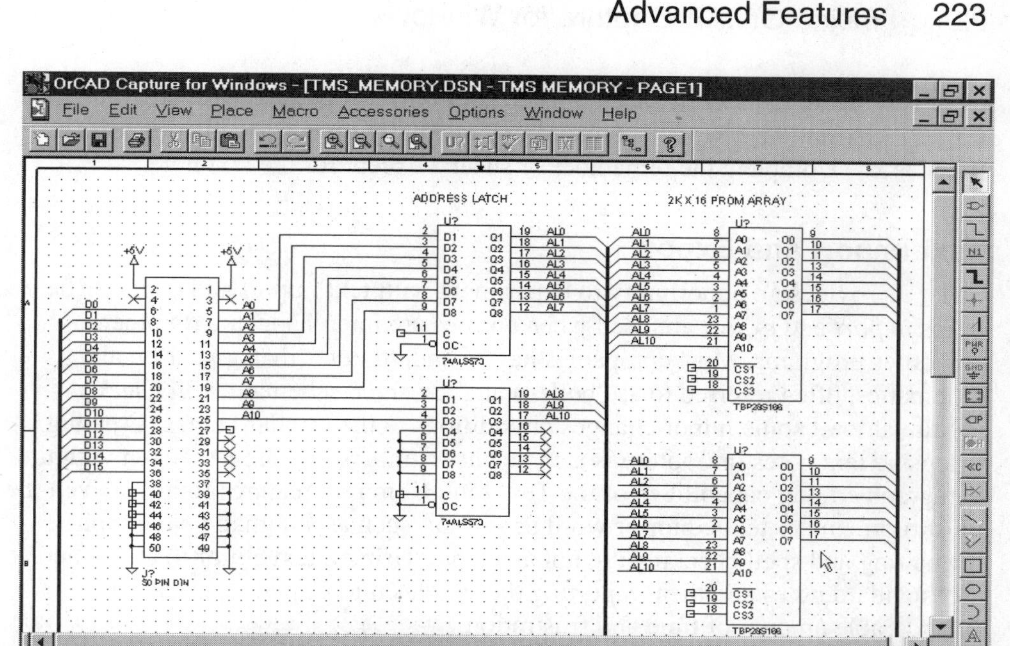

Figure 7-9 **Pasting a Copy of the PROM**

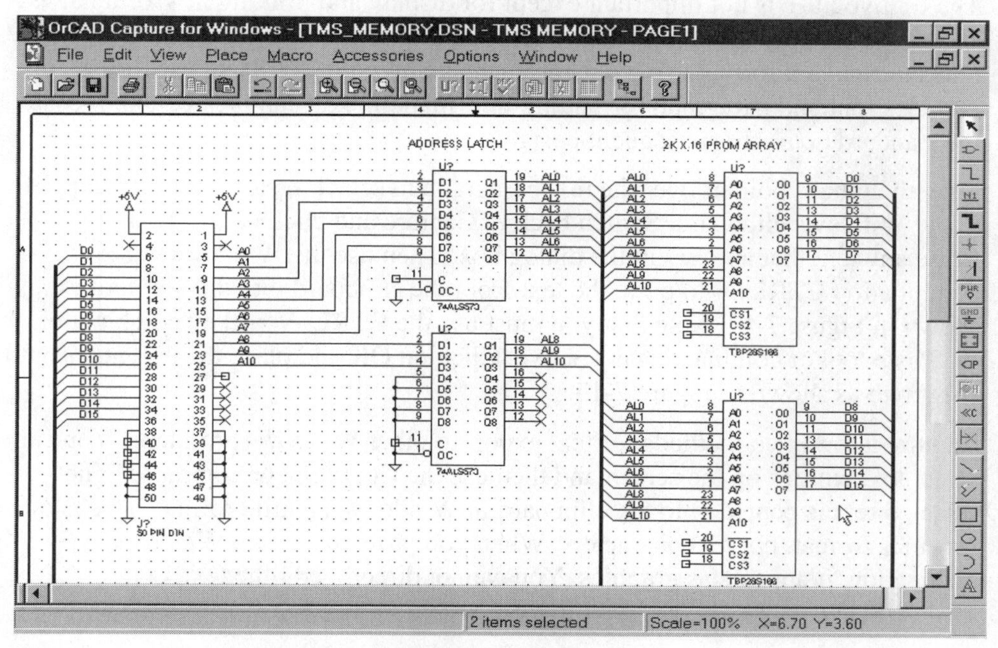

Figure 7-10 **Completed PROM Array**

Complete the remainder of sheet 1. Then start on sheet 2. After drawing the first CY7C128A static RAM, you can use copy and paste operations to complete the RAM array. Complete the remainder of sheet 2 except for the text note.

Text Import and Export

Sheet 2 provides an opportunity to experiment with text import and export. The terms *import* and *export* are used when entities are interchanged with other Windows programs. Most companies have standard notes that appear on all schematics, often referred to as "boilerplate." A given schematic will usually require at least some minor changes or additions to the standard notes. Creating extensive text notes in Capture is a somewhat cumbersome process, even with the paragraph editing capabilities of the Text tool. Common features found in even the most rudimentary text editor or word processor, such as automatic paragraph formatting and spell checking are not available. Using a text editor to create and edit standard notes and then importing these notes into Capture can save a considerable amount of time and frustration.

Use text import to place the notes appearing at the bottom left corner of sheet 2. Open the Windows WordPad. Enter the text for the notes as shown in Figure 7-11. The font you use is not important except for display and printing in WordPad, as Capture will apply the default font when you import the text. You can save these notes in rich text format using a name such as Std_notes.rtf. Rich text format includes font and paragraph formatting information. You can then edit and reuse the standard notes for subsequent designs.

Drag the mouse pointer over the text paragraph to select the text and then copy it to the Windows clipboard. Switch back to Capture and click on the Text tool. Erase any text that appears in the dialog box. Then use the Windows hotkey combination CTRL+V to paste the text from the clipboard into the text tool as shown in Figure 7-12. Note that you must use the CTRL+V hotkey combination, the Paste tool or command is not active. Click on OK and place the text notes onto sheet two as shown in Figure 7-13.

Capture allows two methods for text export to other Windows programs. You can select text entities on the screen and then use the Copy tool or command. The selected text is copied onto the clipboard in Windows metafile format. This is a graphics format, not text characters. When you paste this into a Windows application, it appears as graphics. You cannot directly edit the text.

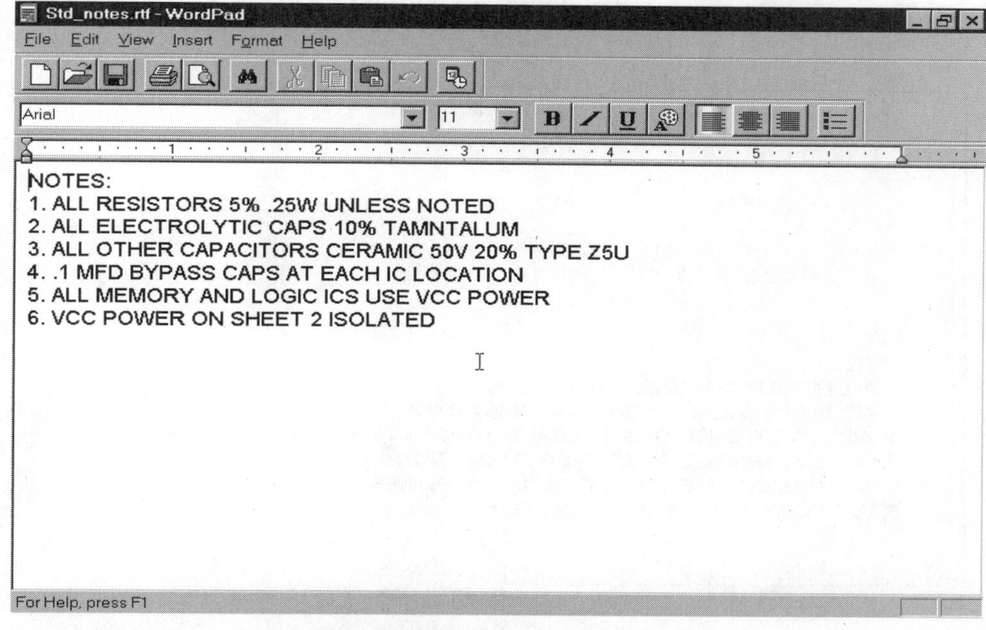

Figure 7-11 WordPad with Text Notes

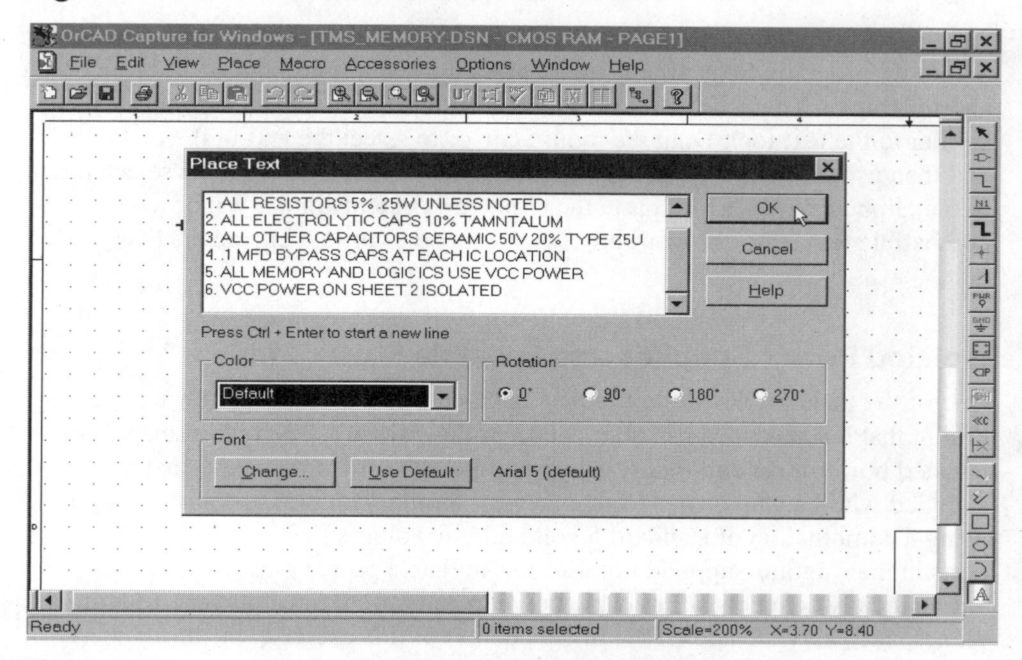

Figure 7-12 Text Notes Imported into Capture

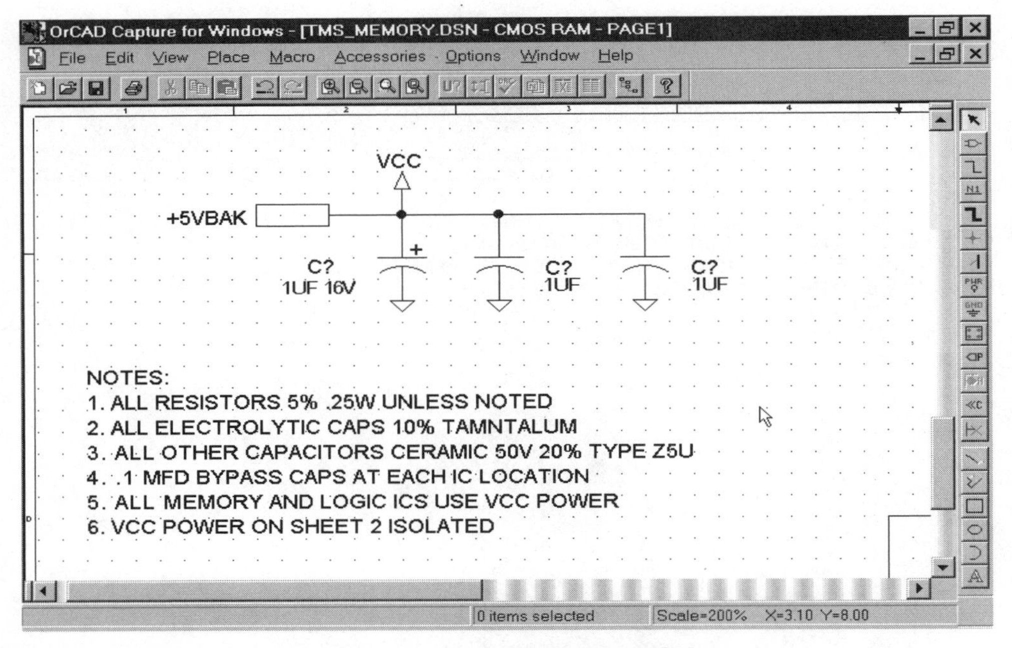

Figure 7-13 Completed Text Notes on Sheet 2

An alternative method is available for exporting text that you can edit in another Windows program. Select the text and press the right mouse button to bring up the shortcut menu. Then click on Edit Properties. This brings up an edit dialog box similar to the text tool. Drag the mouse cursor to select the text in the dialog box and then use the Windows hotkey combination CTRL+C to copy the selected text to the clipboard. When you paste the clipboard contents into a Windows-compatible text editor or word processor, it appears as text characters that you can edit.

Isolated Power Supplies and Invisible Power Pins

One of the main points of the exercise in this chapter is learning how to create a design that isolates power in one section of the circuitry. The requirement for isolated power most commonly occurs with CMOS memory arrays that are supplied with backup power from a battery. Similar situations occur in designs using a combination of standard 5 volt and low voltage (2.7-3.3 volt) logic. Most parts in the Capture-supplied libraries use invisible power pins. Capture automatically connects all invisible power pins to predefined power and ground planes, such as VCC and GND. In the context of this exercise, the terms *power supply* and *power plane* have identical meaning. Engineers think in terms of power supplies; PC designers think in terms of power planes.

Fortunately, Capture provides a simple technique that allows isolating power to a particular circuit block. The only significant limitation is that the circuit block with isolated power must be placed on a separate sheet. In the exercise, sheet 2 (Figure 7-2) uses isolated +5VBAK power from a battery backup supply on sheet 3 (Figure 7-3). The isolated +5VBAK power is brought onto the second sheet via an unspecified type hierarchical port that is tied to a VCC power symbol.

Capture recognizes an unspecified port connected to a power symbol as a special instruction to create an isolated power plane on the particular sheet. All other power symbols and invisible power pins with the same name on that sheet will become part of the isolated power plane. Power symbols and invisible power pins on other sheets are not affected.

The same technique can be used to create multiple isolated power planes on a given sheet. Multiple sheets can have isolated power planes by repeating the technique on every affected sheet.

The same isolation technique can be used with ground symbols and invisible ground pins.

If logic circuitry requiring multiple or isolated power supplies must be shown on a single sheet, one must create custom parts with visible power and ground pins. These pins can then be individually connected to power and ground symbols named after the appropriate supply voltages.

One caveat is that the special technique Capture uses to create isolated power supplies is by no means a universally accepted standard. At the very least, you must include an explanatory text note on the schematic. Even with an appropriate note, the special Capture isolation technique may cause confusion. Ask yourself whether or not all those who will read the schematic during the product's life cycle are likely to fully understand what you had in mind. If you have any doubts, spend the extra time to create new parts with visible power and ground pins.

Creating New User Properties

Capture allows you to create new user-defined properties for many different types of objects. User properties are particularly useful for adding information to parts such as additional description (value) lines, vendor information, or in-house part numbers. You can extract this information using the Create Bill of Materials tool.

Sheet 3 has two components, the battery and the inductor, that require a second description line (see Figure 7-3). You must define a new user property to add this second description line. Start with the battery. Double click on the part to bring up the Edit Part dialog box and then click on User Properties. When the User Properties dialog box appears, click on New. This brings up the New Property dialog box shown in Figure 7-14.

You can enter the name of the new user property and its value. Use the name VALUE_2 for the second parts value description line. Then enter the value as shown and click on OK. Your new user property appears at the bottom of the user properties list below the part value. Note that Capture sorts this list in alphanumeric order based on the property name.

Click on the new property (VALUE_2) to select it and then click on Display. This brings up the Display Properties dialog box shown in Figure 7-15. Change display properties including the color and font selections. Click on Visible to make your new property appear on the screen. Then click on OK.

The User Properties dialog box now appears as shown in Figure 7-16. Note that the "V" in the attributes column for VALUE_2 denotes that the property is visible. Click on OK. Then position the second description line as shown in Figure 7-17.

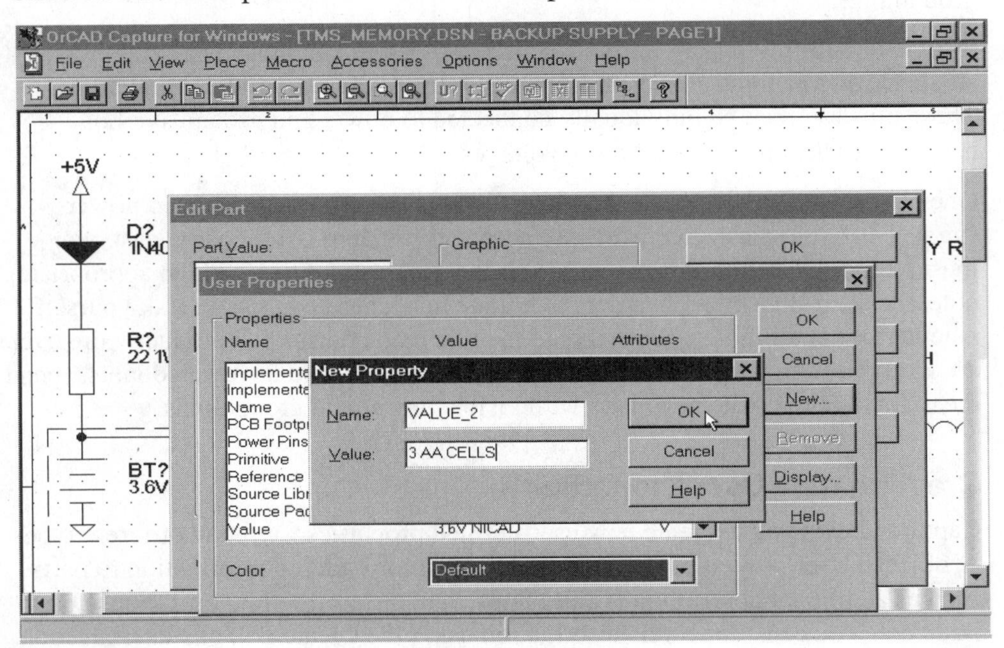

Figure 7-14 Dialog Box for New Property

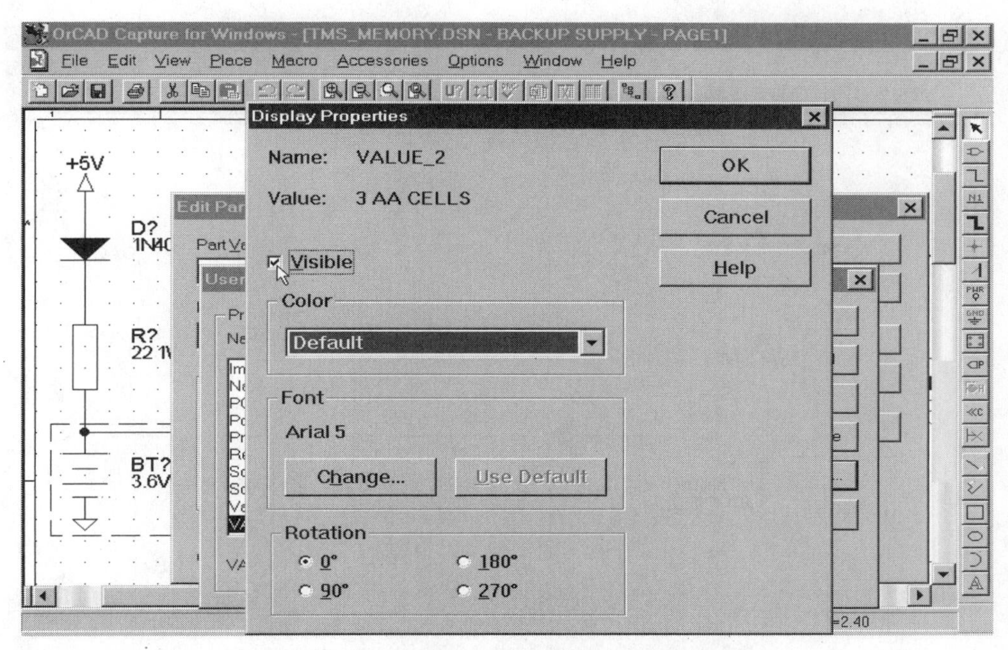

Figure 7-15 Dialog Box for Display Properties

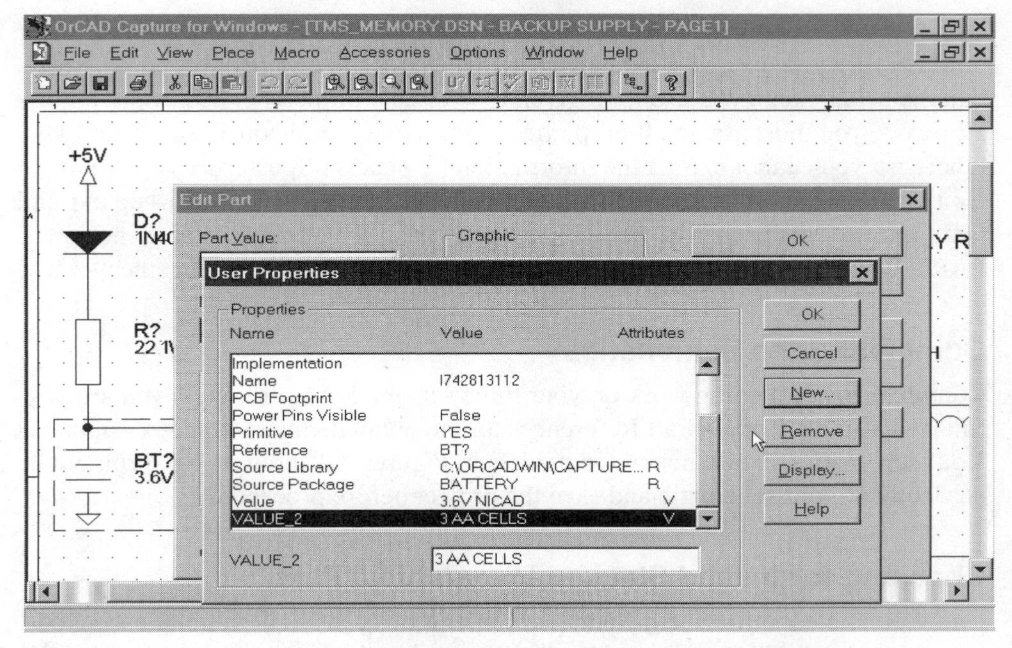

Figure 7-16 Dialog Box Showing New User Property

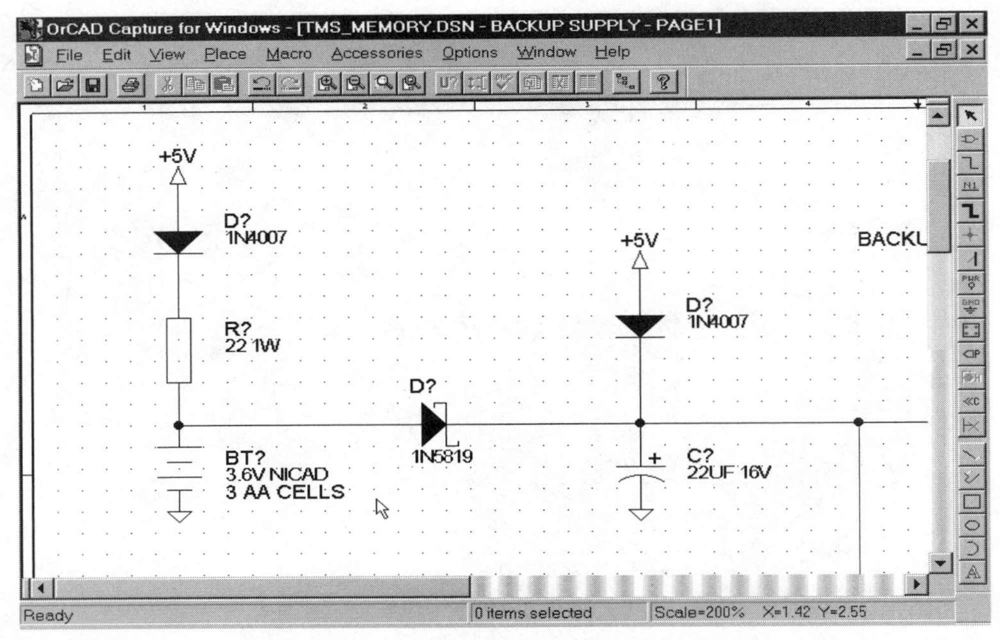

Figure 7-17 New User Property in Final Position

Use the same technique to add the second description line to the inductor on sheet 3. Make sure you use the same name VALUE_2 for the new user property. Note that the actual name you use for a given type of user property is irrelevant. However, you must assign all properties of the same type a consistent name, otherwise you cannot extract the information by means of postprocessing tools. You could just as well have used a name such as VALUE FIELD 2. Keep in mind that Capture sorts properties in alphanumeric order. If you used a name such as PART VALUE 2, the two part-value properties would not appear together.

Completing the Schematics

Complete any remaining work on your three schematic sheets. When you are finished, run the Update Part References tool to annotate the reference designators. Your schematics should match the models in Figures 7-1 through 7-3. Print out a hardcopy of your schematic and save the project before proceeding.

Checking for Invalid Stacked Hierarchical Pins

Under certain conditions, copy and paste operations can result in invalid stacked hierarchical pins. The term *stacked pins* means that the pins are located on top of one another that cannot be deleted. You cannot easily detect or directly delete stacked pins. Capture provides a special tool for this purpose. You can run the Check for Invalid Stacked Hierarchical Pins tool from the Accessories menu to report and fix stacked pins. Good practice is always to run this tool after completing any design containing hierarchical pins.

In Project Manager, click on the design to select it and then run the Check for Invalid Stacked Hierarchical Pins tool. The tool provides options to report or fix invalid pins. Select the report option. After the tool completes, check the results in the session log. If any invalid pins are found, rerun the tool with the fix option. Next, run the Design Rules Check tool. Set the ERC matrix to flag interconnected power supplies with a warning. You will get a warning about a possible conflict between VCC and +5V, which is expected since these supplies are deliberately interconnected. Run the DRC tool again with the delete markers option selected.

Creating a Bill of Materials with User Properties

You can easily create a bill of materials that includes the new VALUE_2 property. From Project Manager, click on the Bill of Materials tool. The dialog box shown in Figure 7-18 appears.

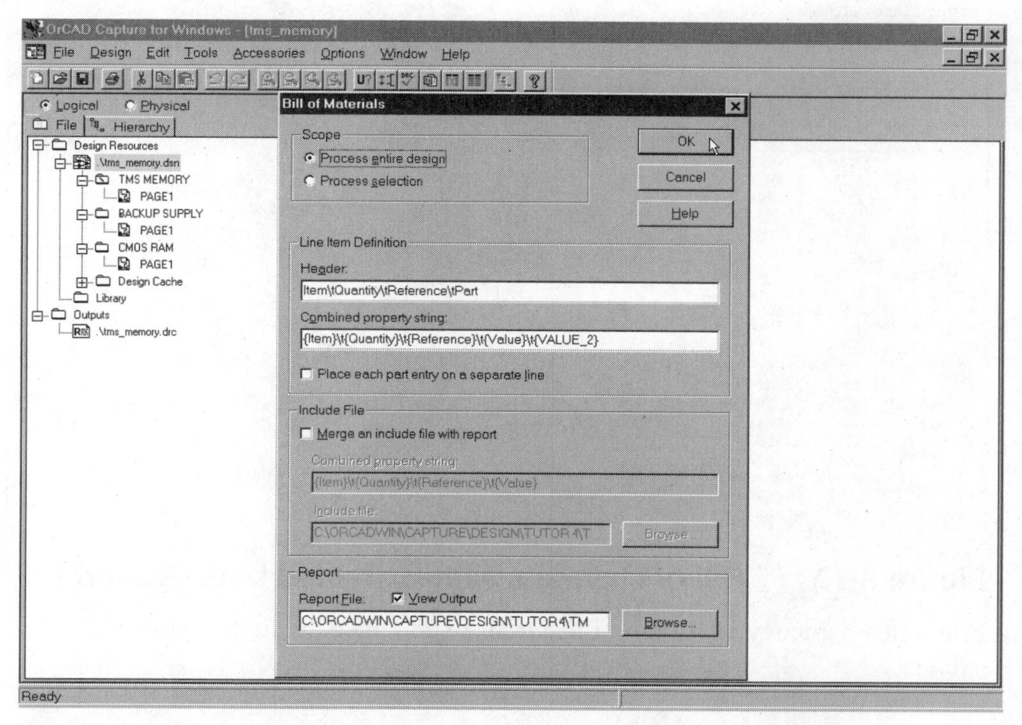

Figure 7-18 Dialog Box for the Bill of Materials Tool

The Line Item Definition option box includes a field for a combined property string. This field is used to format the contents of each line item on the bill of materials. The combined property field also determines which properties must match for parts to be grouped together. The default combined property string includes the item (system-generated line number used for formatting purposes only), quantity, reference, and value properties. Note that each property is enclosed in curly brackets { } and separated by a "\t" sequence, which inserts a tab character between properties.

To extract the VALUE_2 user property, simply add this property to the combined property string as shown in Figure 7-18. Make sure you enclose the VALUE_2 in curly brackets and precede it with the "\t" sequence to insert a tab character. Then click on OK to continue. The bill of materials with the extracted user property appears as shown in Figure 7-19.

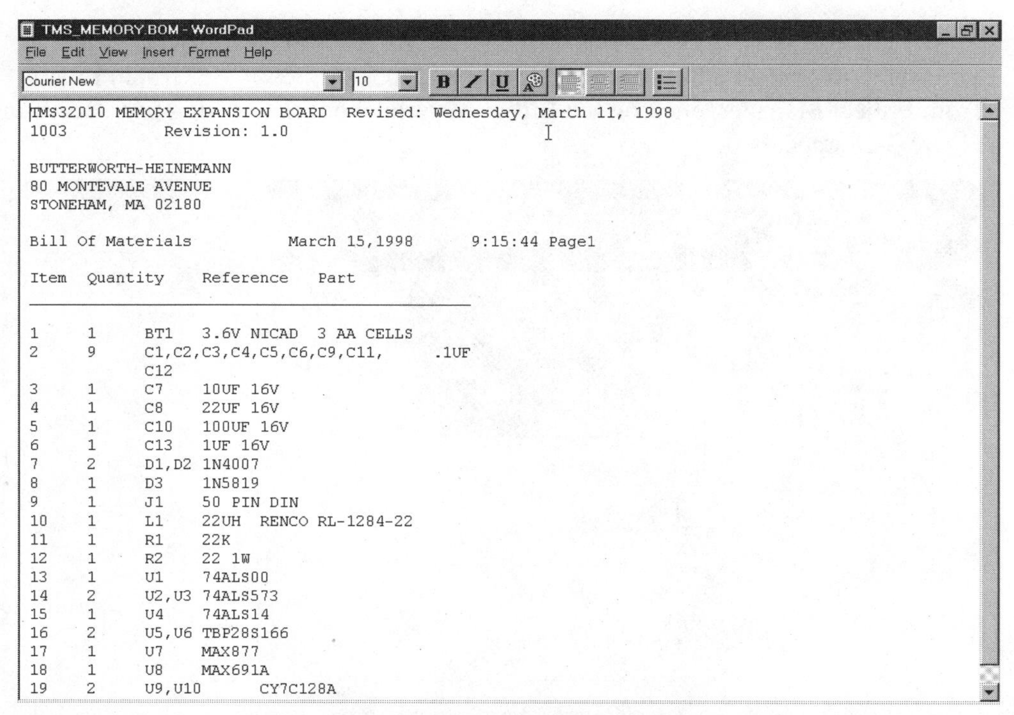


```
TMS32010 MEMORY EXPANSION BOARD  Revised: Wednesday, March 11, 1998
1003          Revision: 1.0

BUTTERWORTH-HEINEMANN
80 MONTEVALE AVENUE
STONEHAM, MA 02180

Bill Of Materials          March 15,1998      9:15:44 Page1

Item  Quantity    Reference    Part
______________________________________________

1     1     BT1   3.6V NICAD  3 AA CELLS
2     9     C1,C2,C3,C4,C5,C6,C9,C11,     .1UF
            C12
3     1     C7     10UF 16V
4     1     C8     22UF 16V
5     1     C10    100UF 16V
6     1     C13    1UF 16V
7     2     D1,D2  1N4007
8     1     D3     1N5819
9     1     J1     50 PIN DIN
10    1     L1     22UH  RENCO RL-1284-22
11    1     R1     22K
12    1     R2     22 1W
13    1     U1     74ALS00
14    2     U2,U3  74ALS573
15    1     U4     74ALS14
16    2     U5,U6  TBP28S166
17    1     U7     MAX877
18    1     U8     MAX691A
19    2     U9,U10      CY7C128A
```

Figure 7-19 Bill of Materials with Extracted User Property

Print out a hardcopy of the bill of materials to complete the first session of the exercise.

Second Session – Using a Flat Design Structure

In the second session you will learn about another type of Capture schematic structure, referred to as a *flat design*, which resembles traditional multiple-page schematics. You will convert the TMS_Memory hierarchical design that you created in the previous session to a flat design. You will also modify the CMOS static RAM array on the second sheet to use visible power pins. This approach clarifies the use of isolated power on the second sheet.

Overview of the Flat Design Structure

Up to now, all the schematics you have created in the exercises have been either single sheet or hierarchical structure. Single-sheet schematics are limited to small, simple designs. With the trend to A size (8.5 × 11 inch) hardcopy generated on laser printers, large and complex designs are best represented with a hierarchical design structure. Capture allows a third option, referred to as a "flat" design structure, which is ideal for moderate-sized designs of three or four pages. Note that a flat design is limited to a single schematic.

The flat structure results in an appearance resembling traditional multiple-page schematics. Hierarchical blocks, ports, and pins do not appear in a flat design. Signals are routed between sheets with off-page connector symbols. No direct means exist for identifying from which sheet a given off-page connector signal originates or to which it is routed. This is a major limitation of the flat design. Workarounds include modifying signal names to include a sheet number suffix or adding text notes. Even with such notations added, following signal paths on flat designs with more than three or four sheets becomes a daunting task.

Project manager is used to navigate between sheets in a flat design. As with hierarchical designs, schematic pages are represented by page icons followed by the page name. Double clicking on a schematic page opens the page and launches the schematic editor.

Converting a Hierarchical Design to a Flat Structure

Figures 7-20 through 7-22 show the memory expansion board schematic converted to a flat structure with off-page connectors used to route signals between sheets. Use these three figures as a model for the second session.

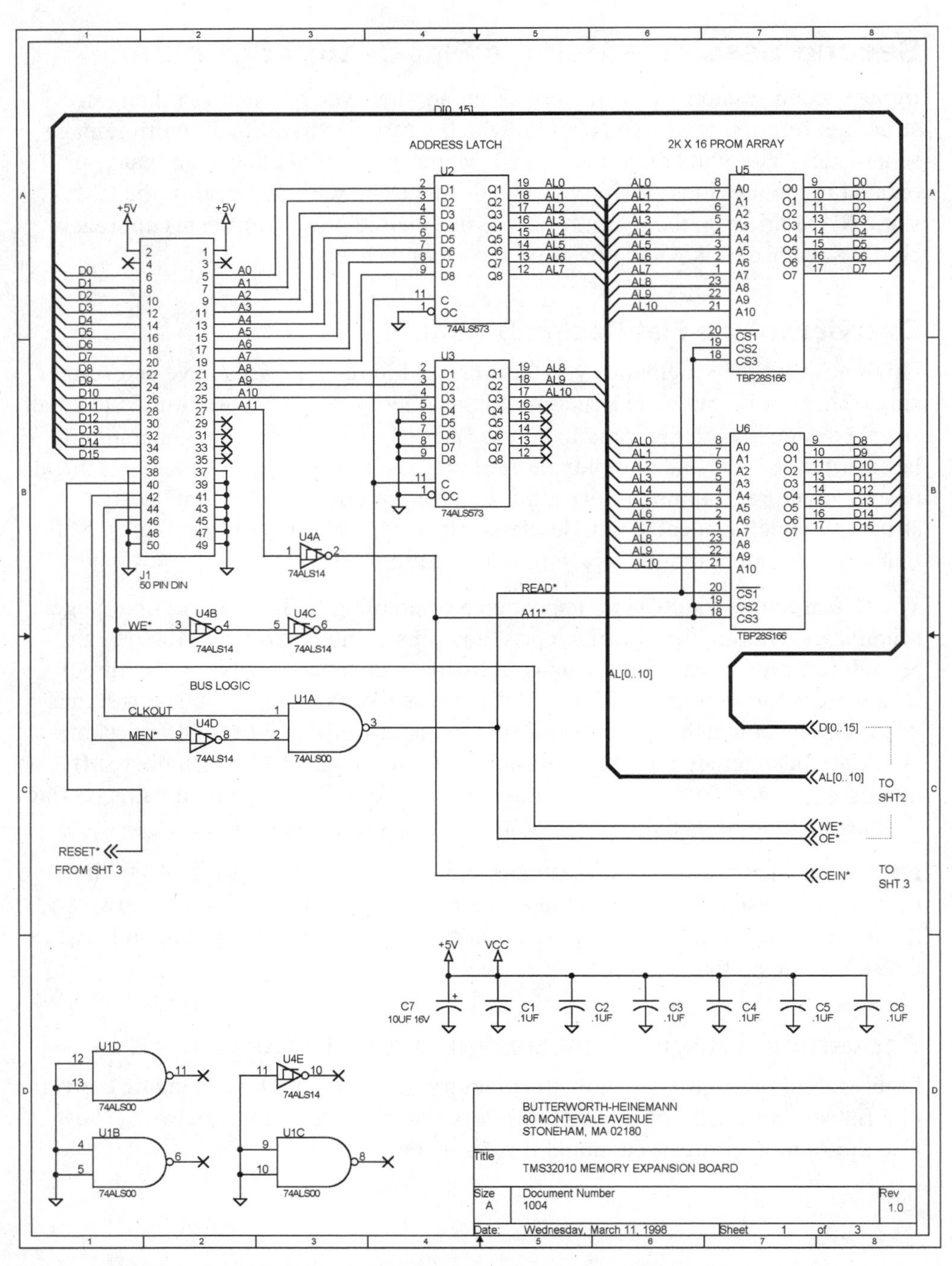

Figure 7-20 Memory Board (Sheet 1 – Flat Design)

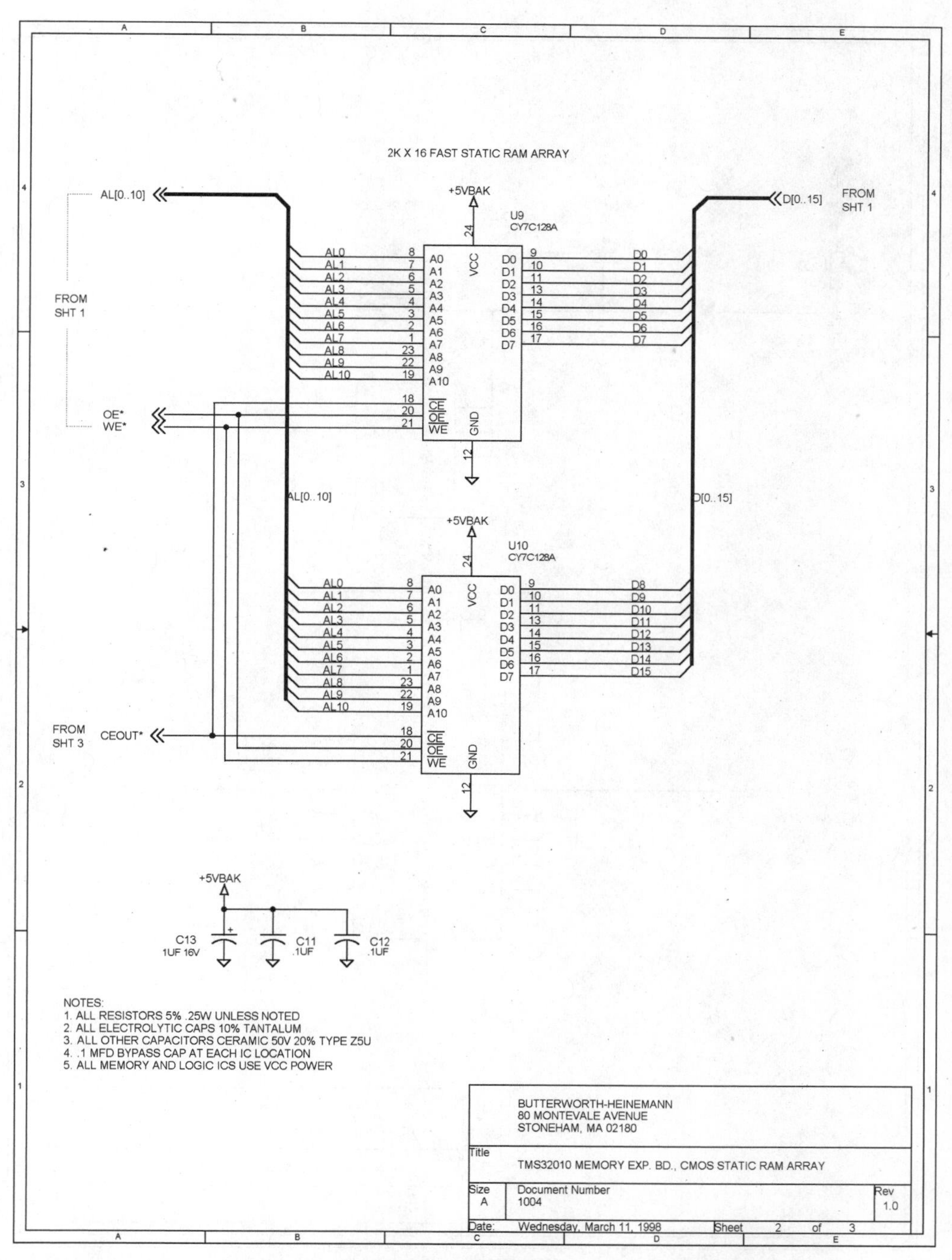

Figure 7-21 Memory Board (Sheet 2 – Flat Design)

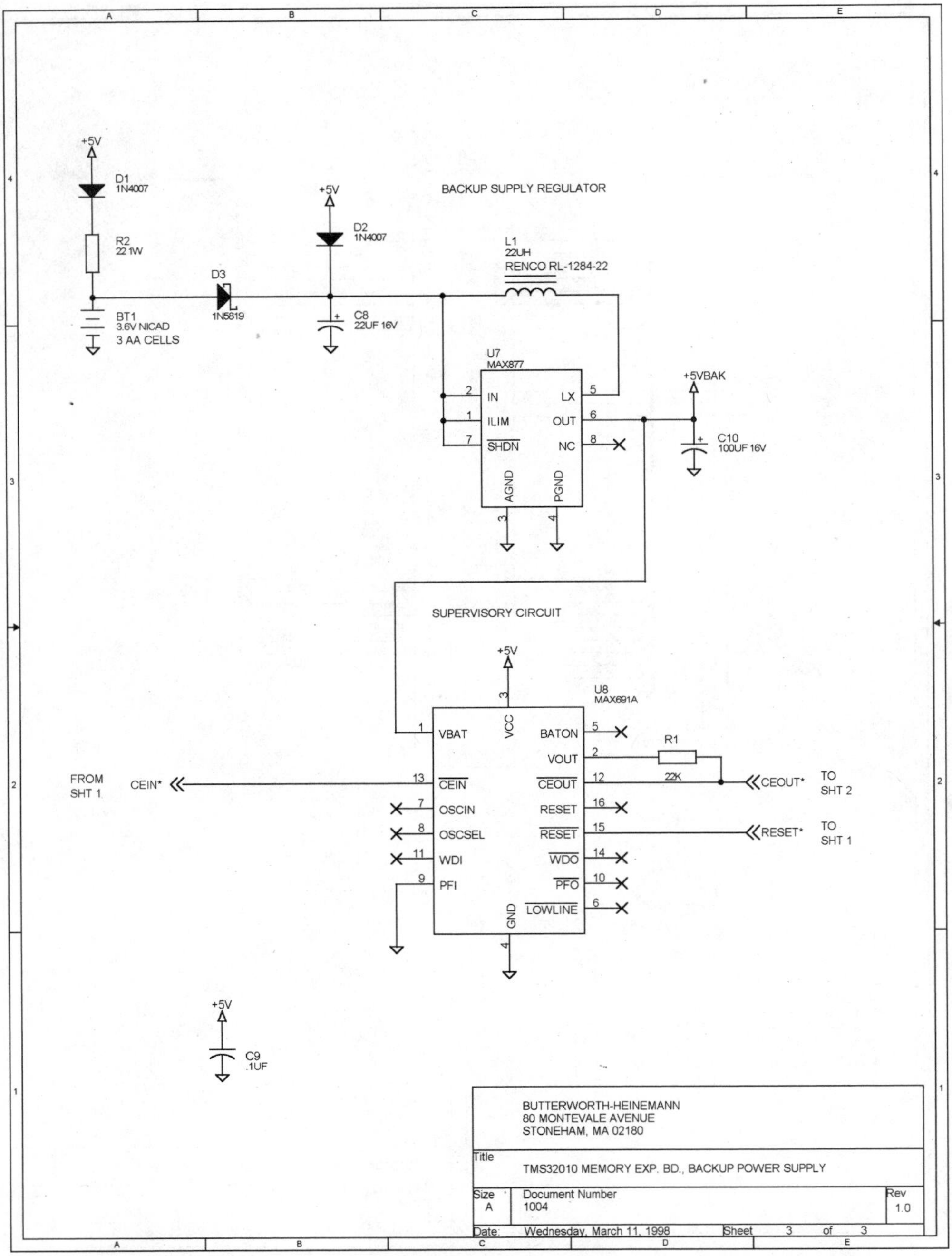

Figure 7-22 Memory Board (Sheet 3 – Flat Design)

Converting the TMS_Memory.dsn to a flat structure is relatively straightforward. First launch Capture and open the original design or project. Note that a project contains the design plus related files. Since you are going to convert the design, you don't care about any other files in the project.

Click on the design icon in Project Manager to select the original design. Then save it as TMS_Memory_Flat.dsn. Next, click on design resources to select the project. Then save it as TMS_Memory_Flat.opj in your existing Tutor4 dierctory. Your screen should now appear as shown in Figure 7-23. At this point you have made a copy of the original design and defined a project using the new name. You are now ready to convert it to a flat design structure.

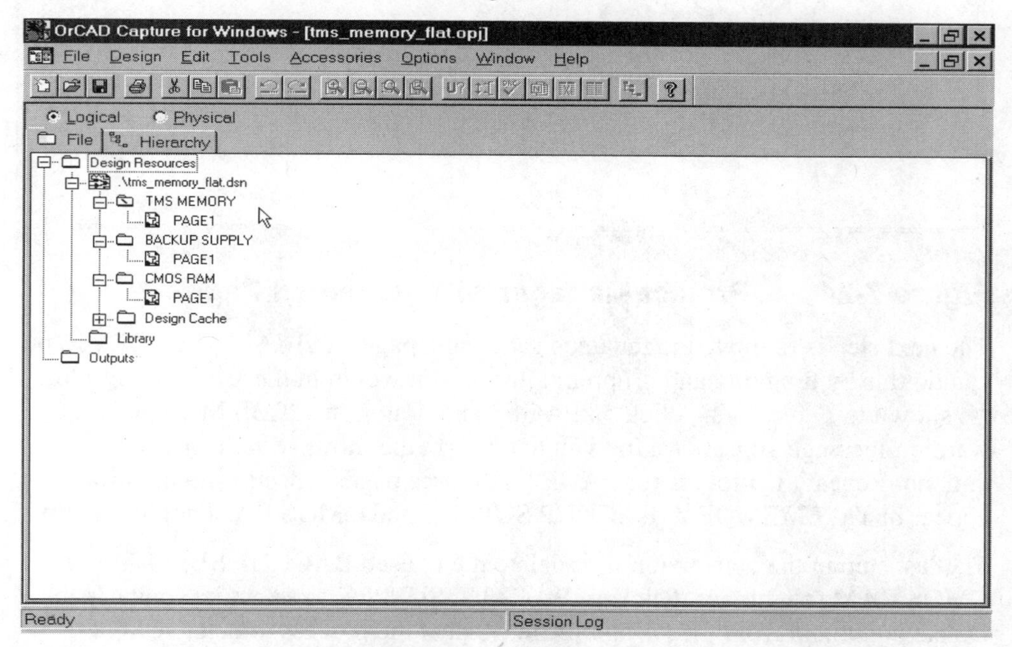

Figure 7-23 Project Manager View before Conversion

A flat design can contain only one schematic, which can have multiple sheets. In this exercise, the TMS Memory schematic will be expanded to three sheets. The CMOS RAM schematic will become sheet 2 in TMS Memory, and the Backup Supply schematic will become sheet 3. Start the conversion by renaming the schematic sheets. Click on CMOS RAM PAGE1 and then use the Rename command from the Design menu to change the sheet name to PAGE2. Use the same technique to rename BACKUP SUPPLY PAGE1 to PAGE3. When you are finished, your project manager screen will appear as shown in Figure 7-24.

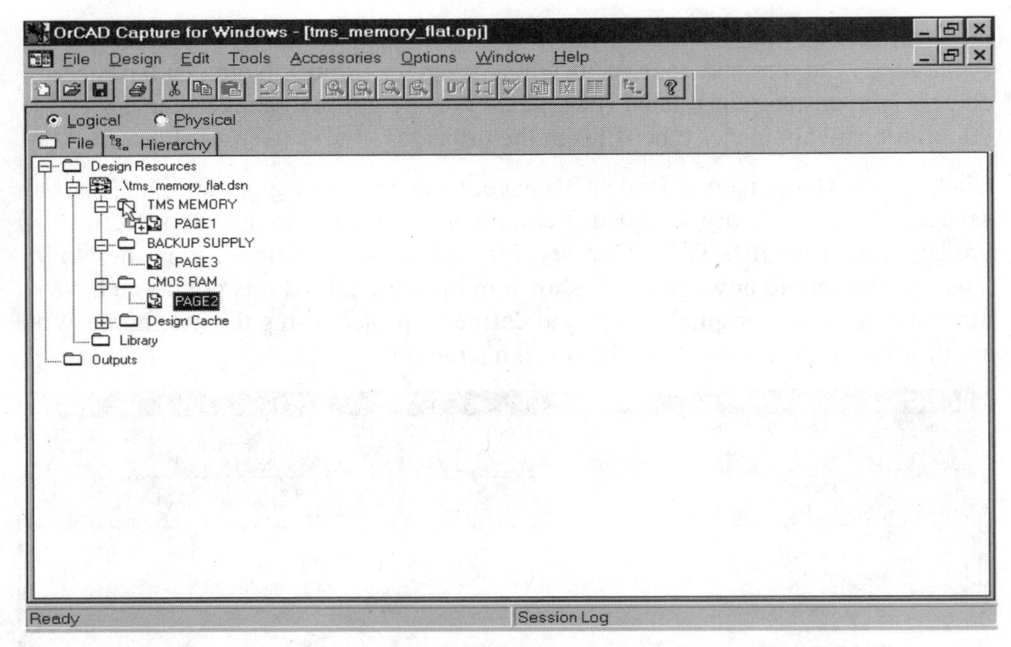

Figure 7-24 Project Manager with Renamed Pages

The next step is to move the renamed schematic pages into TMS MEMORY. You can do this by dragging and dropping, just as you would in the Windows Explorer. As shown in Figure 7-24, click on PAGE3 and drag it into TMS MEMORY. A warning message appears asking you to confirm the move, which cannot be undone. Repeat the process for PAGE2. All three pages (sheets) should now appear in TMS MEMORY. BACKUP SUPPLY and CMOS RAM are left empty.

The last step in the conversion is to delete the unused BACKUP SUPPLY and CMOS RAM schematics. Click on BACKUP SUPPLY to select it. Then use the Delete command from the Design menu. Repeat the process to delete CMOS RAM. Then save the project. You now have a flat design, and your project manager screen should appear as shown in Figure 7-25.

Now that you have converted your design to a flat structure, you can start editing the schematic sheets. Sheet one will require the most extensive edits. Carefully examine Figure 7-1. Start by deleting the hierarchical blocks. Then move the remaining circuitry down to provide proper spacing. The address, data, and control signals that were previously routed to hierarchical blocks now go to off-page connectors. Note that the Place Off-page Connector tool functions exactly the same as the Place Hierarchical Port tool. Off-page connector versions with left- and right-oriented names are available.

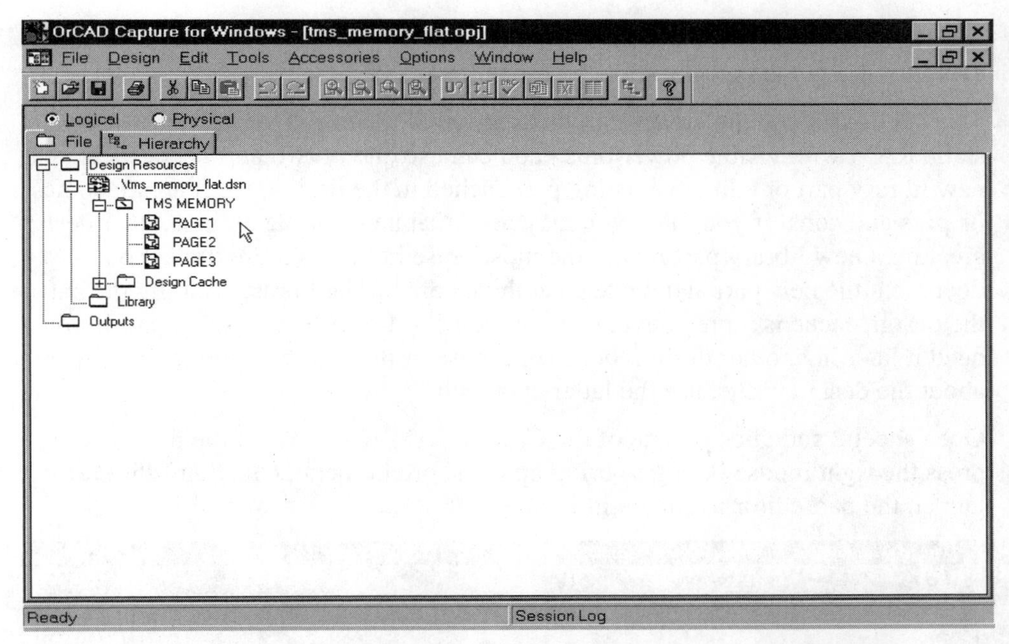

Figure 7-25 Project Manager View of Converted Flat Design

You can use off-page connectors for bus structures as long as you use a proper bus name with the connector. For added clarity, text notes indicating the signal routing are added near the off-page connectors. On the right side of sheet one, dotted lines are used to group the off-page signals that run to sheet 2. Text notes for signal routing have no effect on the electrical connectivity database and are not required by Capture. The only purpose for these notes is to aid the reader in understanding the signal routing.

Complete the required edits to sheet 1. Sheet 2 will require edits to the CMOS static RAM to make the power pins visible. This is explained in the following section.

Editing a Part to Make Power Pins Visible

In the previous chapter, you learned how to use the part editor to edit and create new library parts. Once the part you were working on was completed, it was saved into the Custom library. This allowed the new part to be accessed and placed into any design.

Capture designs include a parts cache. When you initially place a library part, a copy is permanently stored in the design cache. From that point on, whenever you open the design, the part is read in from the cache. You can still open the design even if the original part library is no longer available.

The flat design you are creating in this session of the exercise requires a CMOS static RAM with visible power pins. You can use one of two approaches: create a new library part or edit the existing part cached in the design. Each approach has its pros and cons. If you think you may need that new part again in another design, creating a new library part makes the most sense but takes more time. You must document the new part and make an archive copy of the library. Editing the part in the design cache is more convenient, but you can't readily access the part if you need it later in another design. For the purpose of this exercise and to learn more about the design cache, use the latter approach.

Open sheet 2 and click on one of the CMOS RAM ICs to select the part. Then press the right mouse button to bring up the shortcut menu. Click on Edit Part to launch the part editor as shown in Figure 7-26.

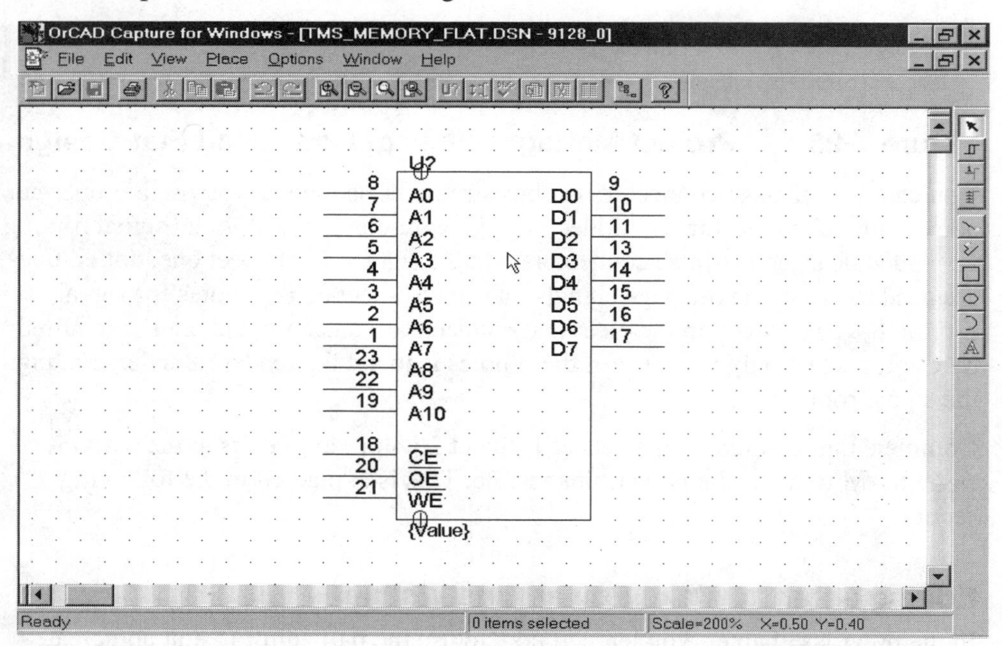

Figure 7-26 Part Editor with Original Version of CMOS RAM

Double click on the invisible power pins (VCC and GND) to bring up the Pin Properties dialog box as shown in Figure 7-27. Change the pins to a line shape and passive type. The pins will become visible. Drag the pins to center them on the part outline as shown in Figure 7-28.

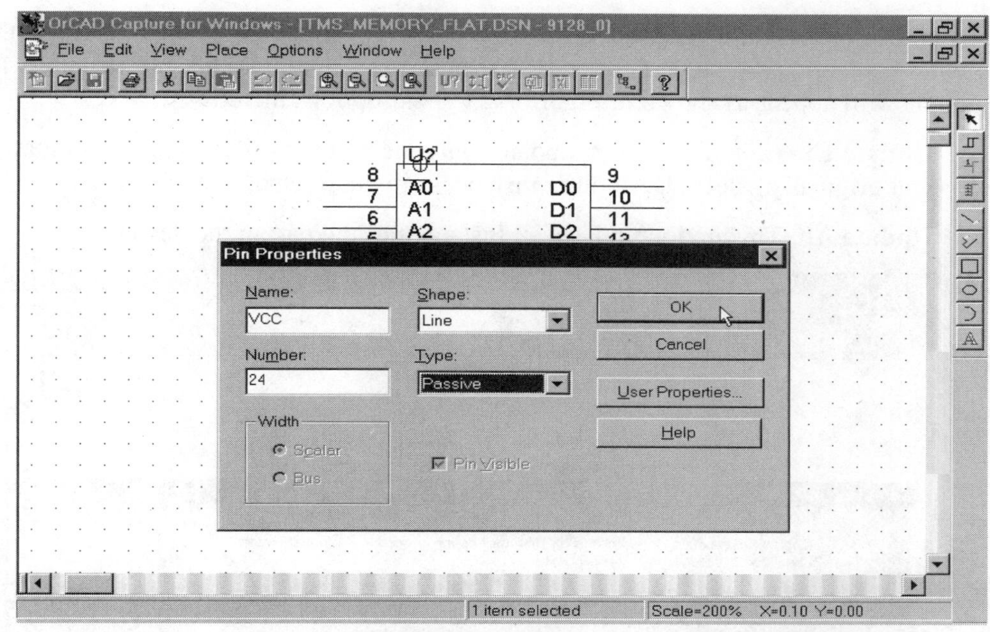

Figure 7-27 Editing CMOS RAM Power Pin Properties

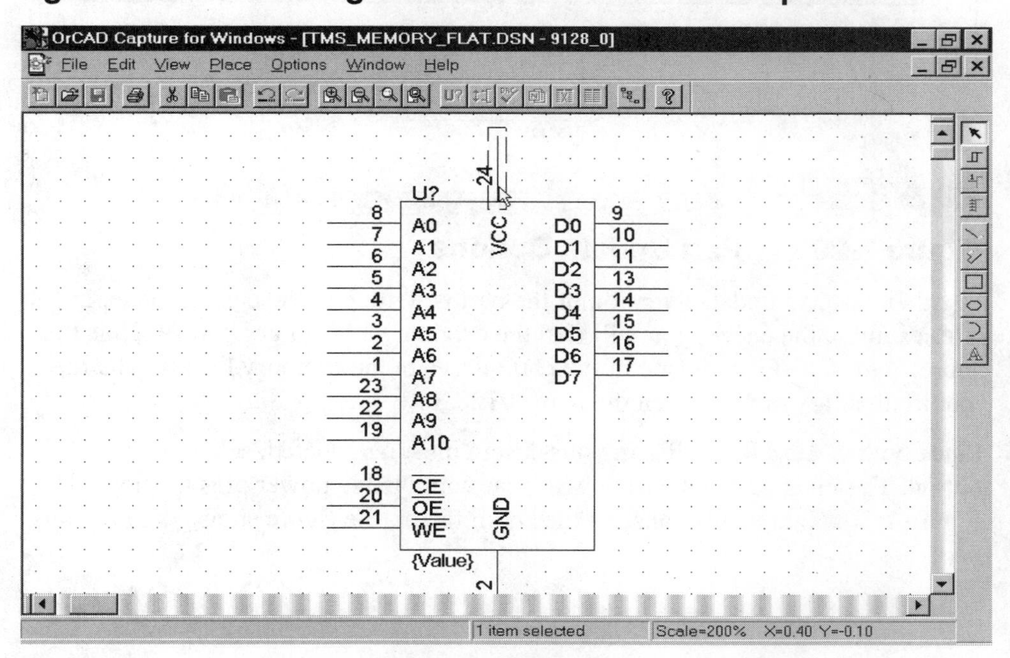

Figure 7-28 Dragging Power Pins into Position

When you have completed editing the CMOS RAM, use the Close command from the File menu to exit the part editor. A screen message with various part update options will appear as shown in Figure 7-29. Your options include:

- **Update Current**. This option updates only the part instance in the design that you originally selected. All other instances are unaffected.
- **Update All**. This option updates all instances of the part in the design.

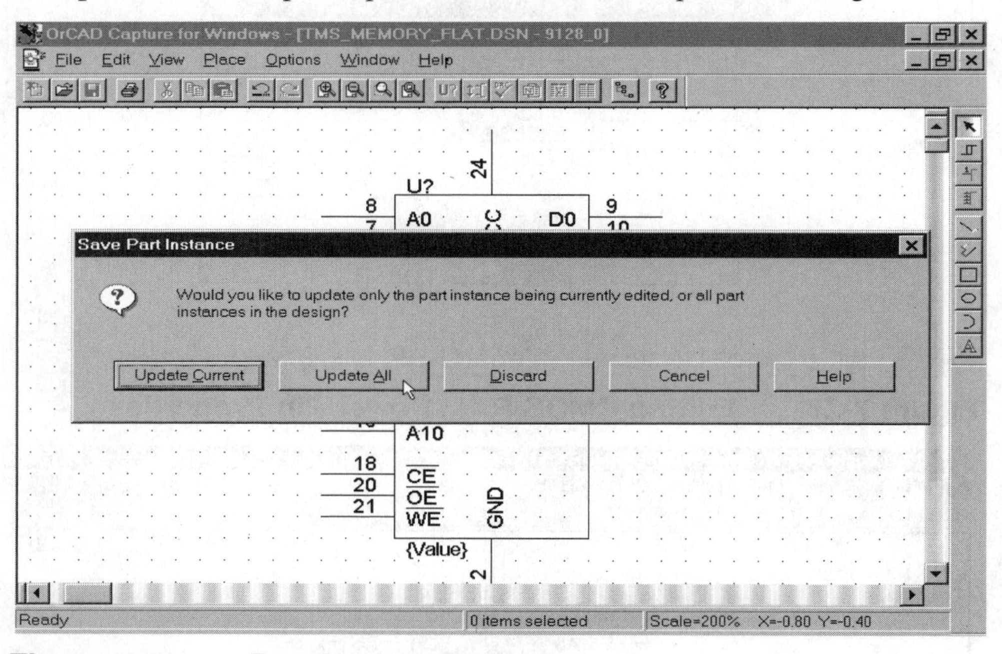

Figure 7-29 Part Update Options

In either case, an updated version of the part is written to the design cache. If you later examine the cache, you will find the original 9128 library part (9128 is the source part, the 7C128 is one of the part aliases in the Memory library) and the modified part, which is given the name 9128_0.

Since both CMOS RAM ICs in your design must be updated, select the Update All option. The updated parts will now appear with visible power pins as shown in Figure 7-30. Note that because of the zoom factor, the figure shows only the top part.

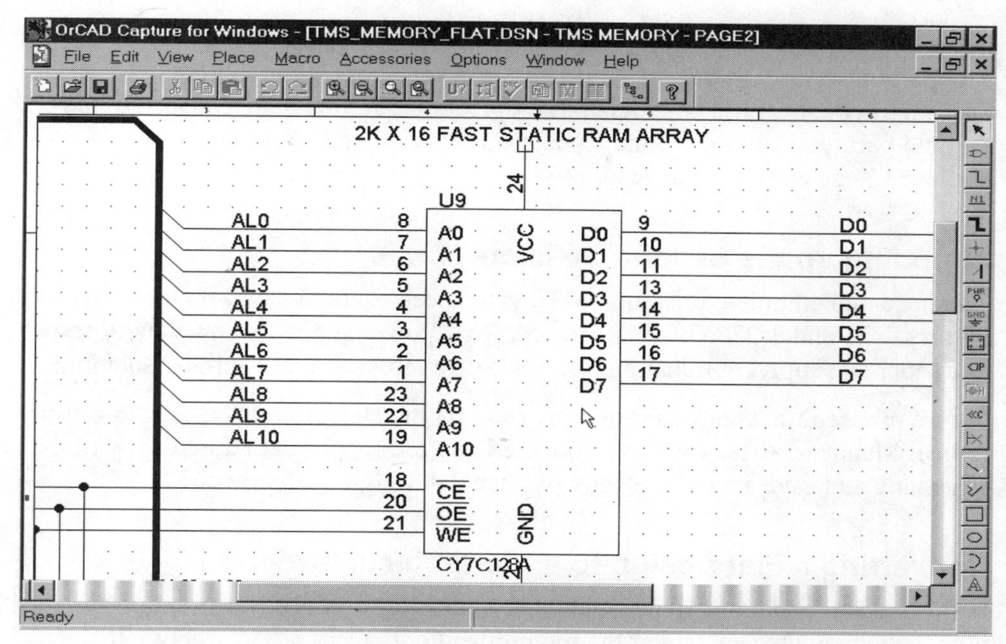

Figure 7-30 Updated Part on Sheet 2

Understanding the Design Cache

When you first place a library part onto a schematic page, a copy of the part is written to the design cache. The part information stored in the design cache includes a link to the original library. If you save and then reopen the schematic, the part is retrieved from the design cache, not the original library.

If you receive an updated library as part of a new Capture release or make edits to your custom library, you may want to update parts in your design. Double click on the design cache in project manager to display the individual parts. Select one or more parts (hold down the CTRL key to select multiple parts). Then use the Update Cache command from the Design menu. When you update parts, user-defined part properties are not affected. However, any user-defined pin properties are lost.

If you want to replace a part in the design cache with a different library part, you can use the Replace Cache command from the Design menu. You can select only a single part before running the Replace Cache command. When you run the command, a dialog box appears with the original part name and library. You can then enter a new part name and part library.

If you edit a part within the design, such as the CMOS RAM in the exercise, the new part exists only in the design cache. To place another copy of the new part in the design you must use copy and paste operations. If you want to return to the original part, you cannot use the Update Cache command. You must use the Replace Cache command instead.

Wrapping up the Second Session

Go ahead and complete the required edits to sheets 2 and 3 based on the models in Figures 7-21 and 7-22. Note that the +5VBAK power supply is now clearly shown as a separate supply, eliminating any possible confusion about supply isolation.

Before proceeding, check for errors by running the Design Rules Check tool from Project Manager. After you have corrected any errors, print out a hardcopy of the schematic and save an archive copy of your flat design to floppy disk.

Converting a Flat Design to a Hierarchical Structure

In this session you converted a small hierarchical design to a flat structure. At times, you may also encounter the requirement to convert a flat structure to a hierarchy. Engineering changes involving extensive addition of circuitry to a small design may necessitate conversion to a hierarchical structure. You can easily perform this conversion. In Project Manager, select the schematic folder that will become the root schematic or create a new schematic folder. Then use the Make Root command from the Design menu. Place the required hierarchical blocks on the root schematic using implementation names that match the names of the corresponding schematic folders.

Third Session— Advanced Postprocessing

The third session covers the use of the Update Properties tool and the use of include files with the Create Bill of Materials tool. You can use the Update Properties tool to import PCB footprint properties or other properties such as part numbers. As their name implies, include files allow you to add information to the bill of materials. Former OrCAD SDT users who abandoned these features as hopelessly confusing will find the Capture implementation greatly improved.

Using the Update Properties Tool to Import PCB Footprints

The Update Properties tool allows one or more properties to be imported into the design on the basis of data in an update file. The most common uses include importing PCB footprints and part numbers. The update file contains ASCII text data. You can create this file with a text editor such as the Windows WordPad. Although you can use other file name extensions, OrCAD suggests using the .upd extension for update files.

A sample update file with PCB footprint values appropriate to this exercise is shown in Figure 7-31. The file name is PCB_Footprint.upd. This file can be found on the disk supplied with this book (see Appendix A for details).

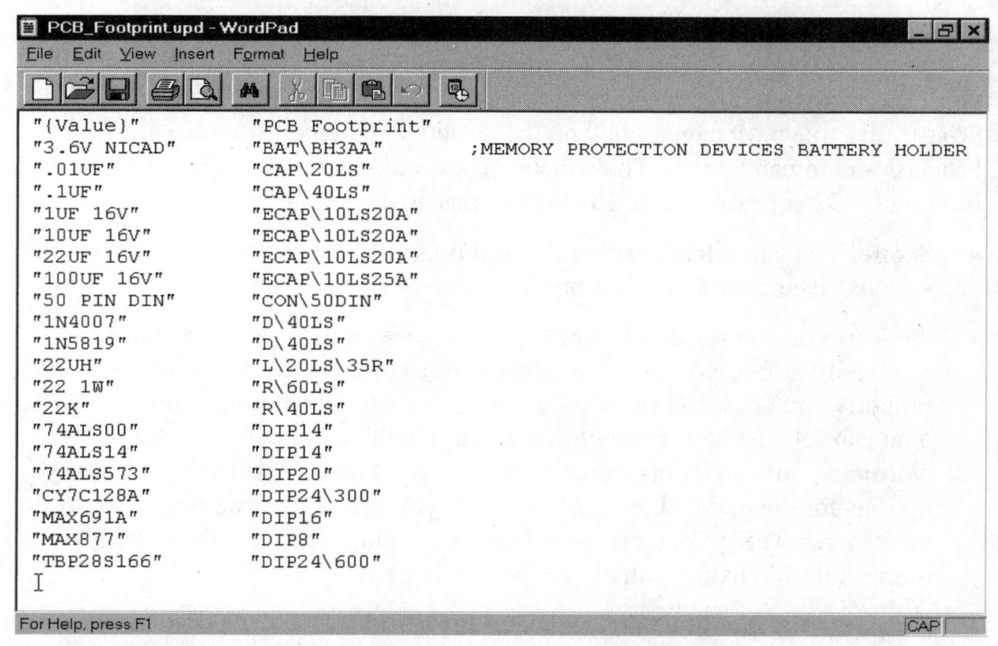

PCB_Footprint.upd - WordPad

File Edit View Insert Format Help

```
"{Value}"          "PCB Footprint"
"3.6V NICAD"       "BAT\BH3AA"        ;MEMORY PROTECTION DEVICES BATTERY HOLDER
".01UF"            "CAP\20LS"
".1UF"             "CAP\40LS"
"1UF 16V"          "ECAP\10LS20A"
"10UF 16V"         "ECAP\10LS20A"
"22UF 16V"         "ECAP\10LS20A"
"100UF 16V"        "ECAP\10LS25A"
"50 PIN DIN"       "CON\50DIN"
"1N4007"           "D\40LS"
"1N5819"           "D\40LS"
"22UH"             "L\20LS\35R"
"22 1W"            "R\60LS"
"22K"              "R\40LS"
"74ALS00"          "DIP14"
"74ALS14"          "DIP14"
"74ALS573"         "DIP20"
"CY7C128A"         "DIP24\300"
"MAX691A"          "DIP16"
"MAX877"           "DIP8"
"TBP28S166"        "DIP24\600"
```

For Help, press F1 CAP

Figure 7-31 PCB_Footprint.upd Update File

The first two lines of text in the update file illustrate the file syntax:

```
"{Value}"        "PCB Footprint"
"3.6V NICAD"     "BAT\BH3AA"    ;MEMORY PROTECTION DEVICES
```

All strings in the update file (except for comments following a semicolon) must be enclosed in quotation marks and cannot exceed 124 characters in length. Strings must be separated by spaces or tabs. The first line lists the combined property string and one or more properties to be updated. Property names are case sensitive. The combined property string at the beginning of the first line contains the properties to be compared to determine whether a match exists. When a match is found, the remaining properties on the first line are updated. The combined match string must be enclosed in curly braces {}.

The second and successive lines contain a match string and the corresponding values for the update properties. For example, for the second line, the Update Properties tool would scan through the design database for any parts with value 3.6V NICAD and then update the PCB footprint property of these matching parts to BAT\BH3AA. Note that the third line of the sample update file in Figure 7-31 contains a .01 UF capacitor not used in the design. Since no match exists, this line will be ignored.

To experiment with the Update Properties tool, create your own update file or use the Windows Explorer to copy the supplied update file (PCB_Footprint.upd) to your Tutor4 design directory. Start Capture and open the TMS_Memory_Flat.opj project. In Project Manager, click on the design and then click on the Update Properties command on the Tools menu. The Update properties dialog box shown in Figure 7-32 appears. The following options are available:

- **Scope**. You can select whether the tool processes the entire design or just the selected schematic folders or pages.
- **Action**. You can select whether to update parts or nets. To help with case sensitivity issues, you can also select various options to convert the combined property and or update property to uppercase (the update file is not changed). You can select whether or not to unconditionally update the property. Normally, only an empty property is updated. You can also select visibility options for the updated property. Finally, you can select whether or not to create a report file. You can specify a report name or accept the default (current design name with an .RPT extension).
- **Property Update File.** You can enter the file name for your update file.

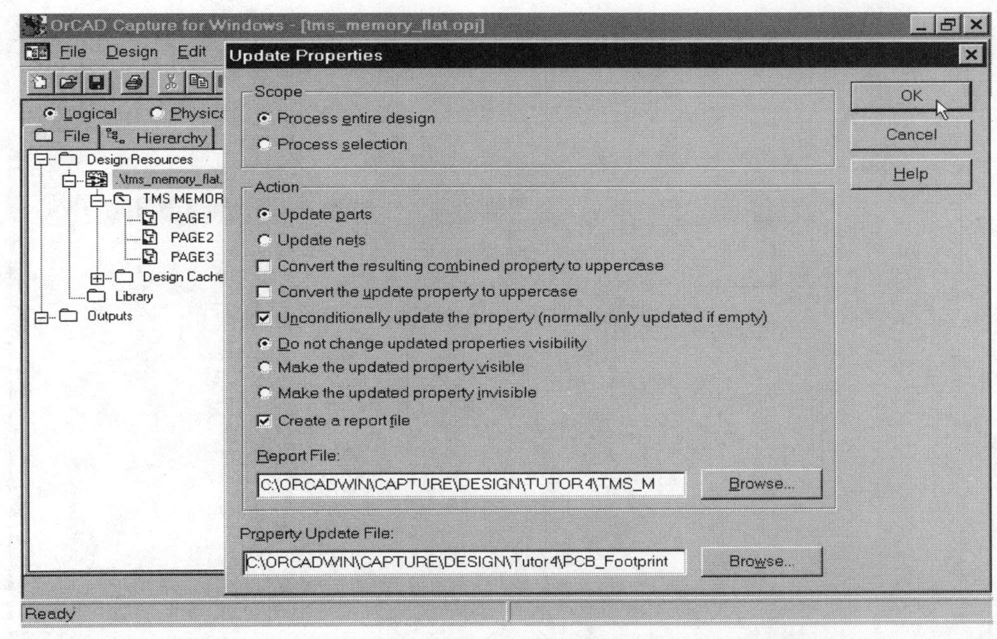

Figure 7-32 Dialog Box for the Update Properties Tool

Select the options shown in Figure 7-32, enter the name of your update file, and click on OK to run the tool. For a small design, such as the exercise in this chapter, the Update Properties tool requires only a few seconds to run. When the tool has completed processing the update, double click on PAGE1. Then double click on one of the parts to examine the updated PCB footprint property as shown in Figure 7-33.

Creating an update file that cross references part values and PCB footprints and then running the Update Properties tool before running the Create Netlist tool can save time in some circumstances. However, the Update Properties tool has limitations. Part values in the update file must precisely match those used in the design. For example, a 1.0UF capacitor in the update file will not match a 1UF or 1.0 UF capacitor in the design. You may find that the trouble of creating and maintaining an update file outweighs the potential time savings. Keep in mind that multiple PCB footprints may be used for some parts. Most users find that the Browse tool as explained in Chapter 5 is usually the quickest method of entering PCB footprints.

Although the example in this section focuses on PCB footprints, you can also use the Update Properties tool to add part numbers to a design.

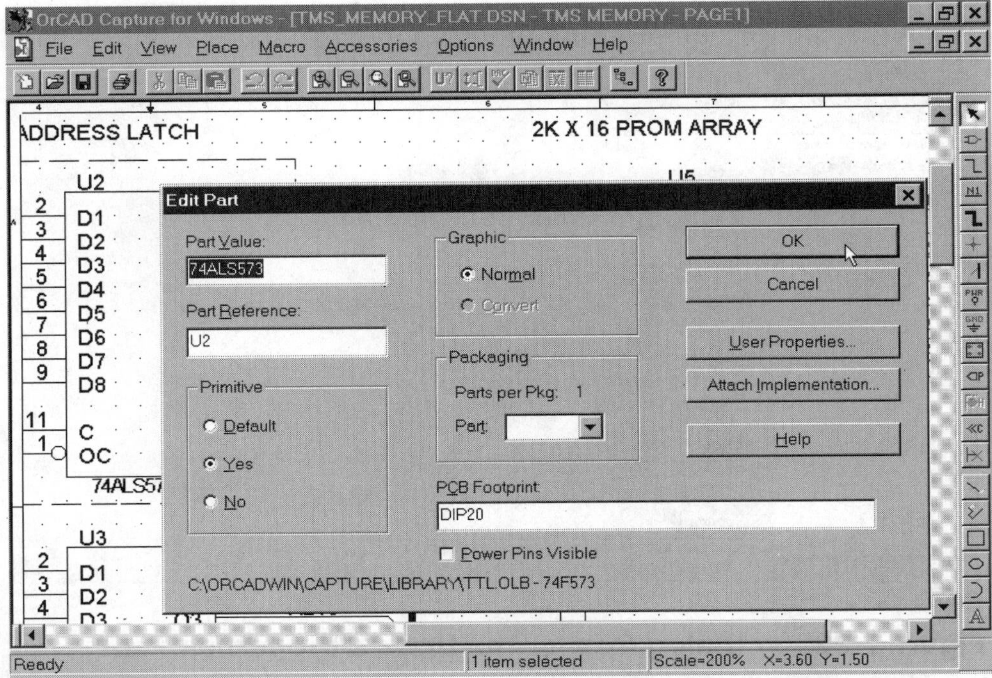

Figure 7-33 Examining an Updated PCB Footprint Property

Using an Include File with the Bill of Materials Tool

An include file is an ASCII text file that contains additional descriptive information to be "included" into the bill of materials report. The effect is somewhat similar to creating a new user property to build a longer part description, except that the information is merged from an external file. You can create the include file with a text editor such as the Windows WordPad. Although you can use other file name extensions, OrCAD suggests using the .inc extension for include files.

A sample include file with additional part description data appropriate to this exercise is shown in Figure 7-34. The file name is BOM_Description.inc. This file can be found on the disk supplied with this book (see Appendix A for details).

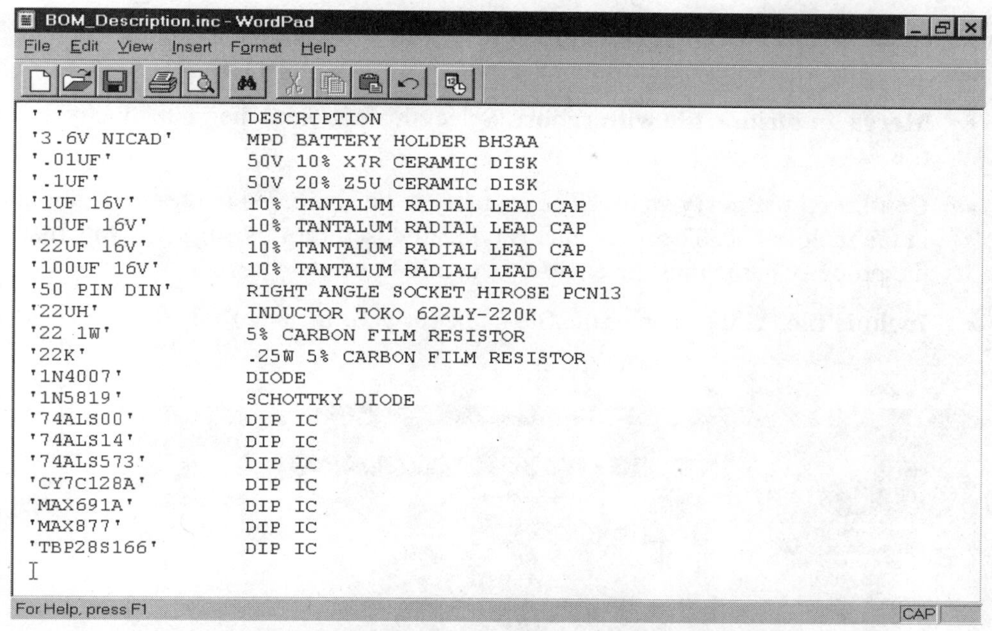
BOM_Description.inc - WordPad

File Edit View Insert Format Help

```
''                  DESCRIPTION
'3.6V NICAD'        MPD BATTERY HOLDER BH3AA
'.01UF'             50V 10% X7R CERAMIC DISK
'.1UF'              50V 20% Z5U CERAMIC DISK
'1UF 16V'           10% TANTALUM RADIAL LEAD CAP
'10UF 16V'          10% TANTALUM RADIAL LEAD CAP
'22UF 16V'          10% TANTALUM RADIAL LEAD CAP
'100UF 16V'         10% TANTALUM RADIAL LEAD CAP
'50 PIN DIN'        RIGHT ANGLE SOCKET HIROSE PCN13
'22UH'              INDUCTOR TOKO 622LY-220K
'22 1W'             5% CARBON FILM RESISTOR
'22K'               .25W 5% CARBON FILM RESISTOR
'1N4007'            DIODE
'1N5819'            SCHOTTKY DIODE
'74ALS00'           DIP IC
'74ALS14'           DIP IC
'74ALS573'          DIP IC
'CY7C128A'          DIP IC
'MAX691A'           DIP IC
'MAX877'            DIP IC
'TBP28S166'         DIP IC
```

For Help, press F1 CAP

Figure 7-34 BOM_Decsription.inc Include File

The first two lines of text in the include file illustrate the file syntax:

‘‘ DESCRIPTION
‘3.6V NICAD’ MPD BATTERY HOLDER BH3AA

The first line of the include file must begin with two single quotation marks without any characters or spaces in between. The remainder of the first line contains a header that will appear at the top of the bill of materials listing to identify the type of information in the include file. The second and successive lines consist of a part value enclosed in single quotes along with the additional descriptive information to be included in the bill of materials.

When Capture creates a bill of materials with the include file option selected, the combined property string (as configured in the Bill of Materials tool dialog box) for each part in the design is compared to each line in the include file. When a match occurs, the additional descriptive text on that line is written to the bill of materials.

To experiment with the use of an include file, create your own version of the file or use the Windows Explorer to copy the supplied file (BOM_Description.inc) to your Tutor4 design directory. Start Capture and open the TMS_Memory_Flat.opj project. In Project Manager, click on the design and then click on the Bill of

Materials tool. The familiar Bill of Materials dialog box shown in Figure 7-35 appears. The following options are available for processing an include file:

- **Merge an include file with report**. Select this option to merge the include file.
- **Combined property string**. The property to be compared to the first column in the include file. The part value is typically used for this purpose. Note that the property name must be enclosed in curly brackets { }.
- **Include file**. You can enter the file name for your include file.

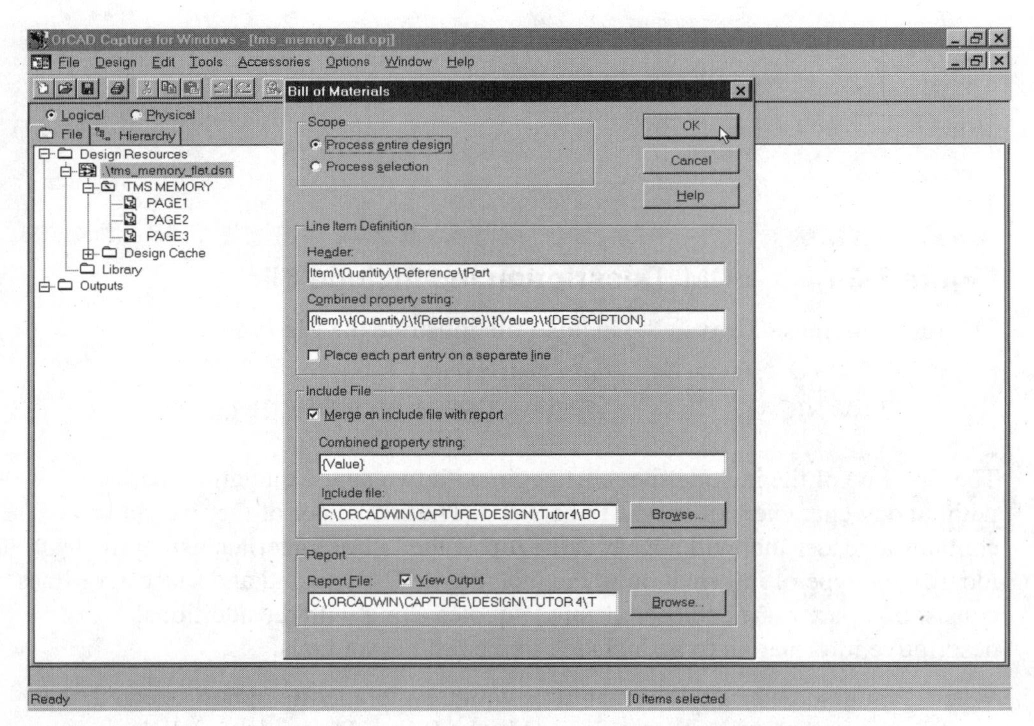

Figure 7-35 Dialog Box for the Bill of Materials Tool

In order for the bill of materials report to format correctly, you must also modify the line-item-definition combined property string. Note that the DESCRIPTION property (preceded by a "/t" to insert a tab character) has been added to the usual combined property string.

Select the options shown in Figure 7-35. Enter the correct combined property strings and the name of your include file. Then click on OK to run the bill of materials. When the tool has completed processing the report, the Windows WordPad is automatically launched and appears as shown in Figure 7-36.

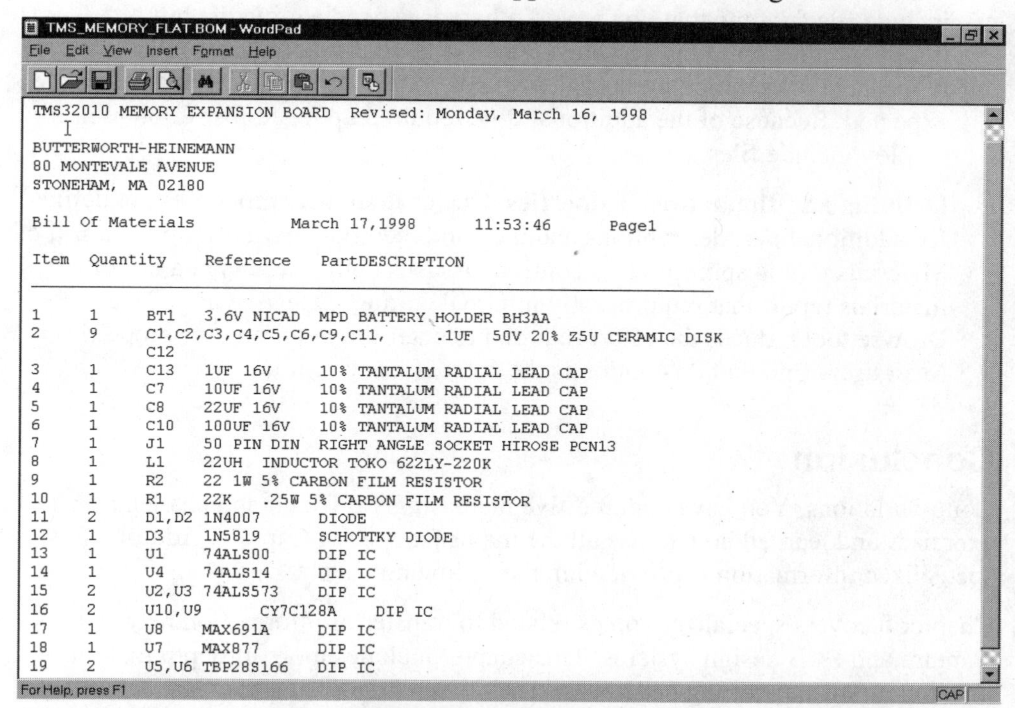

TMS_MEMORY_FLAT.BOM - WordPad

File Edit View Insert Format Help

TMS32010 MEMORY EXPANSION BOARD Revised: Monday, March 16, 1998

BUTTERWORTH-HEINEMANN
80 MONTEVALE AVENUE
STONEHAM, MA 02180

Bill Of Materials March 17,1998 11:53:46 Page1

Item	Quantity	Reference	PartDESCRIPTION
1	1	BT1	3.6V NICAD MPD BATTERY HOLDER BH3AA
2	9	C1,C2,C3,C4,C5,C6,C9,C11, C12	.1UF 50V 20% Z5U CERAMIC DISK
3	1	C13	1UF 16V 10% TANTALUM RADIAL LEAD CAP
4	1	C7	10UF 16V 10% TANTALUM RADIAL LEAD CAP
5	1	C8	22UF 16V 10% TANTALUM RADIAL LEAD CAP
6	1	C10	100UF 16V 10% TANTALUM RADIAL LEAD CAP
7	1	J1	50 PIN DIN RIGHT ANGLE SOCKET HIROSE PCN13
8	1	L1	22UH INDUCTOR TOKO 622LY-220K
9	1	R2	22 1W 5% CARBON FILM RESISTOR
10	1	R1	22K .25W 5% CARBON FILM RESISTOR
11	2	D1,D2	1N4007 DIODE
12	1	D3	1N5819 SCHOTTKY DIODE
13	1	U1	74ALS00 DIP IC
14	1	U4	74ALS14 DIP IC
15	2	U2,U3	74ALS573 DIP IC
16	2	U10,U9	CY7C128A DIP IC
17	1	U8	MAX691A DIP IC
18	1	U7	MAX877 DIP IC
19	2	U5,U6	TBP28S166 DIP IC

For Help, press F1 CAP

Figure 7-36 Bill of Materials Report in Windows WordPad

As with previous bill of materials reports you have run, the body of the report is tab delimited, and the columns do not line up perfectly. Some editing will be required before you can print out a neat hardcopy. Note that tabs are inserted between the part values and the merged part description data from the include file. Also note that the order in which parts occur in the bill of materials report matches that of the include file.

The final bill of materials report generally requires more detailed information than the basic part descriptions that appear on the schematic hard copy. Three options exist for handling this requirement:

- **Editing the bill of materials**. The design database contains only the basic part values, such as .1UF capacitor or 1K resistor. A text editor is used to edit the bill of materials and to add the additional information. This option is the most cost effective if your schematics contain a low percentage of common parts, and if engineering changes throughout the product life cycle are expected to be minimal.

- **Using an include file**. The information in the include file is merged onto the bill of materials. You are trading-off time spent editing and updating the include file for time that you would otherwise spend editing the bill of materials. This option is very cost effective if your schematics contain a high percentage of common parts or if frequent, extensive engineering changes are expected. Because of the additional time required up front, few Capture users employ include files.
- **Defining additional user properties**. One or more user properties are defined for additional part descriptions and the combined property string for the Bill of Materials tool is appropriately configured. This option results in a bill of materials report that requires only minimal editing. Effective use of the Browse tool reduces the effort required to enter the required user properties. Most users find that this option is the most effective choice.

Conclusion

Congratulations. You have endured five increasingly difficult and challenging exercises and learned how to use all the major features of Capture. Additional specialized information is provided in the following chapters:

Chapter 8 covers specialized topics related to transfer of information between Capture and PCB design systems. This chapter includes more in-depth material about creating and editing netlists and back annotation (using the Gate and Pin Swap tool to update the schematic after PCB design).

Chapter 9 covers using Capture to create netlists for SPICE circuit simulation. If you do not use SPICE, you can skip this specialized material.

Chapter 10 discusses techniques related to bills of materials. This chapter includes material on editing bills of materials, the use of a special sort utility to sort parts by part value, and importing bills of materials into spreadsheet programs such as Microsoft Excel.

Chapter 11 covers translation of designs from OrCAD SDT into Capture. If you do not have old designs in SDT format that require translation, you can skip this chapter.

Chapter 12 covers image and data transfer between Capture and other Windows programs.

Review Exercises

1. Describe the suggested guidelines for schematic structure based on the complexity and expected total number of sheets in the design.
2. How does autoincrement improve productivity when placing net aliases on bus structures?
3. How can you import text from a text editor such as the Windows WordPad and paste the text onto a schematic sheet?
4. Describe the two methods of text export from Capture. Which method would be more appropriate if you wanted to edit the text after exporting it?
5. What are the Windows hotkey combinations for copy and paste?
6. What technique can you use to isolate power on a particular sheet that contains ICs with invisible power pins?
7. Describe the technique for creating new user properties. List some possible applications for user properties.
8. How can you create a bill of materials report with user properties?
9. Compare the relative merits of hierarchical versus flat schematic structures.
10. How can you convert a hierarchical design to a flat schematic?
11. What information should accompany off-page connectors?
12. Describe the function of the design cache.
13. Describe the difference between updating and replacing a part in the design cache.
14. Compare the relative merits of importing PCB footprints via the Update Properties tool versus using the Browse tool.
15. Using one of the design exercises, create a new user property for a part number. Create a short update file with several part numbers and use the Update Properties tool to import the part numbers into your design. Then create a bill of materials report that includes the new part number property.
16. Describe several possible applications for using an include file with the Bill of Materials tool.

8

PCB Netlists and Back Annotation

Netlists were introduced in Chapter 5 with a discussion of netlist concepts and an exercise using the Create Netlist tool to generate a Tango format netlist. Additional material was introduced in Chapter 7, including the use of the Update Properties command to import PCB footprints. Recall from the discussion in Chapter 5 that two primary issues arise when preparing a netlist for PCB design:

- PCB design software requires PCB footprint values to identify the physical layout of parts on the schematic.
- The pin number sequence used by Capture may differ from that used by the PCB design software to represent the physical layout of actual devices. Problems are most prevalent with discrete parts, especially transistors.

The author's experience has shown that using Capture's Browse tool to add PCB footprints and then using a text editor to clean up the pin numbers is usually the most efficient approach to preparing a netlist for PCB design.

This chapter consists of two sessions. In the first session, you will prepare a netlist suitable for input into PowerPCB by Pads Software. PowerPCB is widely used and considered by many to be the leading-edge PCB design software for the Windows environment. PCB footprints, pin numbers, and pin arrangements used in the exercise are based on the PowerPCB parts library. Although PowerPCB imposes some specific requirements, the techniques you will learn are applicable to most PCB design software.

The second session introduces the subject of back annotation and the Gate and Pin Swap tool. As you work on a PCB design, the software may swap device gates and pins in order to improve trace routing. You may also renumber reference designators to correspond to the spatial sequence of parts on the board. Updating the schematic to incorporate these changes is referred to as *back annotation*. The Gate and Pin Swap tool uses the information in a swap file to automatically back annotate the schematic.

First Session – Creating a PCB Netlist

The model for this exercise is the Ignition_Sys design that you created in Chapter 3. Recall that Ignition_Sys is the four-sheet hierarchical schematic shown in Figures 1-18A through 1-18D in Chapter 1. For your convenience, a completed design file that you can use for this exercise is included on the disk supplied with this book (see Appendix A for details).

Create a new design directory called Tutor5. Use the Windows Explorer to copy either your own or the supplied Ignition_Sys.dsn file to this directory. Start Capture and then open the Ignition_Sys design. Print out a hardcopy of the design for reference.

First, check the design for errors. In Project Manger, click on the design to select it. Then run the Check for Invalid Stacked Hierarchical Pins tool from the Accessories menu to report any invalid stacked pins. If the tool reports any invalid stacked pins, rerun the tool with the fix option. Next, run the Design Rules Check tool with the usual options selected. The tool will report numerous warnings about bidirectional pins. You can ignore these warnings. Fix any errors and then rerun the tool with the option selected to delete existing DRC markers. Save your design. If you made any major changes to fix errors, you should also print out a new hardcopy.

Entering PCB Footprint Properties

After you have checked the design for errors, the next step is to add the PCB footprint properties. Use the Browse tool as explained in Chapter 5 and the PCB footprint values in Table 8-1. PCB footprint values will depend on the PCB design software parts libraries. PCB design software packages generally come with a standard library of common parts. The user then defines additional parts in a custom library. The situation exactly parallels Capture. The PCB footprints in Table 8-1 come from the author's PowerPCB part libraries.

Table 8-1 PCB Footprint Cross Reference

Reference Designator	Part Value	PCB Footprint
C1	22UF 35V	ECAP\30SQ\SMD
C2,C4	1UF 16V TANT	ECAP\3216\SMD
C3	.01UF	C\0805\SMD
D1	1N4007	D\40LS
PL1	+12V	CONN\08PAD

Table 8-1 Module Value Cross Reference (Cont'd)

Reference Designator	Part Value	PCB Footprint
PL2	GROUND	CONN\08PAD
PL3	VACUUM	CONN\08PAD
PL4	TACH OUTPUT	CONN\08PAD
PL5	COIL OUTPUT	CONN\08PAD
Q1,Q2	2N4401	TO-92A
Q3	FUJI ET365	TO-220\UPA
R1	3.3K	R\0805\SMD
R2	100K	R\0805\SMD
R3	39K	R\0805\SMD
R4,R7,R9	2.2K	R\0805\SMD
R5,R6	10K	VRES\T\ADJ\6MM
R8	1K .25W	R\1210\SMD
R10	470 1W	R\2512\SMD
RV1	ERZ-CF1MK270	ZNR\3224\SMD
U1	HA-640	HALL\HA640\1
U2	MIC2951	DIP8\SO
U3	PIC16C71	DIP18\SOL
Y1	8.00 MHZ	XTAL\RES\SMD

At first glance, these PCB footprint value names might appear somewhat strange. The names can be interpreted as follows: ECAP\3216\SMD for 3216 size SMD tantalum capacitors, C\0805\SMD for 0805 size SMD capacitors, D\40LS for axial lead diodes with .40 inch lead spacing, CONN\08PAD for .080 inch diameter pads used for connections to signal wires, VRES\T\ADJ\6MM for 6 mm top adjust trimpots, and DIP8\SO for 8-pin SOICs.

Creating a Bill of Materials with PCB Footprint Properties

Many users find that a bill of materials report with PCB footprint properties is useful during the PCB design process. You can easily create such a report by adding the PCB footprint property to the header and combined property string fields in the Bill of Materials dialog box. From Project Manager, click on the Bill of Materials tool. The dialog box shown in Figure 8-1 appears. Add the PCB footprint property as shown. Recall that the "\t" sequence inserts a tab character and that property names in the combined property string must be enclosed in curly brackets { }.

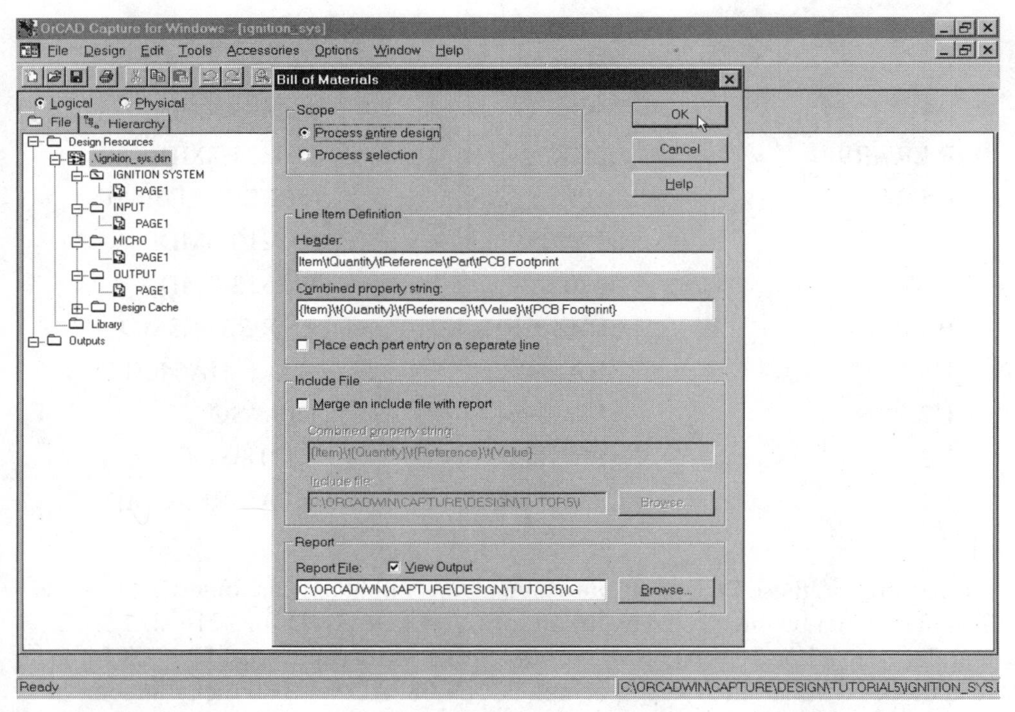

Figure 8-1 Dialog Box for the Bill of Materials Tool

The bill of materials report with PCB footprints appears as shown in Figure 8-2. Your report should be similar. Note that the report in the figure has been edited by adding tab characters in order to make the columns line up neatly. Check your bill of materials report against Table 8-1 to make sure that you entered the correct PCB footprint for each component. Print out a hardcopy of the report. You are now ready to create the netlist.

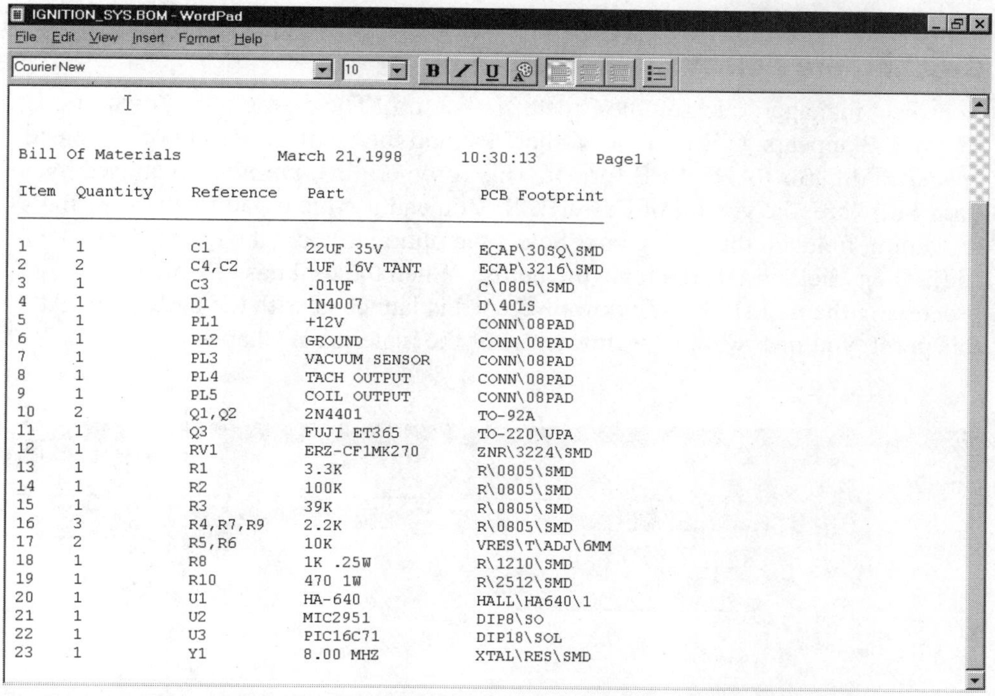

IGNITION_SYS.BOM - WordPad

File Edit View Insert Format Help

Courier New 10

Bill Of Materials March 21,1998 10:30:13 Page1

Item	Quantity	Reference	Part	PCB Footprint
1	1	C1	22UF 35V	ECAP\30SQ\SMD
2	2	C4,C2	1UF 16V TANT	ECAP\3216\SMD
3	1	C3	.01UF	C\0805\SMD
4	1	D1	1N4007	D\40LS
5	1	PL1	+12V	CONN\08PAD
6	1	PL2	GROUND	CONN\08PAD
7	1	PL3	VACUUM SENSOR	CONN\08PAD
8	1	PL4	TACH OUTPUT	CONN\08PAD
9	1	PL5	COIL OUTPUT	CONN\08PAD
10	2	Q1,Q2	2N4401	TO-92A
11	1	Q3	FUJI ET365	TO-220\UPA
12	1	RV1	ERZ-CF1MK270	ZNR\3224\SMD
13	1	R1	3.3K	R\0805\SMD
14	1	R2	100K	R\0805\SMD
15	1	R3	39K	R\0805\SMD
16	3	R4,R7,R9	2.2K	R\0805\SMD
17	2	R5,R6	10K	VRES\T\ADJ\6MM
18	1	R8	1K .25W	R\1210\SMD
19	1	R10	470 1W	R\2512\SMD
20	1	U1	HA-640	HALL\HA640\1
21	1	U2	MIC2951	DIP8\SO
22	1	U3	PIC16C71	DIP18\SOL
23	1	Y1	8.00 MHZ	XTAL\RES\SMD

Figure 8-2 Bill of Materials Report with PCB Footprints

Creating the Netlist

In project manager, click on the Create Netlist tool. The dialog box shown in Figure 8-3 appears. Click on the "Other" tab and then use the scroll box to select "padspcb.dll" for PADS-PCB format. This is the original netlist format used by Pads Software, the vendor of PowerPCB. You can use the default values for the remaining fields in the dialog box. Select the option to view the output (netlist file). Then click on OK to create the netlist. When the tool has completed processing the netlist, the Windows Notepad is launched with the netlist file. At this point, you may want to review some of the material in Chapter 5.

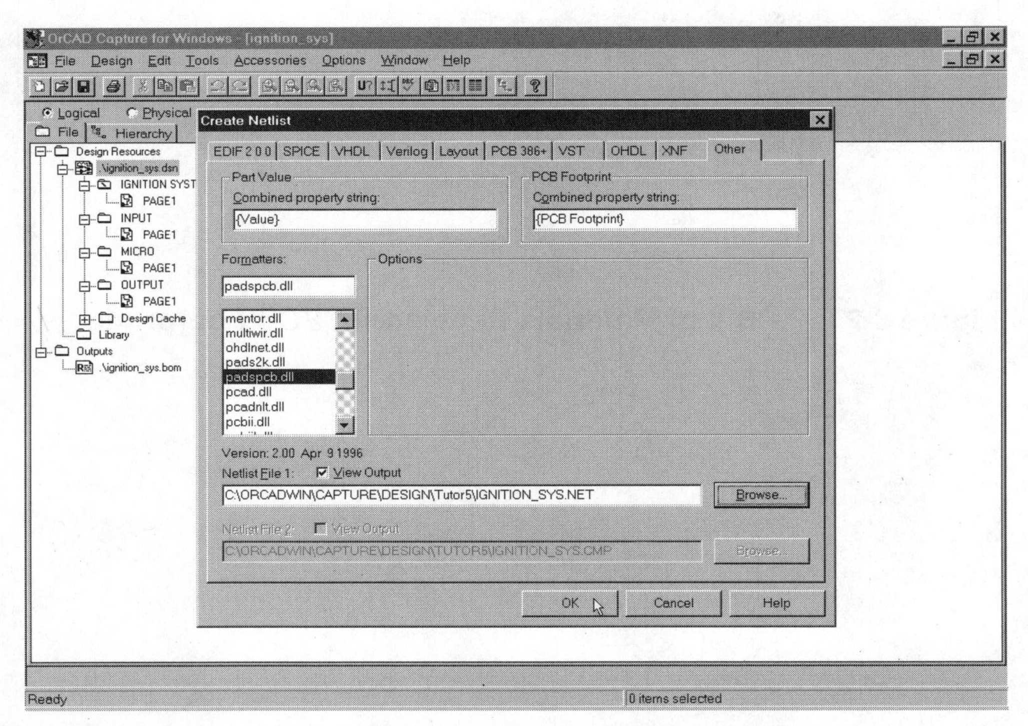

Figure 8-3 Dialog Box for the Create Netlist Tool

PADS-PCB Netlist Format

The PADS-PCB format is somewhat similar to the Tango format introduced in Chapter 5. The PADS-PCB netlist is an ASCII file. Unlike Tango format, there is no design identification header, just the netlist identification header *PADS-PCB*. The PowerPCB netlist then continues with the parts section shown in Figure 8-4. The parts section is identified by the section header *PART*. Each line following the section header contains the reference designator and PCB footprint for one part.

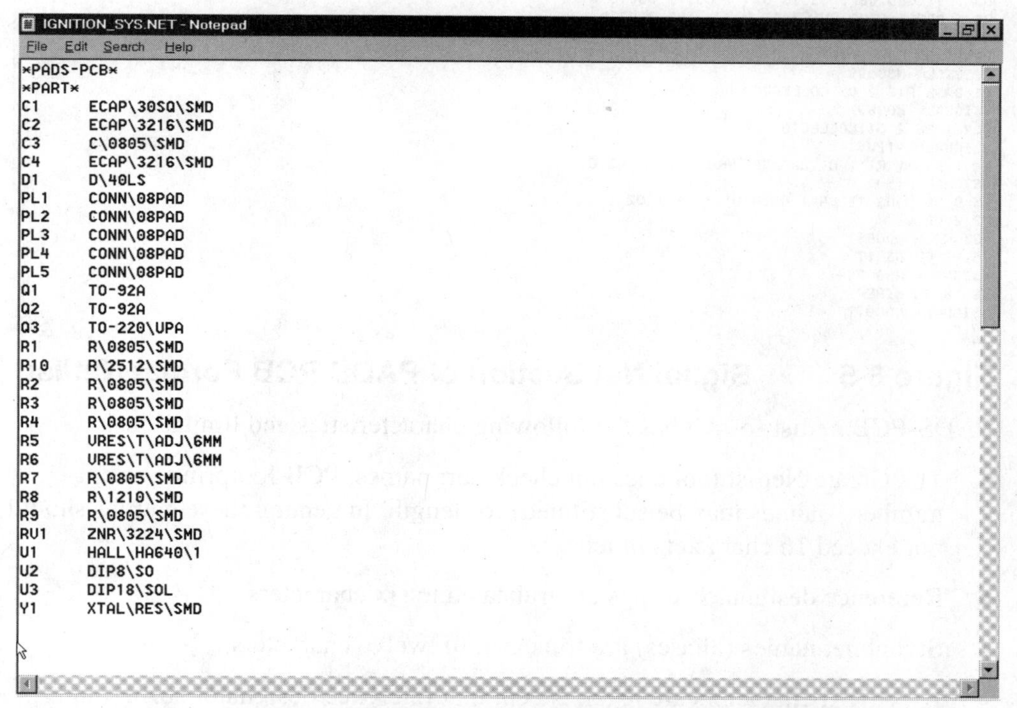
IGNITION_SYS.NET - Notepad

```
*PADS-PCB*
*PART*
C1      ECAP\30SQ\SMD
C2      ECAP\3216\SMD
C3      C\0805\SMD
C4      ECAP\3216\SMD
D1      D\40LS
PL1     CONN\08PAD
PL2     CONN\08PAD
PL3     CONN\08PAD
PL4     CONN\08PAD
PL5     CONN\08PAD
Q1      TO-92A
Q2      TO-92A
Q3      TO-220\UPA
R1      R\0805\SMD
R10     R\2512\SMD
R2      R\0805\SMD
R3      R\0805\SMD
R4      R\0805\SMD
R5      VRES\T\ADJ\6MM
R6      VRES\T\ADJ\6MM
R7      R\0805\SMD
R8      R\1210\SMD
R9      R\0805\SMD
RU1     ZNR\3224\SMD
U1      HALL\HA640\1
U2      DIP8\SO
U3      DIP18\SOL
Y1      XTAL\RES\SMD
```

Figure 8-4 Parts Section of PADS-PCB Format Netlist

The signal net section follows the parts section and is shown in Figure 8-5. The signal net section is identified by the section header *NET*. Each signal net data block is preceded by the string *SIGNAL*. The signal net data block consists of the signal name followed by one or more lines of node data. For example, the first signal net data block in Figure 8-5 is for signal N00043 with nodes R9.2 and Q2.BASE. R9.2 refers to pin 2 on R9. N00043 is a system-generated name.

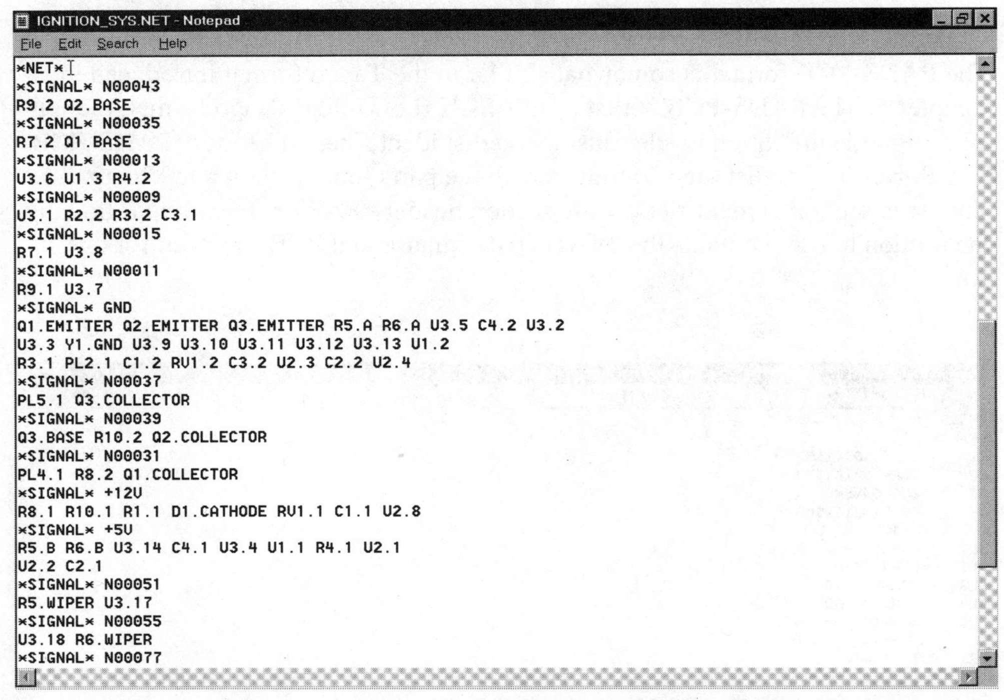

IGNITION_SYS.NET - Notepad

File Edit Search Help

```
*NET*
*SIGNAL* N00043
R9.2 Q2.BASE
*SIGNAL* N00035
R7.2 Q1.BASE
*SIGNAL* N00013
U3.6 U1.3 R4.2
*SIGNAL* N00009
U3.1 R2.2 R3.2 C3.1
*SIGNAL* N00015
R7.1 U3.8
*SIGNAL* N00011
R9.1 U3.7
*SIGNAL* GND
Q1.EMITTER Q2.EMITTER Q3.EMITTER R5.A R6.A U3.5 C4.2 U3.2
U3.3 Y1.GND U3.9 U3.10 U3.11 U3.12 U3.13 U1.2
R3.1 PL2.1 C1.2 RU1.2 C3.2 U2.3 C2.2 U2.4
*SIGNAL* N00037
PL5.1 Q3.COLLECTOR
*SIGNAL* N00039
Q3.BASE R10.2 Q2.COLLECTOR
*SIGNAL* N00031
PL4.1 R8.2 Q1.COLLECTOR
*SIGNAL* +12V
R8.1 R10.1 R1.1 D1.CATHODE RV1.1 C1.1 U2.8
*SIGNAL* +5V
R5.B R6.B U3.14 C4.1 U3.4 U1.1 R4.1 U2.1
U2.2 C2.1
*SIGNAL* N00051
R5.WIPER U3.17
*SIGNAL* N00055
U3.18 R6.WIPER
*SIGNAL* N00077
```

Figure 8-5 Signal Net Section of PADS-PCB Format Netlist

PADS-PCB netlist format has the following characteristics and limitations:

- The Create Netlist tool does not check part names, PCB footprints, or pin numbers (names may be substituted) for length. In general these entities should not exceed 16 characters in length.
- Reference designator strings are truncated to six characters.
- Signal net names (aliases) are truncated to twelve characters.
- All ASCII characters are legal, except that reference designator strings and signal net names are restricted to alphanumeric characters and the following special characters:

  ~ ! # $ % _ - =

  + | / . : ; < >

- System-generated net names are truncated to five digits following the "N" prefix.

You will note an immediate problem with the limitation imposed on legal ASCII characters for signal net names. PADS-PCB format does not allow the asterisk (*) character commonly used to identify negative or active low signals. If the Create Netlist tool encounters a signal net name with an asterisk character, it will generate a warning message and substitute a system-generated name. Details about the substitutions will appear in the session log.

After running the Create Netlist tool and examining the netlist file, print out a hardcopy of the netlist. You should always keep a hardcopy of the original netlist. If you later make a mistake editing the netlist and delete or overwrite something, you can refer to your original netlist hardcopy to make the required correction. Note that your netlist may differ slightly from the example in Figures 8-4 and 8-5 as to the sequence of parts and nets. System-generated net names will also differ. Part names, pin numbers, and signal interconnections should be the same.

Note that diode D1, transistors Q1-3, and trimpots R5-6 all have pin names instead of pin numbers. These parts, and many others in the DEVICE.LIB library, are defined without pin numbers. If the Create Netlist tool encounters a part without pin numbers, the pin names are used instead. The PCB design software uses layout parts defined with pin numbers, so any alphanumeric pin names such as ANODE or BASE must be replaced with appropriate pin numbers. You can make the required changes to the netlist file with a text editor.

Pin Numbers and Pin Arrangements

Before you can edit pin data on the netlist, you need to know what pin numbers and pin arrangements are required by your PCB design software. These requirements are determined by part definitions in the parts library used with the PCB design software (not the Capture libraries).

Two-lead nonpolarized components such as ceramic capacitors, inductors, and resistors generally do not cause any problems or require special attention. The same applies to ICs because industry standards exists for pin numbering on most common packages. Polarized capacitors, diodes, LEDs, transistors, and other three-lead discrete semiconductor devices, trimpots, switches, and transformers are a different situation. You must examine these parts in detail and make sure that the netlist matches the PCB parts library.

Start with the documentation for the PCB parts library. Most PCB design software vendors provide graphic illustrations of standard library parts giving dimensions and pin names. You will also need the manufacturer's data sheets for all parts with unique pin arrangements.

The Ignition_Sys design includes the following parts that require close examination of pin numbers and pin arrangement. Pin information abstracted from manufacturers' data sheets is included for your convenience:

- **C1** (22UF 35V electrolytic capacitor): Refer to Figure 8-6 for the ECAP\30SQ\SMD pin arrangement. Note that Capture assigns pin 1 to the positive (+) terminal on polarized capacitors. This is not an industry standard and matches the PCB part only by coincidence. No editing is required.
- **C2,C3** (1UF 16V tantalum capacitor): Refer to Figure 8-7 for the ECAP\3216\SMD pin arrangement. Same note about pin numbers as for C1. No editing is required.
- **D1** (1N4007 diode): Refer to Figure 8-8 for the D\40LS pin arrangement. Capture uses the pin names CATHODE and ANODE for all diodes and LEDs. Change the pin names to pin numbers 1 and 2, respectively.
- **Q1,Q2** (2N4401 transistor): Refer to Figure 8-9 for the TO-92A pin arrangement. Capture uses the pin names EMITTER, BASE, and COLLECTOR for all bipolar transistors. The data sheet for a 2N4401 shows pin sequence EBC corresponding to pin numbers 1, 2, and 3, respectively. Note that not all TO-92 bipolar transistors use this same pinout.
- **Q3** (Fuji ET365 power transistor): Refer to Figure 8-10 for the TO-220\UPA pin arrangement. The Fuji data sheet for the ET365 shows pin sequence BCE corresponding to pin numbers 1, 2, and 3, respectively. Note that most TO-220 bipolar transistors use this same pinout.
- **R5,R6** (10K trimpot, Mepcopal CT6P series): Refer to Figure 8-11 for the VRES\T\ADJ\6MM pin arrangement. Capture uses the pin names A, WIPER, and B for all trimpots. Examination of the schematic and netlist signal section shows that pin B is the CW (clockwise) terminal. The part data sheet shows that this corresponds to pin 1 of the PCB layout part in Figure 8-11. Thus assign A, WIPER, and B pin numbers 3, 2, and 1, respectively.
- **Y1** (8.00 MHZ ceramic resonator with built-in capacitors): Refer to Figure 8-12 for the XTAL\RES\SMD pin arrangement. Capture uses pin numbers 1, 2, and GND. The PCB layout part uses pin numbers 1 to 6. The part data sheet shows that pins are connected in parallel across the two rows and that pins 2 and 5 are ground. Thus change netlist pins 1, 2, and GND to pin numbers 1, 3, and 2, respectively.

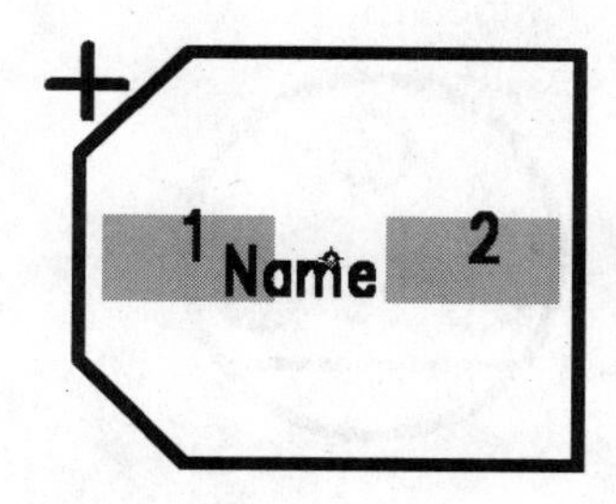

Figure 8-6 **ECAP\30SQ\SMD Part Layout**

Figure 8-7 **ECAP\3216\SMD Part Layout**

Figure 8-8 **D\40LS Part Layout**

Figure 8-9 TO-92A Part Layout

Figure 8-10 TO-220\UPA Part Layout

Figure 8-11 VRES\T\ADJ\6MM Part Layout

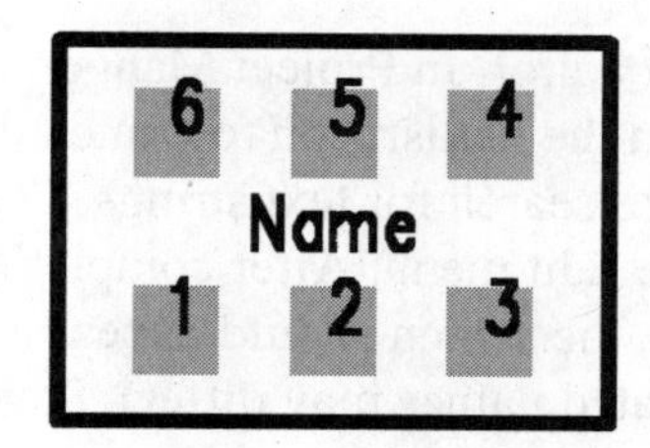

Figure 8-12 XTAL\RES\SMD Part Layout

Netlist Editing

You can use most text editors to edit ASCII netlists such as the PADS-PCB netlist in this exercise. The text editor should include search and replace capability. Capture includes a text editor that has search and replace capability. The Capture text editor is readily accessible by double clicking on any report in the Outputs folder. You can also use the Windows WordPad or Microsoft Word. Remember to save your file in "text only" format. Use a fixed-space font such as Courier to help maintain neat columns. The Windows Notepad is not recommended as a netlist editor, as it lacks search and replace capability.

Editing Pin Numbers and Signal Names

Each of the parts listed on page 264, with the exception of C1 through C3, require pin number edits. Search and replace techniques can greatly reduce the editing time for large, complex designs containing many common components. For example, if the design used several dozen diodes, you could replace every occurrence of the strings CATHODE and ANODE with the numbers 1 and 2, respectively.

The other parts require more detailed edits because there are no unique one-to-one correspondences between Capture and PCB design pin numbers. Q1, Q2, and Q3 are a good example. The BASE pins on Q1 and Q2 must be changed to pin number 2, whereas the BASE pin on Q3 must be changed to pin number 1. In this case, you have to search for every occurrence of Q1, Q2, and Q3 in the net signal section of the netlist and make the appropriate edits. The same applies to the remaining parts on page 264. You must search for every occurrence of these parts and make the required edits.

Long system-generated signal names can prove to be a real nuisance in the PCB design process. Shorten any long system-generated names. Try to keep signal names to no more than six characters.

You can use the Capture text editor. In Project Manger, double click on the outputs icon and then double click on the netlist icon to launch the text editor. Use the Find command on the Edit menu to search for text strings. For search and replace, use the Replace command on the Edit menu. After completing the pin number edits and shortening signal names, the screen should appear as shown in Figure 8-13 (sequence and system-generated names may differ). Save and then print out a hardcopy of your edited netlist.

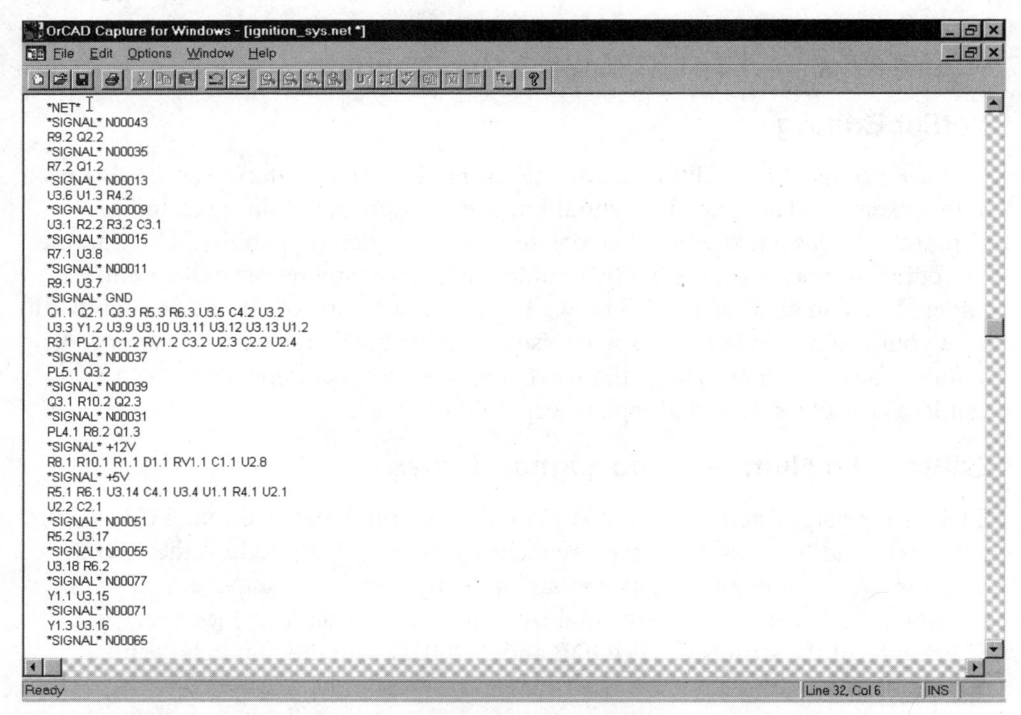

Figure 8-13 Edited Netlist in the Capture Text Editor

The edited netlist is now ready for import to the PCB design system. Figure 8-14 shows the results of loading the Ignition_Sys netlist into PowerPCB. The figure shows the PCB design after creation of the board outline and placement of all parts. "Rubberbanded" interconnections appear between pins. The PCB design shown is ready for trace routing.

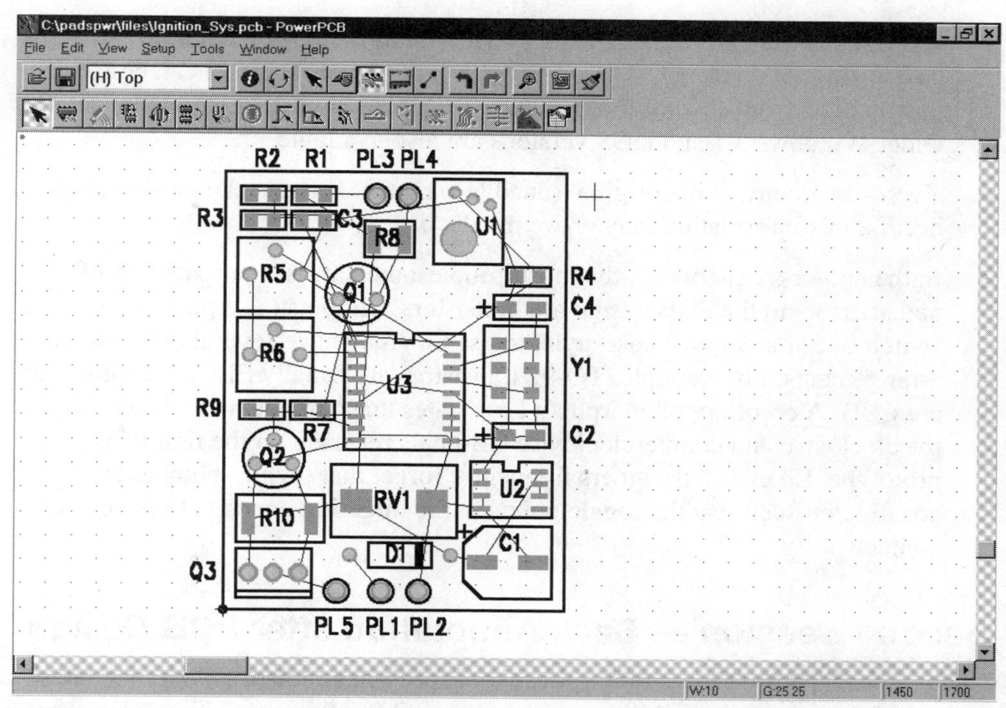

Figure 8-14 Netlist Loaded into PowerPCB

PCB Design Netlist Tips and Techniques

The following is a list of suggested tips and techniques related to netlists for PCB design:

- Before starting on netlist edits, make a file with all the available documentation: schematics, bill of materials and unedited netlist printouts, data sheets for all devices that may require edits, and any available data from the PCB layout part libraries. Keep copies of all this information in a permanent file for the design.
- Capture and most PCB design software packages share a common limitation. The programs do not include sufficient documentation for library parts, such as pin names for discrete devices. Often the only solution is to use the library editor to examine parts in question. The author uses a screen capture utility to generate hardcopy that is organized in a binder containing information on all the parts for future reference. The part footprints in Figures 8-6 through 8-12 were captured as TIFF (tagged image file format) files and then printed out. A recommended screen capture utility that appears compatible with most EDA

and CAD software is Screen Thief 95 (for Windows 95) by Nildram Software (www.nildram.co.uk). A shareware version of Screen Thief 95 can be downloaded from several Internet sites, including www.rocketdownload.com. Older Windows 3.1 and DOS versions are also available.

- Keep a hardcopy of the original (unedited) netlist. If a mistake or discrepancy occurs, this information may prove invaluable.
- In the author's experience, the most troublesome parts for propagation of netlist errors in the PCB design are transistors, trimpots (also panel pots), and switches. For any given case style, transistors may have several different pin arrangements. For example, TO-92 transistors are usually EBC, but some parts are CEB. A corollary of Murphy's Law states that trimpots will always have the clockwise and counterclockwise terminals reversed on the first PCB prototype. Likewise, designers frequently forget that toggle switches are constructed such that the toggle points in the direction opposite to the closed contacts.

Second Session — Back Annotation after PCB Design

Capture provides the Gate and Pin Swap tool for back annotation. With the latest release of Capture, the capabilities of the Gate and Pin Swap tool have been greatly enhanced. The tool now allows properties to be updated as well, similar to the Update Properties tool.

The model for this exercise is the TMS_Memory_Flat design that you created in Chapter 7. Recall that TMS_Memory_Flat is the three-sheet flat schematic shown in Figures 7-20 through 7-22. The gate and pin swap in this exercise is specific to certain reference designators and pins. Since the design you completed in the previous chapter may differ, use the sample design file included on the disk supplied with this book (see Appendix A for details).

Use the Tutor5 directory that you created for the netlist session. Use the Windows Explorer to copy the sample TMS_Memory_Flat.dsn file to this directory. Start Capture and then open the TMS_Memory_Flat design. Print out a hardcopy for your reference.

Using the Gate and Pin Swap Tool for Back Annotation

The Gate and Pin Swap tool allows you to back annotate the schematic with changes that occurred during the PCB design phase. The tool uses commands and data in a swap file. The swap file contains ASCII text. You can create this file with a text editor such as the Windows WordPad or Notepad. Although you can use other filename extensions, OrCAD suggests using the .swp extension for swap files.

A sample swap file with back annotation data appropriate to this exercise is shown in Figure 8-15. The file name is TMS_Memory_Flat.swp. This file can be found on the disk supplied with this book (see Appendix A for details).

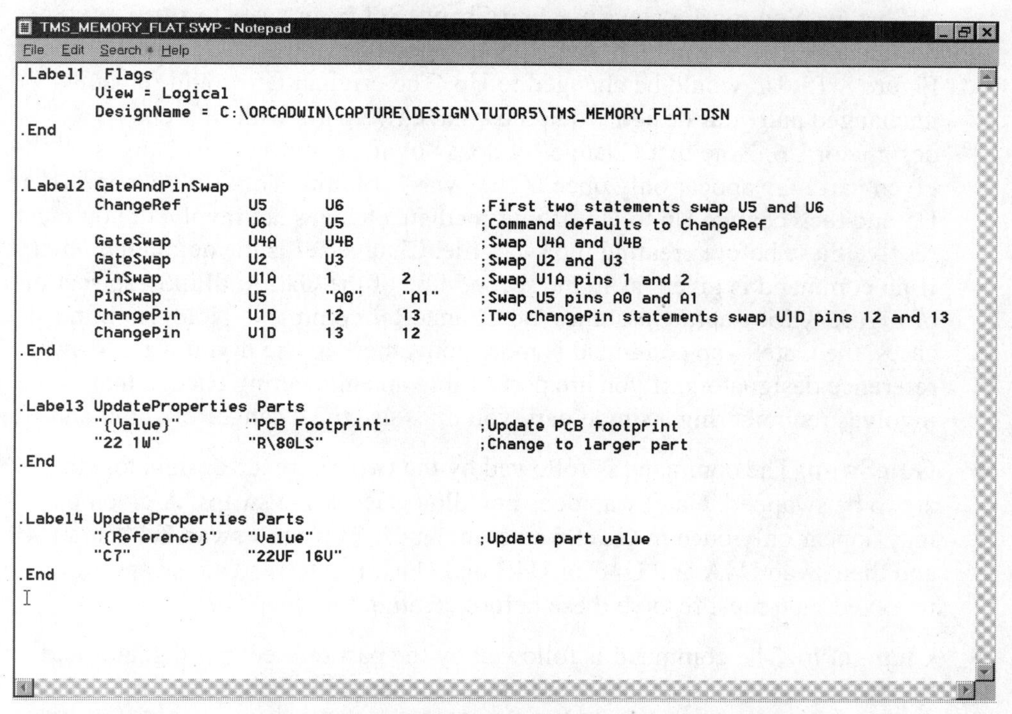

TMS_MEMORY_FLAT.SWP - Notepad

File Edit Search Help

```
.Label1  Flags
        View = Logical
        DesignName = C:\ORCADWIN\CAPTURE\DESIGN\TUTOR5\TMS_MEMORY_FLAT.DSN
.End


.Label2 GateAndPinSwap
        ChangeRef       U5      U6                      ;First two statements swap U5 and U6
                        U6      U5                      ;Command defaults to ChangeRef
        GateSwap        U4A     U4B                     ;Swap U4A and U4B
        GateSwap        U2      U3                      ;Swap U2 and U3
        PinSwap         U1A     1       2               ;Swap U1A pins 1 and 2
        PinSwap         U5      "A0"    "A1"            ;Swap U5 pins A0 and A1
        ChangePin       U1D     12      13              ;Two ChangePin statements swap U1D pins 12 and 13
        ChangePin       U1D     13      12
.End


.Label3 UpdateProperties Parts
        "{Value}"       "PCB Footprint"                 ;Update PCB footprint
        "22 1W"         "R\80LS"                        ;Change to larger part
.End


.Label4 UpdateProperties Parts
        "{Reference}"   "Value"                         ;Update part value
        "C7"            "22UF 16V"
.End
```

Figure 8-15 TMS_Memory_Flat.swp Swap File

The swap file shown in Figure 8-15 combines gate and pin swap and properties update information. The file is divided into sections. Each section starts with a label. For the purposes of the exercise, the strings Label1, Label2, Label3, and Label4 are used as labels. Any alphanumeric string can be used as a label. A period (.) must precede all labels. Immediately following the label is the name of the utility program applicable to the section. Valid utility names include: Flags, GateAndPinSwap and UpdateProperties. The flags section is used to specify the applicable view and the design file name. Logical view is used for all schematic exercises in this book. Commands and data follow the label line. Tabs or spaces are used to separate commands and data elements. Indentation is not a requirement but improves readability. Each section terminates with a .End string.

The Label2 section in the sample swap file contains gate and pin swap information. Each line contains a separate command. Four commands are available for gate and pin swap purposes:

- **ChangeRef**. The command is followed by the "was" and "is" reference designators. It changes the reference designator from the "was" value to the "is" value. You must always use **two** ChangeRef commands to swap reference designators. For example, if your file included only the first ChangeRef line in Figure 8-15, U5 would be changed to U6. The original U6 would remain unchanged and your design would now have two parts with reference designator U6. Note that ChangeRef does not allow successive changes. A given part may appear only once in the "was" column. You can't change U6 to U5 and then change U6 to U7. If intermediate changes are involved, you must resolve these before creating the swap file. ChangeRef is the default command. If no command is given, as in the second line of the GateAndPinSwap section of Figure 8-15, Capture assumes the ChangeRef command. Note that in most cases, the GateSwap command is more convenient to use if you want to swap reference designators. If you are performing an engineering change that involves renumbering a single part, you must use the ChangeRef command.

- **GateSwap**. The command is followed by the two reference designators that are to be swapped. GateSwap does not allow successive swaps. A given part may appear only once in GateSwap commands. You can't swap U4A and U4B and then swap U4A and U4C or U4B and U4C. If intermediate swaps are involved, you must resolve these before creating the swap file.

- **ChangePin.** The command is followed by the part reference designator and the "was" and "is" pin name or numbers. It changes the pin name or number from the "was" value to the "is" value. You must always use **two** ChangePin commands to swap pins. For example, if your file included only the first ChangePin line in Figure 8-15, U1D pin 12 would be changed to pin 13. The original pin 13 would remain unchanged, and your design would now have a gate with two pins numbered 13 – definitely not a lucky occurrence. As with the other commands listed, ChangePin does not allow successive changes. A given pin may appear only once in the "was" column. If intermediate changes are involved, you must resolve these before creating the swap file. Note that in most cases, the PinSwap command is more convenient to use if you want to swap pins. Pin names must be enclosed in double quotation marks (""). Many passive devices, such as capacitors and transistors, have **numeric** pin names — these must still be enclosed in double quotation marks.

- **PinSwap**. The command is followed by the part reference designator and the two pin names or numbers that are to be swapped. Unlike the other tools, PinSwap **does** allow successive swaps. You can swap pins 1 and 2 and then pins 2 and 3 on a part. Consequently, the ordering of PinSwap commands in the swap file effects the final result. Pin names must be enclosed in double quotation marks (""). Many passive devices, such as capacitors and transistors, have **numeric** pin names – these also must be enclosed in double quotation marks.

The Label3 and Label4 sections in the sample swap file contain property update information. A separate section is used for each type of property to be updated. The label lines include the utility program name, i.e., UpdateProperties, and the entity to be updated, i.e., parts. You can also update nets. Capture allows some additional parameters; refer to online help for the Gate and Pin Swap tool for more detailed information. The format of the information following the label line is identical to that of a standard update file. The first line lists the combined property string and the property to be updated. Subsequent lines contain the match string and the corresponding values for the update properties. Note that property names are case sensitive. For additional format information about update files, see pages 245-247 in Chapter 7.

Run the Gate and Pin Swap tool to back annotate the TMS_Memory_Flat design. In Project Manager, click on the design and then click on the Gate and Pin Swap tool on the main toolbar. The Gate and Pin Swap dialog box shown in Figure 8-16 appears.

The dialog box includes options to select the scope of the tool (process the entire design or just the selected schematic folders or pages). You can also enter the file name for your swap file.

Select the options shown in Figure 8-16, enter the name of your swap file and click on OK to run the tool. For a small design and simple swap file, such as the exercise in this chapter, the Gate and Pin Swap tool requires only a few seconds to run. When the tool has completed processing the swap file, double click on PAGE 1. Examine the page and, if desired, print out a hardcopy. Your schematic should appear the same as that shown in Figure 8-17.

Note that pins A0 and A1 are swapped on U6. This is a consequence of the "was" "is" nature of the swap file. The program swapped pins while swapping gates (parts). Before the gate swap, U6 was U5.

You can also examine the PCB footprint of R2 on sheet 3. This is the 22 ohm 1 watt resistor that had its PCB footprint changed to R\80LS.

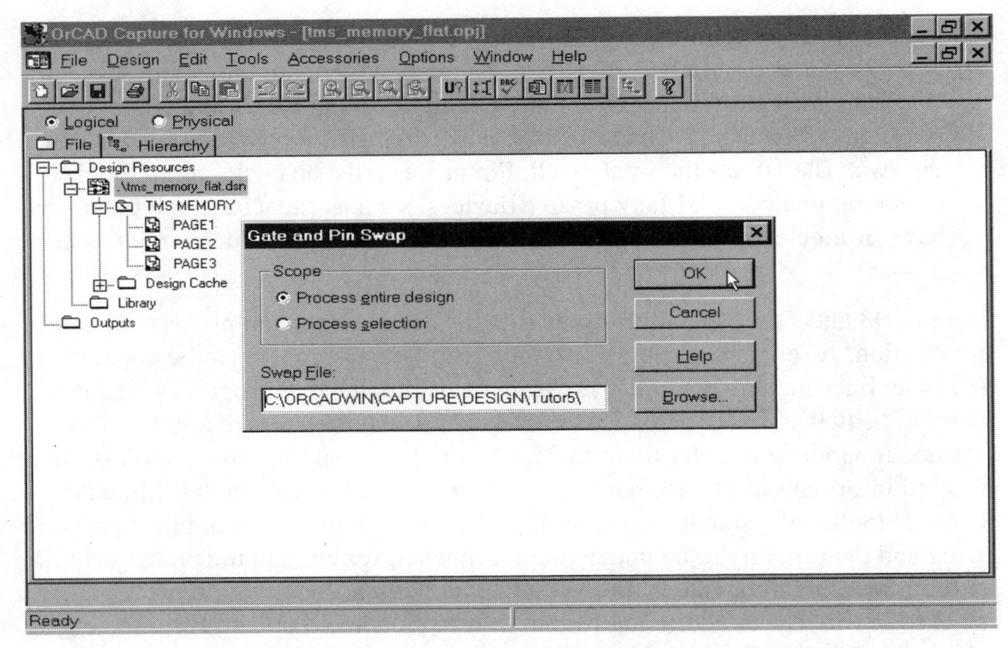

Figure 8-16 Dialog Box for the Gate and Pin Swap Tool

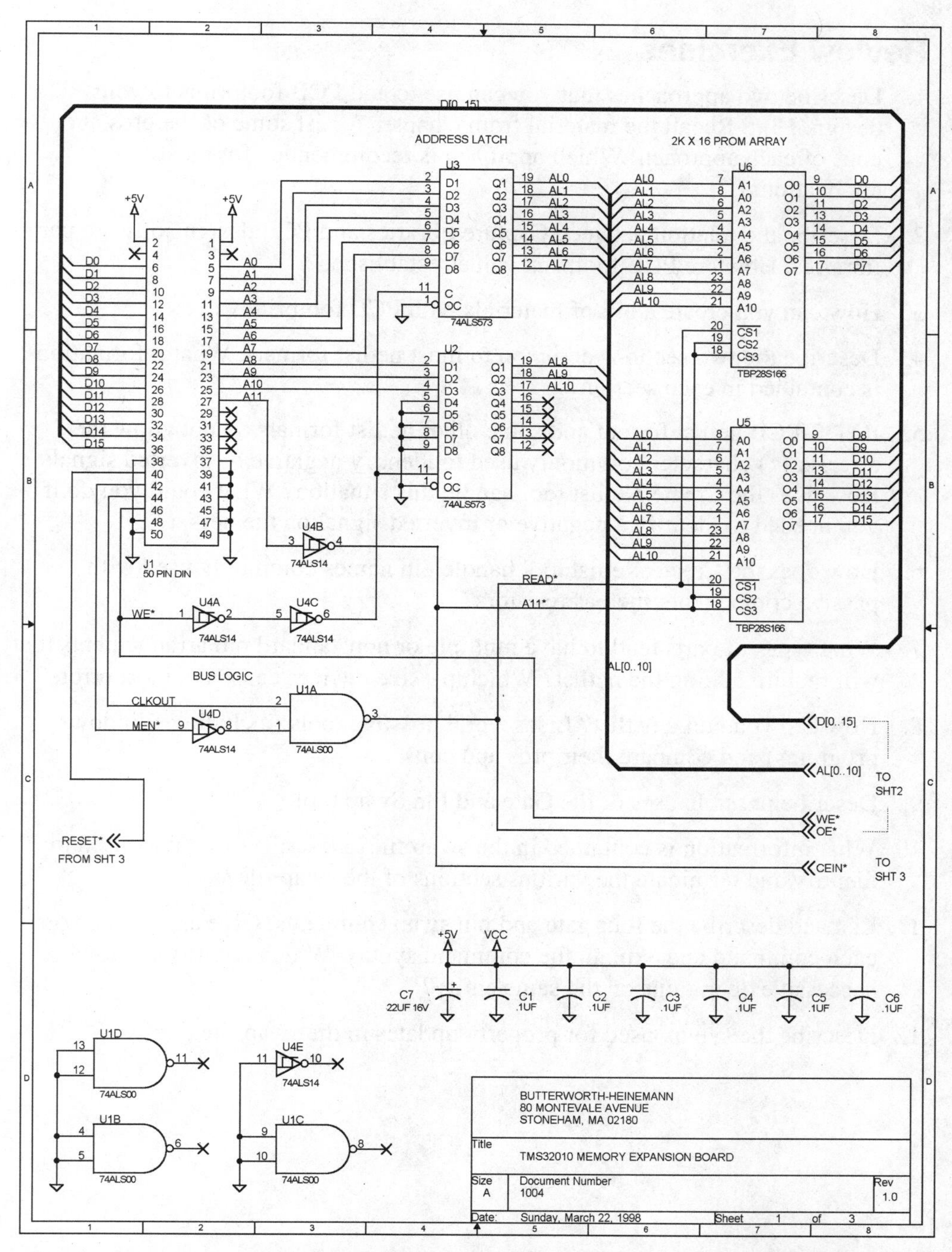

Figure 8-17 Memory Board Schematic after Gate and Pin Swap

Review Exercises

1. Describe two approaches that you can use to add PCB footprints to your design. Hint: Recall the material from Chapter 7. List some of the pros and cons of each approach. Which approach is recommended for most applications?
2. Describe the relation between Capture libraries and PCB design software parts libraries. How are PCB footprint values established?
3. How can you create a bill of materials with PCB footprint values?
4. Describe the two sections common to most netlist formats. What information is contained in each section?
5. PADS-PCB netlist format and some other netlist formats do not allow the asterisk (*) character commonly used to identify negative or inverted signals. How does the Create Netlist tool handle this situation? What could you do if you needed to identify a negative or inverted signal on the netlist?
6. How does the Create Netlist tool handle pin names commonly used with passive components such as resistors?
7. What types of parts tend to have multiple or nonstandard pin arrangements that will require editing the netlist? Which passive devices cause the most errors?
8. How can you edit a netlist? List several possible tools (including Windows programs) and compare their pros and cons.
9. Describe possible uses of the Gate and Pin Swap tool.
10. What information is contained in the swap file? Describe the syntax used to identify and terminate the various sections of the swap file?
11. List and describe the four gate and pin swap comments. Give an example of each command and explain the command syntax. Which command allows successive operations on the same entity?
12. Describe the syntax used for property updates in the swap file.

9

SPICE Netlists

The chapter consists of an exercise that shows you how to capture a schematic and create a netlist for SPICE circuit simulation. SPICE stands for simulation program with integrated circuit emphasis. The University of California, Berkeley, developed the original SPICE program during the mid 1970s. Since then, many commercial versions have been developed. One of the most popular for the Windows 95 environment is PSpice. OrCAD recently acquired the rights to PSpice. This may result in a closer integration of Capture and PSpice in the near future. For now, Capture supports PSpice with a customized component library and netlist format. With some limitations, the same library and netlist format are compatible with most other SPICE programs.

SPICE evolved as a result of the problems associated with designing ICs. By the 1970s, ICs had reached the state of miniaturization and complexity in which breadboarding a prototype was no longer feasible. SPICE allowed the designers to simulate the circuit before running a wafer through a fab line and testing an actual part. Although SPICE simulation had its roots in IC design, the concept is now widely used throughout the electronics industry.

In this exercise, you will draw a schematic and then create a SPICE netlist for a small analog circuit, a magnetic pickup phonograph amplifier. The assumption is made that the reader has some level of familiarity with SPICE.

SPICE Netlist Overview

SPICE simulation programs, such as PSpice, require a circuit file as input. The circuit file consists of:

- **Title line and comments**. PSpice expects the first line to be the title line. Additional comment lines starting with an asterisk (*) character usually appear after the title line. The Create Netlist tool automatically extracts information from the title block and creates the title line and additional comment lines.
- **Circuit device declarations list**. Each device declaration appears on a separate line. Device declarations include device name (reference designator),

node names, and part/parameter value(s). Capture extracts data for circuit devices directly from the schematic. An example of a device declaration is

R1 INPUT 0 47K ; COMMENTS

where R1 is the device name, INPUT and 0 are the two node names, and 47K is the part value. PSpice accepts alphanumeric node names up to 131 characters in length. Ground is always assigned node 0. The semicolon is treated as an end-of-line character, and any text after the semicolon is treated as a comment. Node names are established by net aliases. Capture uses system-generated names for nodes without associated net aliases.

- **SPICE commands**. These are program commands for the SPICE simulator. All SPICE commands start with a period (.) character. The exact syntax and SPICE-supported commands vary somewhat between the different commercial versions.

Capture can extract SPICE commands, comments, and device declarations from special text strings placed on the schematic. SPICE text strings must be preceded by the pipe character (|). On most PC AT style keyboards, the pipe character is above the backslash (\) and appears as a broken vertical bar. For example:

```
|SPICE
|* ADDITIONAL COMMENT LINE
|.DC LIN I2 5MA 2MA 1MA ; DC SWEEP STATEMENT
|KXFORM L1 L2 .99 ; COUPLED INDUCTOR DECLARATION
```

SPICE strings should be grouped together on the schematic. The |SPICE statement on the first line instructs Capture to place the remaining lines of text at the top of the netlist. The pipe characters are stripped from the text placed on the netlist.

The order of comments, device declarations, and commands in the circuit file is generally not important, except that the first line must be the title and the .END statement must appear on the last line.

SPICE programs use the term *circuit input file*. Capture uses the term *SPICE netlist*. The term *SPICE netlist* is used in a very broad sense, because unlike a netlist for PCB design, no actual listing of circuit nets is created.

Most of the PC-based SPICE circuit simulation programs, including the present version of PSpice, offer schematic capture. So why use Capture? If simulating a circuit is the only requirement, using the simulation program's built-in schematic capture capability may be the best approach. However, if PCB design is a requirement after the simulation is completed, using Capture from the start is more effective.

PSpice Parts Library

OrCAD provides a special PSpice parts library. To avoid problems, you must use parts exclusively from this library. PSpice, and other SPICE versions as well, have certain special requirements and considerations regarding parts:

- **Special reference designators**. Capacitors, inductors, resistors, diodes, and bipolar transistors use standard reference designators. JFETS use the letter J and MOSFETS use M. Various controlled current and voltage sources use the letters E, F, G, and H. Transformers are modeled as coupled inductors; SPICE uses T to represent a transmission line.
- **Device values and units**. SPICE uses units of volts, amperes, seconds, meters, ohms, henries, and farads with the multiplier prefixes in Table 9-1.

Table 9-1 SPICE Multiplier Prefixes

SPICE PREFIX	UNIT	MULTIPLIER
F	femto	10^{-15}
P	pico	10^{-12}
N	nano	10^{-9}
U	micro	10^{-6}
M	milli	10^{-3}
K	kilo	10^{3}
MEG	mega	10^{6}
G	giga	10^{9}
T	tera	10^{12}
MIL	.001 inch	2.54×10^{-6}

Note that a 10-megohm resistor would become 10MEG not 10M.

- **Device models**. Other than simple passive devices such as capacitors, inductors, and resistors, most semiconductor devices are based on models with specified parameters. PSpice comes with its own parts library that contains device models used during simulation (not to be confused with Capture's PSpice library). You can use only parts for which models exist in the simulator's library. All the parts in Capture's PSpice library are supported. PSpice does not support parts in other Capture libraries.
- **Device pin names.** SPICE device declarations require nodes to be in a specified sequence. For example, the sequence for bipolar transistor device nodes is collector, base, and emitter. The Create Netlist tool generates the node sequence based on the numerical order of the pin names. Bipolar transistor pins are named 1, 2, and 3 corresponding to collector, base, and emitter. All the parts in the PSpice library have pin names with the proper numerical order for SPICE.

Circuit Considerations for SPICE Simulation

A detailed discussion of circuit considerations for SPICE simulation is beyond the scope of this book. Some basic considerations relevant to Capture and the exercise in this chapter are listed below. For additional information, the reader should consult the reference given at the end of this chapter.

- **Circuit simplification**. Circuits for SPICE simulation can generally be simplified. For example, SPICE voltage sources for power supplies are ideal, with zero source impedance. In most cases, you can delete bypass capacitors without affecting simulation results.
- **Ideal circuit devices**. Simple passive circuit devices such as resistors and capacitors are modeled as ideal devices without parasitic elements. Parameters such as resistor wattage or capacitor polarity and rated voltage are not applicable to the ideal devices modeled in SPICE.
- **Initial conditions**. In addition to part value, initial conditions can be specified for capacitor voltage and inductor current. The initial condition statement follows the device value. For example:

  1U IC=5V (1 microfarad capacitor charged to 5 volts)

  5MH IC=1A (5 millihenry inductor with 1 amp current)

- **Ground**. The Capture PSpice library includes a special ground symbol. Note that SPICE considers ground as a node, not as a circuit device. SPICE requires the name "0" for the ground node. The Create Netlist tool correctly interprets the standard Capture ground symbols, even though these are named GND.
- **Floating nodes prohibited**. SPICE does not allow isolated terminals or floating nodes. All devices must have at least two terminal nodes. Every terminal must be connected to another terminal or ground. Every SPICE circuit must have a ground node, and every other node in the circuit must have a DC path to ground. If necessary, high value resistors must be added in parallel with circuit devices such as capacitors to meet the DC path to ground requirement.
- **Power objects**. You can use Capture power symbols if the value is correctly named. The Create Netlist tool translates power symbols into SPICE voltage source device declarations. For example:

VCC +5V	(power symbol name on schematic)
VCC VCC 0 +5V	(SPICE voltage source declaration)

  Note that the name of the voltage source, VCC, is also used as the name for the first node. Capture automatically assigns the second node to ground. The first letter of the name must be V, because this is the prefix used by SPICE for voltage sources. The "+" and "V" characters associated with the +5V value are optional.
- **Inputs and outputs**. SPICE circuits must be completely self-contained. Inputs are excited via a voltage or current source with an associated waveform definition. Outputs are measured across a load impedance, such as R9 in Figure 9-1.

Starting the SPICE Netlist Exercise

Use the circuit shown in Figure 9-1 as the model for this exercise. Start the exercise by creating a new folder called Tutor6. Then create a new design called Phono_Amp.dsn using the techniques you learned in previous chapters. Use the name PHONO AMP for the single-sheet schematic. You will need to add the PSpice library to your list of configured libraries before you can start placing parts.

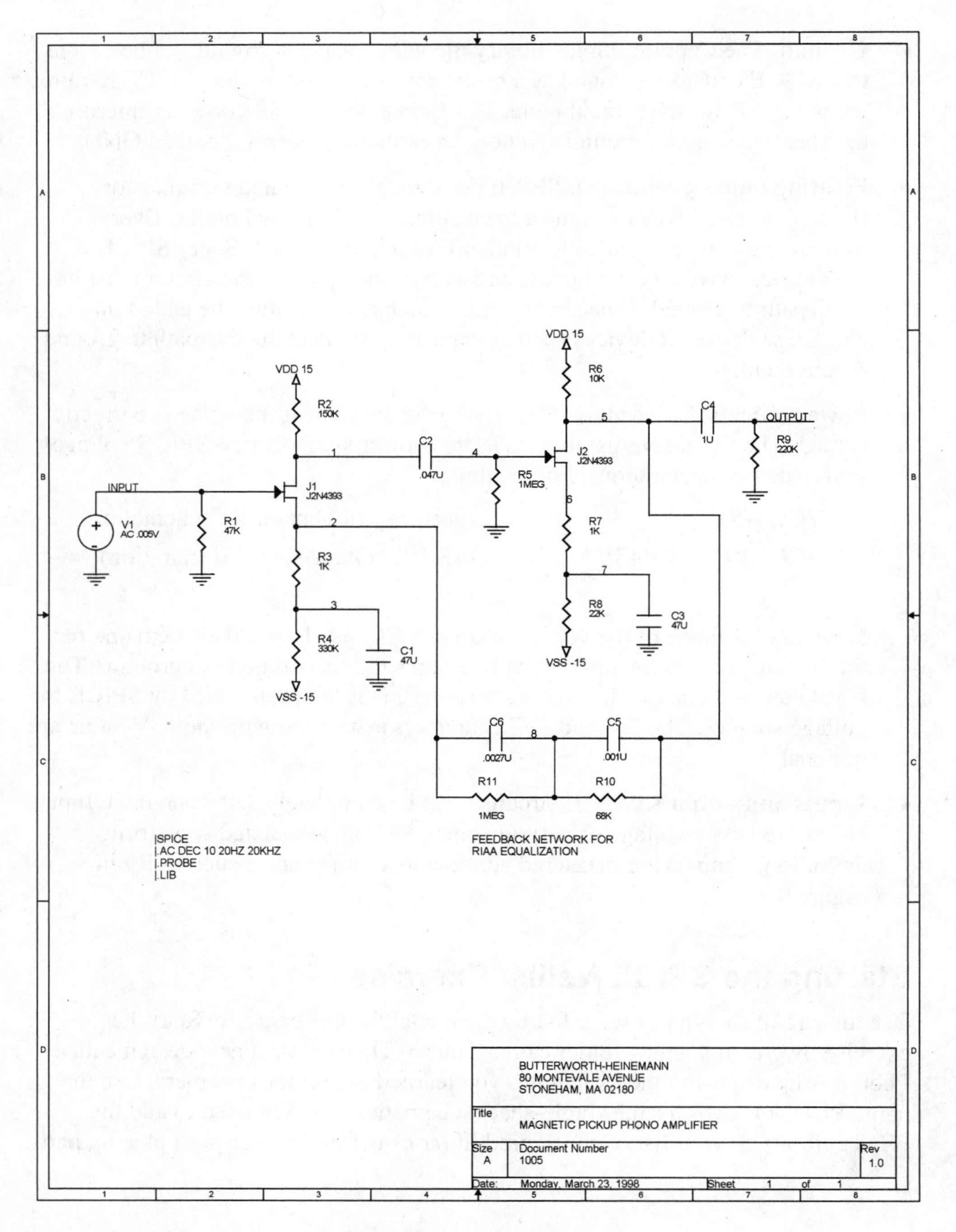

Figure 9-1 Phono Amplifier Schematic

Drafting the Schematic

Complete the schematic as shown in Figure 9-1. All the parts that you place must come from the PSpice library. To avoid possible errors, you may want to delete all other libraries except for Capsym. Listed below are hints on the parts and symbols used in the phono amplifier design:

- **Power symbols**: use the standard Capsym power symbol and edit the name as required.
- **Ground**: use GND from the PSpice library.
- **Capacitors**: use CAP from the PSpice library. All capacitors in a SPICE simulation are nonpolarized.
- **Resistors**: use RES from the PSpice library.
- **J2N4393**: use JFET N from the PSpice library. Edit the name. Note that SPICE requires the reference designator J for JFETS. J2N4393 is the PSpice name for a 2N4393 JFET.
- **Voltage source V1**: use V SRC from the PSpice library. Edit the name.

Additional objects that appear on the schematic include the net aliases 1 through 8, text comments, and the SPICE pipe commands.

Because the schematic has few components, manually edit the reference designators. This will assure that your schematic matches the model in Figure 9-1 and will facilitate checking the netlist.

After completing the schematic, run the Design Rules Check tool. You will get an error message about power connected to an output at V1. You can ignore this error. Correct any other errors. Rerun the Design Rule Check tool to remove DRC markers. Save your design and print out a hardcopy of the schematic.

Creating the SPICE Netlist

In Project Manager, click on the Create Netlist tool. The dialog box shown in Figure 9-2 appears. Click on the Other tab and then use the scroll box to select "spice.dll" for PSpice format. Note that the Create Netlist tool has a separate SPICE tab. The netlist format generated by the SPICE tab appears always to use system-generated names and ignore net aliases. Use the SPICE tab only if you require a hierarchical netlist (refer to online help in Capture for more details). Use spice.dll from the Other tab for this exercise. This selection will generate a flat netlist suitable for most simulations. Note that the terms *hierarchical* and *flat* refer to the netlist, not the design. A simple hierarchical design is converted into a flat netlist when you select spice.dll from the Other tab.

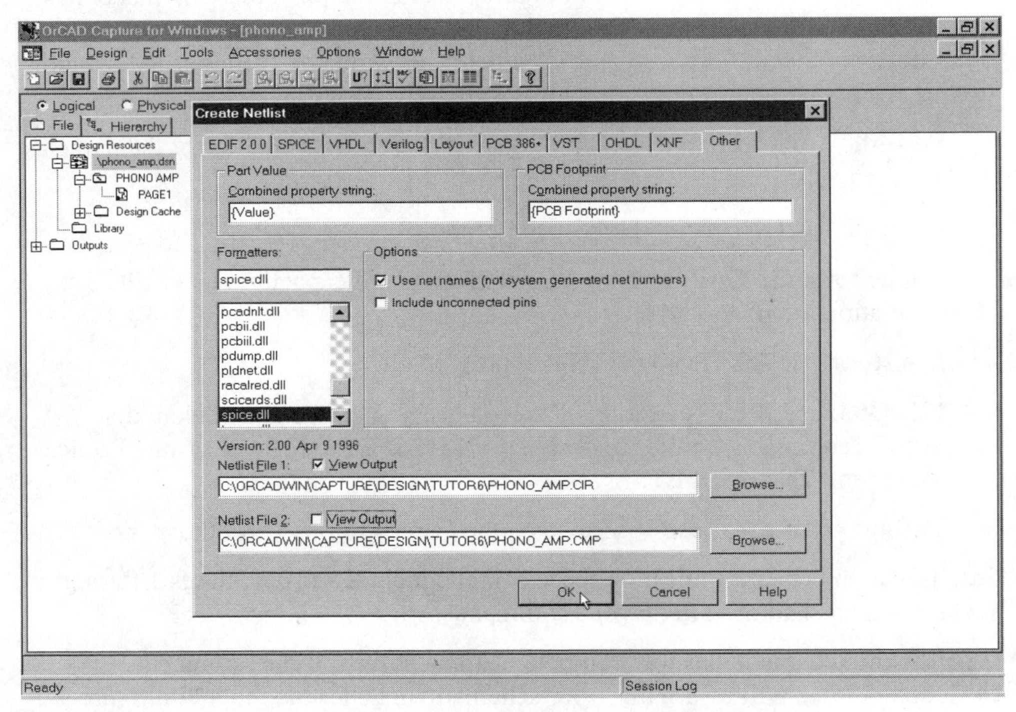

Figure 9-2 Dialog Box for the Create Netlist Tool

The following options are available for creating a SPICE netlist:

Use net names (not system-generated net numbers). Causes net aliases to be used as node names. Some SPICE versions do not accept alphanumeric node nodes. If this option is not selected, Capture assigns arbitrary numerical node names starting with 10000. The map file (see Netlist File 2) cross references net aliases on the schematic to the assigned node names.

Include unconnected pins. SPICE never allows unconnected (floating) pins. This option will cause SPICE to generate an error message if unconnected pins appear on the schematic.

Netlist File 1. This is the actual SPICE netlist file. The file path defaults to the current design directory, and the file name defaults to the design name with a .NET extension. You can change these defaults. Note that PSpice expects the netlist to have a .CIR extension. If you select the option to view the output, Windows launches the appropriate file viewer or text editor. If

PSpice is loaded on your system, Windows will launch the PSpice (MicroSim) text editor with the netlist.

Netlist File 2. This is the map file that cross references net aliases on the schematic to the assigned node names. The data in this file is invalid if you select the Use net names option. The file path defaults to the current design directory, and the file name defaults to the design name with a .CMP extension. If you select the option to view the output, Capture launches the Windows Notepad with the map file.

Select the Use net names option and change the netlist file extension to .CIR as shown in Figure 9-2. Then click on OK to create the netlist. When the tool has completed processing the netlist, Windows launches an appropriate viewer with the netlist file. Your file should appear similar to that shown in Figure 9-3. Statements may appear in a different order, depending on how you placed parts onto the schematic.

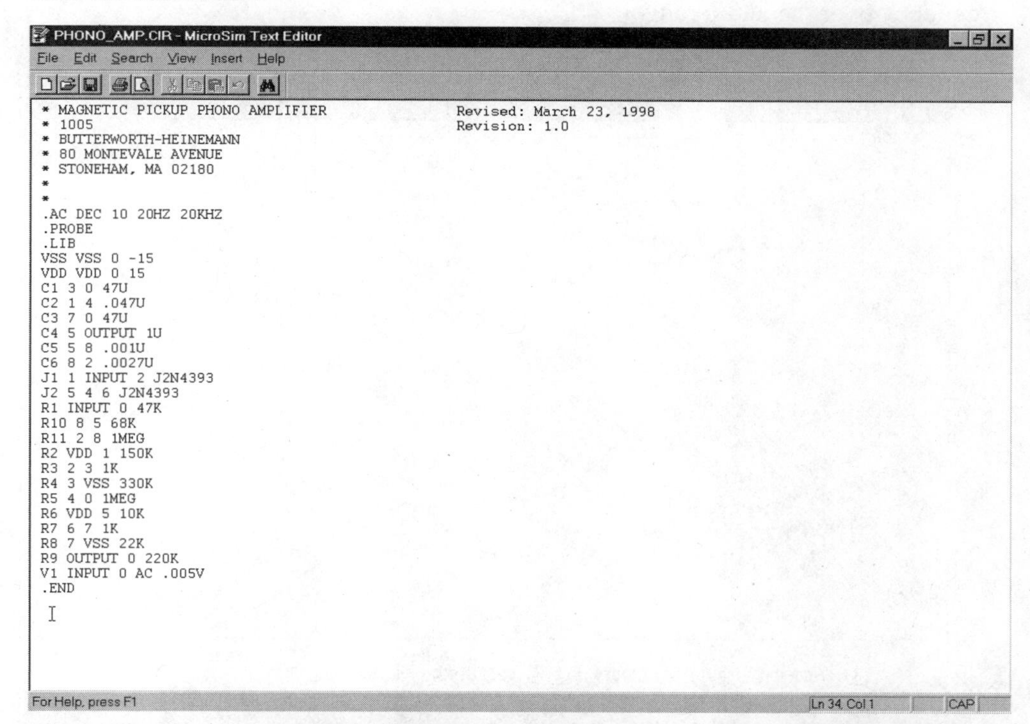

Figure 9-3 SPICE Netlist for the Phono Amplifier Circuit

SPICE netlists created by Capture have the following characteristics and limitations:

- The Create Netlist tool does not check part names, net aliases, port names, and pin numbers for length.
- Node numbers are limited to five characters (however, longer system-generated node names may be output in some cases).
- If you select the Use net names option, node names are restricted to alphanumeric characters, the dollar sign ($), and the underscore (_).
- All ASCII characters are legal for part (device) names, values, comments, and commands.

PSpice runs the netlist shown in Figure 9-3 without any requirement for editing. Figure 9-4 shows a Bode plot of the amplifier gain and phase shift as a function of frequency over the audio range.

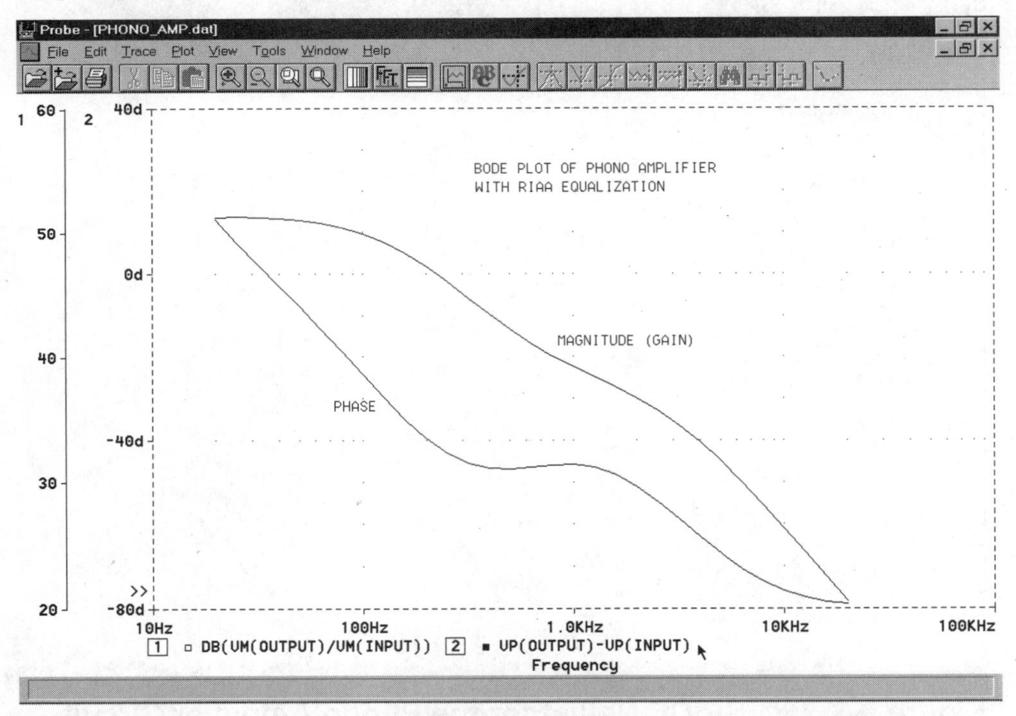

Figure 9-4 Bode Plot from Phono Amplifier Simulation

Conclusion

Several excellent reference sources for readers who are interested in learning more about SPICE include:

Connelly, J. *Macromodeling with SPICE*, Englewood Cliffs: Prentice-Hall, 1992.

Kielkowski, R. *Inside SPICE: Overcoming the Obstacles of Circuit Simulation*, New York: McGraw-Hill, 1994.

Rashid, M. *SPICE for Power Electronics and Electric Power*, Englewood Cliffs: Prentice-Hall, 1993.

Sandler, S. *SMPS Simulation with SPICE*, New York: McGraw-Hill, 1997.

Tuinega, P. *SPICE: A Guide to Circuit Simulation and Analysis Using PSpice*, 2d ed. Englewood Cliffs: Prentice-Hall, 1992.

The SPICE guide by Tuinega is an introductory book that is available with a student version of PSpice on disk, which is suitable for running the phono amplifier circuit model used as an example in this chapter. The book by Connelly covers the modeling of complex circuit functions via macromodels. The books by Rashid and Sandler cover power electronics and SMPS (switch mode power supply) systems. The book by Kielkowski examines the internals of SPICE and contains useful information for more advanced users.

Review Exercises

1. Describe the contents of a typical circuit file used as input for SPICE simulation.
2. What character must you use to begin SPICE comment lines?
3. What character must you use to begin SPICE commands?
4. How does Capture extract text strings that contain SPICE commands, comments and circuit device declarations?
5. What character must precede all SPICE text strings placed on the schematic?
6. What is the purpose of the ISPICE statement?
7. Why should you use the special SPICE libraries supplied with Capture? What netlist edits are required if you use conventional Capture library parts?

8. List the SPICE multiplier prefixes. How do these differ from the multiplier prefixes used with conventional schematics? Hint: Refer to the material in Chapter 1.
9. List some of the basic circuit considerations for SPICE simulation.
10. Run the SPICE netlist exercise again without selecting the Use net names option. What node names are output by Capture? Examine the map file. Do the cross references match the schematic?
11. What characters are allowed in node names?

10

Bill of Materials Techniques

Bills of materials were first discussed in Chapter 4 when the Bill of Materials tool was introduced. Additional bill of materials concepts, such as extracting user properties and the use of include files, were covered in Chapter 7. Recall that the include file provides a means of automatically merging parts information into bills of materials.

Most users find that simply editing the basic Capture-generated bills of materials is the most effective approach. Even with the use of an include file or user properties to complete the part descriptions, bills of materials require further editing in part because of limitations of the Bill of Materials tool. Capture generates a somewhat disorganized header and does not sort parts by value. In this chapter, you will learn simple techniques for editing and sorting bills of materials. Sorting is accomplished via a simple sort utility included on the disk supplied with this book. You will also learn how to import bills of materials into spreadsheet programs such as Microsoft Excel.

Starting the Bill of Materials Exercise

The model for this exercise is the Ignition_Sys design that you created in Chapter 3. Recall that Ignition_Sys is the four-sheet hierarchical schematic shown in Figures 1-18A through 1-18D in Chapter 1. For your convenience, a completed design file that you can use for this exercise is included on the disk supplied with this book (see Appendix A for details). The same design was also used for the exercise in Chapter 8.

Create a new design directory called Tutor7. Use the Windows Explorer to copy either your own or the supplied Ignition_Sys.dsn file to this directory. Start Capture and then open the Ignition_Sys design. Print out a hardcopy of the design for reference.

First, check the design for errors. In Project Manger, click on the design to select it. Then run the Check for Invalid Stacked Hierarchical Pins tool from the Accessories menu to report any invalid stacked pins. If the tool reports any invalid stacked pins, rerun the tool with the fix option. Next, run the Design Rules Check tool with the usual options selected. The tool will report numerous warnings about bidirectional pins. You can ignore these warnings. Fix any errors and then rerun the tool with the option selected to delete existing DRC markers. Save your design. If you made any major changes to fix errors, you should also print out a new hardcopy.

Next, create a bill of materials for the design using the techniques you learned in Chapter 4. From Project Manager, click on the Bill of Materials tool. You can accept the default values in the dialog box. Select the option to view the output. The unedited bill of materials report appears as shown in Figure 10-1 (the top part of the header is not shown on the screen). Print out a hardcopy of the report for future reference.

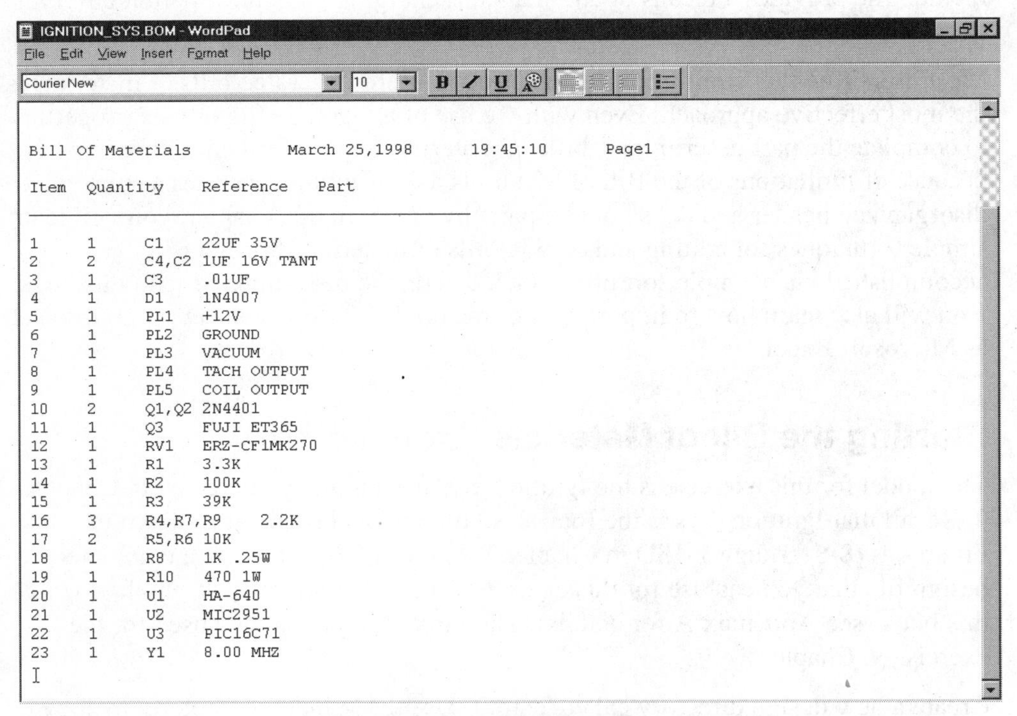

Bill Of Materials March 25,1998 19:45:10 Page1

Item	Quantity	Reference	Part
1	1	C1	22UF 35V
2	2	C4,C2	1UF 16V TANT
3	1	C3	.01UF
4	1	D1	1N4007
5	1	PL1	+12V
6	1	PL2	GROUND
7	1	PL3	VACUUM
8	1	PL4	TACH OUTPUT
9	1	PL5	COIL OUTPUT
10	2	Q1,Q2	2N4401
11	1	Q3	FUJI ET365
12	1	RV1	ERZ-CF1MK270
13	1	R1	3.3K
14	1	R2	100K
15	1	R3	39K
16	3	R4,R7,R9	2.2K
17	2	R5,R6	10K
18	1	R8	1K .25W
19	1	R10	470 1W
20	1	U1	HA-640
21	1	U2	MIC2951
22	1	U3	PIC16C71
23	1	Y1	8.00 MHZ

Figure 10-1 Unedited Bill of Materials Report

Preliminary Bill of Materials Editing

Use the Windows WordPad for preliminary editing. Do not use the Windows Notepad or DOS EDIT, as these programs do not support tab characters. Note that Capture inserts tab characters between fields in the bill of materials. The sort utility uses these tab characters to delimit fields. Do not remove or insert any tab characters or add any extra spaces. The sort utility will clean up the file for you and align the fields in neat columns.

Before editing, save the file under a new name. This allows you to retain the original file without modifications, in case you need it later. Save the file using a name such as Ign_Bom.txt. You must use a short "8.3 file name," i.e., with no more than eight characters in the file name and three characters in the extension. This is for compatibility with the sort utility, which does not accept Windows 95 long file names.

Once you have saved the file under a new name, the first editing step is to remove the header. The reason for removing the header is that it will interfere with the sort utility. In most cases, you will use only a few items of information from the header, such as the design name and revision code. You can easily retype this information or cut and paste from the original file after the sorting process.

The next step is to make preliminary edits to the bill of materials. Two areas that generally require special attention are as follows:

- Capture schematics may contain PCB layout parts that do not correspond to actual physical parts. In this example, PL1 through PL5 are pads used to solder wire harness connections to the PCB. Because they are not physical parts, they should be deleted from the bill of materials.
- Different types of capacitors or resistors can appear together on the same line if they are assigned the same part value on the schematic. For example, if the design has both 10K resistors and trimpots, these parts may all appear together. You may not notice this problem unless you closely examine the bill of materials. You can avoid this situation by always using distinctive part values. If it does occur, just edit the bill of materials and place the affected parts on separate lines.

Complete the preliminary edits by removing the lines containing PL1 through PL5. The file should appear as shown in Figure 10-2. The file is now ready for sorting.

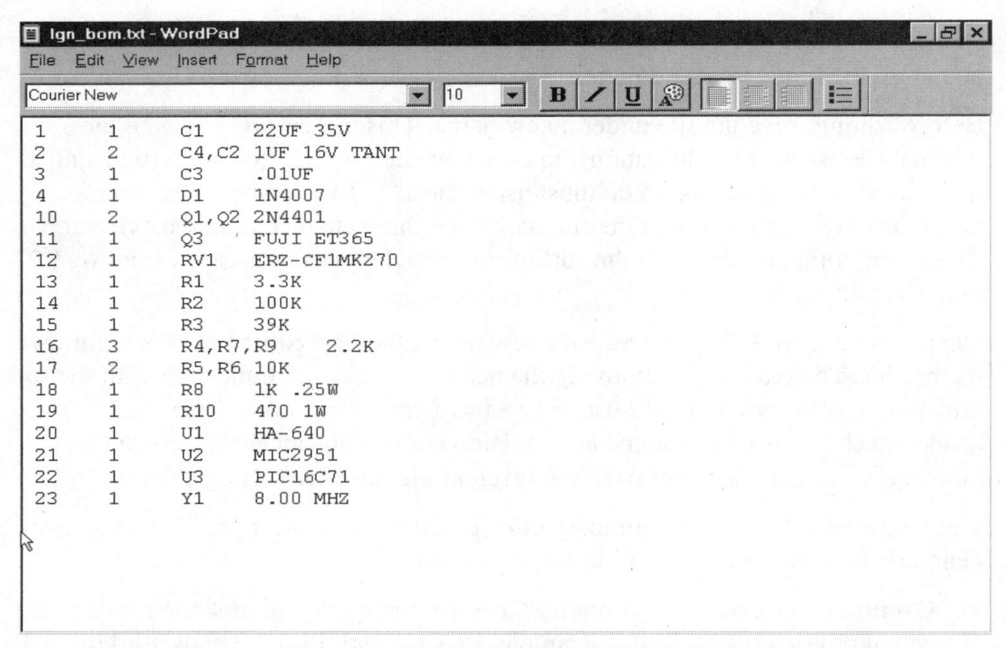

1	1	C1	22UF 35V
2	2	C4,C2	1UF 16V TANT
3	1	C3	.01UF
4	1	D1	1N4007
10	2	Q1,Q2	2N4401
11	1	Q3	FUJI ET365
12	1	RV1	ERZ-CF1MK270
13	1	R1	3.3K
14	1	R2	100K
15	1	R3	39K
16	3	R4,R7,R9	2.2K
17	2	R5,R6	10K
18	1	R8	1K .25W
19	1	R10	470 1W
20	1	U1	HA-640
21	1	U2	MIC2951
22	1	U3	PIC16C71
23	1	Y1	8.00 MHZ

Figure 10-2 Bill of Materials after Preliminary Edits

Sorting the Bill of Materials

Standard industry practice is to sort bills of materials by reference designator and part value. The Bill of Materials tool sorts only by reference designator. As you can see in Figure 10-2, capacitor and resistor part values appear in random order. Although you could easily use a text editor manually to sort the parts in a small design, manual sorting would be very time consuming and prone to errors in a real-world design with dozens of different parts.

A handy sort utility called BOMSORT is included on the disk supplied with this book. The BOMSORT utility automatically sorts parts by value. BOMSORT recognizes the common electrical value multiplier prefixes listed in Table 1-2 in Chapter 1. BOMSORT makes the following assumptions about a bill of materials file that is to be sorted:

- **No header**. The file must contain only parts information, without any blank lines. The file shown in Figure 10-2 has a suitable input format. Each line contains item number, quantity, reference designator, and parts description fields separated by a single tab character (not spaces). The line length must not exceed 79 characters. Note that some text editors, such as the Windows Notepad insert spaces in place of tab characters. This is not acceptable.

- **Maximum of 399 lines**. Note that the limitation is 399 lines, not 399 parts. This should suffice for most practical designs.
- **Reference designator prefix.** BOMSORT accepts reference designators with no more than a two-letter prefix. Note that none of the recommended prefixes listed in Table 1-1 in Chapter 1 exceeds two characters.
- **Part description (value)**. BOMSORT sorts based on the first five characters of the part value.

Installing and Running the BOMSORT Utility

Load the BOMSORT utility onto your system. Use the C:\Orcadwin\Capture\Vendor directory for Capture-related utilities. This directory should have been created during the Capture installation process. Use the Windows Explorer to copy the Bomsort.exe file from the disk supplied with this book to your Vendor directory. Next create a Windows Start menu entry for BOMSORT. Click on the Start button. Then click on Settings, Taskbar, and Start Menu Programs. Click on the Add button. Browse for the Bomsort.exe file in the Vendor directory. In the Select Program Folder dialog box, select New Folder. Use a name such as OrCAD Utilities for the new program folder. Then enter a name for the shortcut such as BOM Sort Utility. The last step is to pick an icon to associate with the shortcut, such as the MS-DOS box.

Launch the BOMSORT utility from the Windows 95 Start menu as shown in Figure 10-3. The first time you run the program, click on the MS-DOS icon in the upper left corner of the program window and then click on Properties. Select the options to Run Maximized and Close on Exit. This will automatically maximize the program window when you launch BOMSORT and return to the Windows desktop when the program terminates. The BOMSORT file name entry and options setup screen appears as shown in Figure 10-4.

BOMSORT prompts for the names of the bill of materials input file and output file. You must enter the complete path and file name. Long Windows 95 file names are not accepted. Enter C:\Orcadwin\Capture\Design\Tutor7\Ign_Bom.txt for the input file name and use a file name such as C:\Orcadwin\Capture\Design\Tutor7\Ign_Bom1.txt for the output file name. You can select options to include item numbers in the output and to insert blank lines between records. For this exercise, enter Y (yes) for item numbers and N (no) for blank lines. BOMSORT requires about one second to sort the bill of materials for the exercise and then returns to the Windows 95 desktop. Note that BOMSORT does not modify the input file.

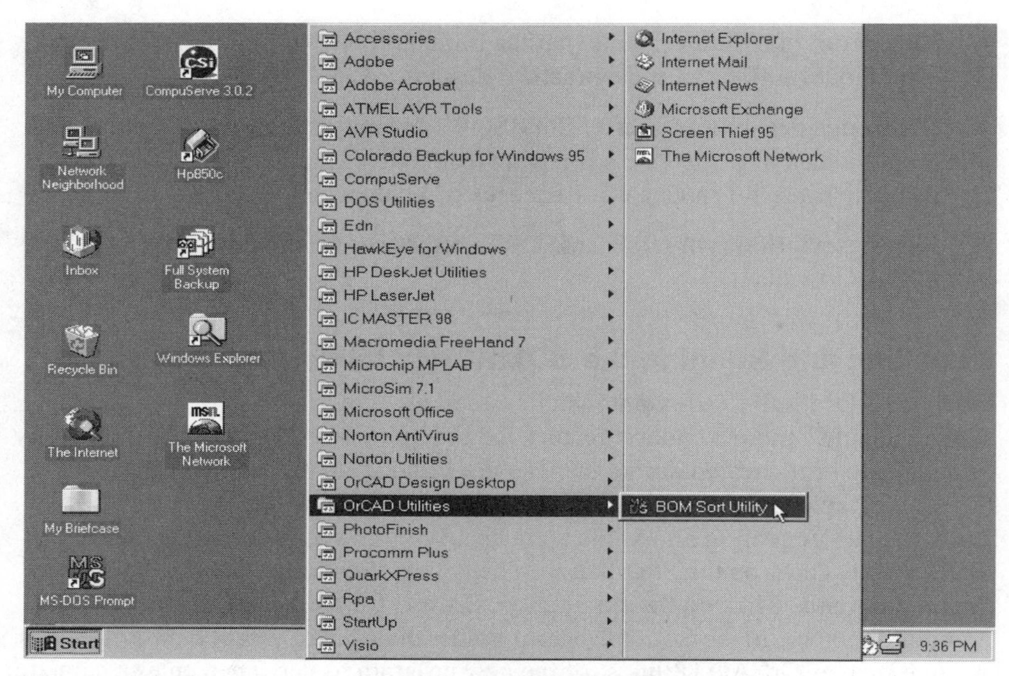

Figure 10-3 Launching the BOMSORT Utility

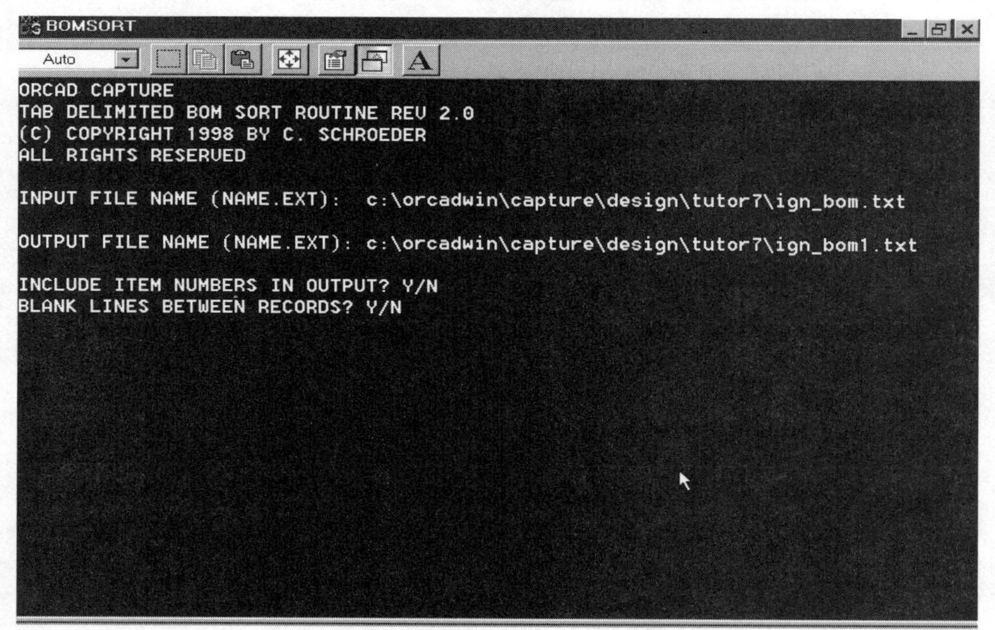

Figure 10-4 Entering File Names and Options for BOMSORT

Completing the Bill of Materials

Use the Windows WordPad or other text editor to examine and print out the sorted bill of materials file Ign_Bom1.txt. Verify that your sorted file appears similar to that shown in Figure 10-5. Note that capacitors and resistors are now listed in order of parts value.

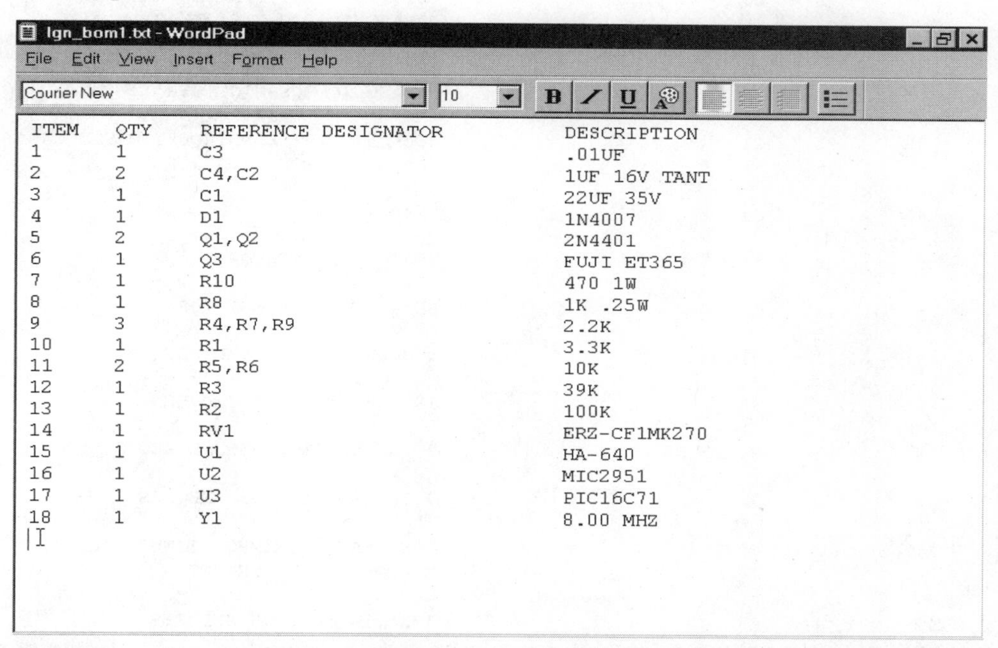

ITEM	QTY	REFERENCE DESIGNATOR	DESCRIPTION
1	1	C3	.01UF
2	2	C4,C2	1UF 16V TANT
3	1	C1	22UF 35V
4	1	D1	1N4007
5	2	Q1,Q2	2N4401
6	1	Q3	FUJI ET365
7	1	R10	470 1W
8	1	R8	1K .25W
9	3	R4,R7,R9	2.2K
10	1	R1	3.3K
11	2	R5,R6	10K
12	1	R3	39K
13	1	R2	100K
14	1	RV1	ERZ-CF1MK270
15	1	U1	HA-640
16	1	U2	MIC2951
17	1	U3	PIC16C71
18	1	Y1	8.00 MHZ

Figure 10-5 Sorted Bill of Materials Processed by BOMSORT

The final steps in completing a bill of materials include inserting some of the header information, such as the design name and revision code, back into the sorted file. You will also need to clean up some of the parts descriptions and add information such as vendor names and vendor part numbers. Abbreviated parts descriptions used on schematics are usually rewritten in a more detailed and formal manner on the bill of materials. Schematics usually include notes to the effect that all resistors and capacitors are a certain wattage, voltage, and tolerance. Recommended practice is to write out such information in detail for each part on the bill of materials. The completed bill of materials must include any parts that are not on the schematic but that are required to build the product such as the PCB and hardware.

An example of a completed bill of materials is shown in Figure 10-6.

```
Ignition System                              Revised:  March 25, 1998
Drawing Number: 1001                         Revision: 1.0
Bill Of Materials                            Page:     1
```

ITEM	QTY	REFERENCE DESIGNATOR	DESCRIPTION
1	1	C3	.01UF 50V 10% X7R 805 SMD CAP PANASONIC ECU-V1H103KBG
2	2	C2,C4	1UF 16V 20% TANTALUM 3216 SMD CAP PANASONIC ECS-H1CY105R
3	1	C1	22UF 35V 20% ELECTROLYTIC SMD CAP PANASONIC ECE-V1VA220P
4	1	D1	1N4007 DIODE
5	2	Q1,Q2	2N4401 TRANSISTOR
6	1	Q3	FUJI ET365 TRANSISTOR
7	1	R10	470 1W 5% 2512 SMD RES
8	1	R8	1K .25W 5% 1210 SMD RES
9	3	R4,R7,R9	2.2K .1W 5% 0805 SMD RES
10	1	R1	3.3K .1W 5% 0805 SMD RES
11	2	R5,R6	10K CERMET TRIMPOT 1 TURN TOP ADJ MEPCOPAL CT6P SERIES
12	1	R3	39K .1W 5% 0805 SMD RES
13	1	R2	100K .1W 5% 0805 SMD RES
14	1	RV1	ERZ-CF1MK270 SMD SURGE ABSORBER PANASONIC
15	1	U1	HA-640 HALL EFFECT SENSOR CLAROSTAT
16	1	U2	MIC2951 SOIC MICREL
17	1	U3	PIC16C71 SOIC MICROCHIP WITH REV 1.0 FIRMWARE
18	1	Y1	8.00 MHZ SMD RESONATOR PANASONIC EFO-V8004E5
19	1		PCB REV 1.0

Figure 10-6 Completed Bill of Materials

Recommended practice is to make a bill of materials for each separable assembly. For the design used in this exercise, the bill of materials represents the PCB assembly. A separate bill of materials would be generated for the final assembly, which would include the PCB assembly, housing, wire harness, mounting hardware, labels, and any other required miscellaneous items.

Importing Bill of Materials Files into Microsoft Excel

Microsoft Excel is an excellent tool for managing and printing bills of materials. Other spreadsheet programs can be used for the same purpose. Using a spreadsheet offers easy organization of data in columns and the capability of inserting and totaling cost information.

Use the following steps to import a Capture-generated bill of materials file into Excel (the assumption is made that the file has been sorted and edited as previously described in this chapter):

- Edit the file so that all parts descriptions are on a single line. Note that the example shown in Figure 10-6 uses multiple line parts descriptions, with the vendor name and part number on the second line. For better readability, leave a blank line between successive parts (you can select the option for the BOMSORT utility to automatically insert blank lines). You can eliminate the header at this stage or during the file import process.
- Launch Excel and open the file. The Excel Text Import Wizard will appear because the file is not in Excel format. The Text Import Wizard consists of three steps, described in detail below.
- Text Import Wizard Step 1 — Select Data Type. This step selects the type of data to be imported. Select fixed-width data, since the sorted bill of materials file is organized into columns. You can also select the starting row to eliminate any header information. You must also select a file origin. Select DOS or Windows.
- Text Import Wizard Step 2 — Set Field Widths. This step sets the column breaks. Set the column breaks to make separate columns for item number, quantity, reference designator, part value, and vendor name and part number.
- Text Import Wizard Step 3 — Set Data Format. This is the final step. You can set the data format for each column or skip particular columns. Skip the first column (item number) and set the remaining columns to text format. Note that the BOMSORT utility includes an option to eliminate item numbers. You can select this option if you are going to import the sorted file in Excel.

Excel now imports the bill of materials data in neat columns. You can format the column width and insert a row at the top with bold column labels. The finished spreadsheet appears as shown in Figure 10-7. Note that Excel does not automatically save the file in Excel format. Make sure you select Excel format when you save your spreadsheet, otherwise some of the formatting information may be lost.

QTY	**REFERENCE DESIGNATOR**	**DESCRIPTION**	**VENDOR PART NUMBER**
1	C3	.01UF 50V 10% X7R 805 SMD CAP	PANASONIC ECU-V1H103KBG
2	C2,C4	1UF 16V 20% TANTALUM 3216 SMD CAP	PANASONIC ECS-H1CY105R
1	C1	22UF 35V 20% ELECTROLYTIC SMD CAP	PANASONIC ECE-V1VA220P
1	D1	1N4007 DIODE	
2	Q1,Q2	2N4401 TRANSISTOR	
1	Q3	FUJI ET365 TRANSISTOR	
1	R10	470 1W 5% 2512 SMD RES	
1	R8	1K .25W 5% 1210 SMD RES	
3	R4,R7,R9	2.2K .1W 5% 0805 SMD RES	
1	R1	3.3K .1W 5% 0805 SMD RES	
2	R5,R6	10K CERMET TRIMPOT 1 TURN TOP ADJ	MEPCOPAL CT6P SERIES
1	R3	39K .1W 5% 0805 SMD RES	
1	R2	100K .1W 5% 0805 SMD RES	
1	RV1	ERZ-CF1MK270 SMD SURGE ABSORBER	PANASONIC
1	U1	HA-640 HALL EFFECT SENSOR	CLAROSTAT
1	U2	MIC2951 SOIC	MICREL
1	U3	PIC16C71 SOIC	MICROCHIP
		WITH REV 1.0 FIRMWARE	
1	Y1	8.00 MHZ SMD RESONATOR	PANASONIC EFO-V8004E5
1		PCB REV 1.0	

Figure 10-7 Bill of Materials Imported into Excel

Figure 10-7 shows the bill of materials spreadsheet as it appears within Excel. Before you print out the spreadsheet, you can use the Header/Footer tab in Page Setup on the File menu to define a custom header with information such as the design name, revision code, and date.

Review Exercises

1. Why are tab characters important when editing Capture-generated bills of materials. What Windows text editing program supports tab characters?
2. Describe the preliminary editing steps required before sorting a bill of materials with the BOMSORT utility.
3. List the assumptions BOMSORT makes about bill of materials files.
4. Does BOMSORT support Windows 95 long file names? What is the maximum number of characters BOMSORT allows in the file name?
5. Describe possible uses for the BOMSORT processing options?
6. Describe the steps required to finish a bill of materials after sorting if you are using a word processor or text editor.
7. Describe the steps required to load a sorted bill of materials file into Microsoft Excel.
8. What advantages does Microsoft Excel offer for handling bills of materials?

11

Translating Designs from OrCAD SDT

Since OrCAD Capture running under Windows is still relatively new, many users have designs and custom part libraries created with the older DOS version, OrCAD SDT. The translation from OrCAD SDT is straightforward if you use the techniques shown in the exercises in this chapter.

Starting the Design Translation Exercise

An OrCAD SDT design that you can use for this exercise is included on the disk supplied with this book (see Appendix A for details). The design consists of a three-sheet hierarchical schematic that represents a development board for the new AVR family of RISC (reduced instruction set computer) microcontrollers from Atmel.

Use the Windows Explorer to create a new directory folder called Tutor8 in your Capture Design directory. Copy all the files from the Tutor8 directory on the disk supplied with this book to this new directory.

Files Required for Design Translation

Take a moment to examine the contents of Tutor8. The files in this directory represent all the files associated with an actual OrCAD SDT design. You will not need all these files for the translation process. In general, SDT files required for the translation include:

- Schematic files with a .sch extension.
- Library files with a .lib extension.
- SDT configuration file Sdt.cfg.

Note that the name of an OrCAD SDT design is the name of the design directory. OrCAD SDT also requires that the root schematic have the same name as the design directory. Thus the root schematic of the Tutor8 design is Tutor8.sch.

The most important issue in translation from OrCAD SDT relates to part libraries. Unlike Capture, OrCAD SDT schematic files do not contain an internal parts cache. Part information is always loaded from one or more external part library files when a schematic file is opened. If a required library file is no longer available, the affected parts are deleted from the schematic.

If you still have the OrCAD SDT program and part libraries loaded onto your system, library-related problems should not occur during the translation process as Capture can access the required SDT libraries. A different situation exists if you no longer have OrCAD SDT loaded or receive a design originally created on another system. If recommended practices were followed and the OrCAD SDT design includes an archive library file with all parts that appear on the schematics, Capture can use this archive library during the translation process. If no archive library is available, a last resort is to translate your Capture libraries back to OrCAD SDT format and then use these libraries for the design translation process.

During the translation process, Capture reads the library list in the OrCAD Sdt.cfg file to determine what libraries to use. The Sdt.cfg file is an ASCII file that you can examine with any text editor. The library list in the Sdt.cfg file for this exercise is shown in Figure 11-1.

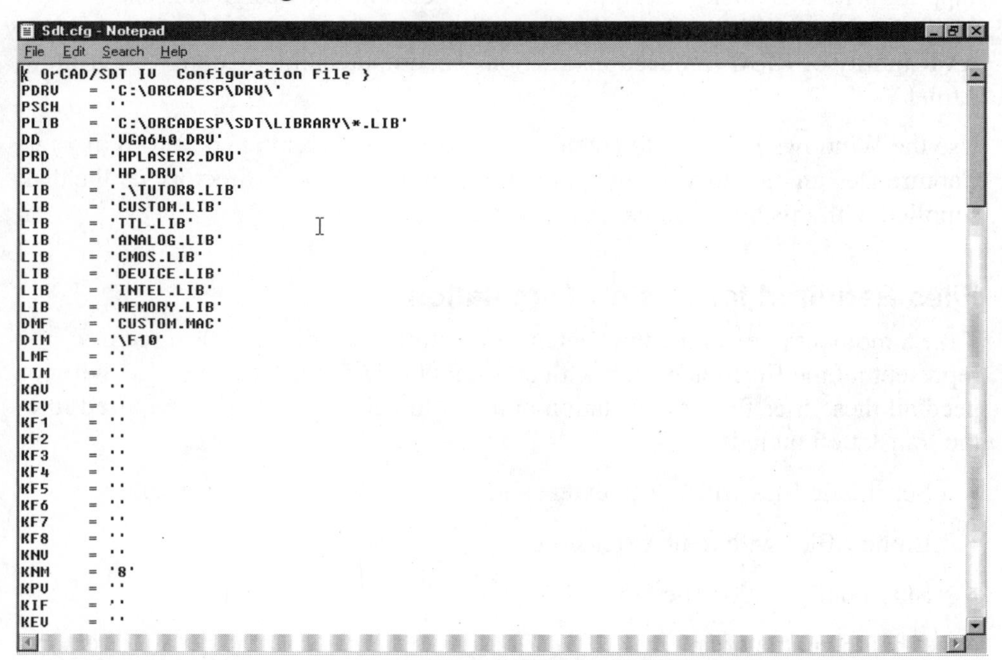


```
{ OrCAD/SDT IV  Configuration File }
PDRV    = 'C:\ORCADESP\DRV\'
PSCH    = ''
PLIB    = 'C:\ORCADESP\SDT\LIBRARY\*.LIB'
DD      = 'VGA640.DRV'
PRD     = 'HPLASER2.DRV'
PLD     = 'HP.DRV'
LIB     = '.\TUTOR8.LIB'
LIB     = 'CUSTOM.LIB'
LIB     = 'TTL.LIB'
LIB     = 'ANALOG.LIB'
LIB     = 'CMOS.LIB'
LIB     = 'DEVICE.LIB'
LIB     = 'INTEL.LIB'
LIB     = 'MEMORY.LIB'
DMF     = 'CUSTOM.MAC'
DIM     = '\F10'
LMF     = ''
LIM     = ''
KAV     = ''
KFV     = ''
KF1     = ''
KF2     = ''
KF3     = ''
KF4     = ''
KF5     = ''
KF6     = ''
KF7     = ''
KF8     = ''
KNV     = ''
KNM     = '8'
KPV     = ''
KIF     = ''
KEV     = ''
```

Figure 11-1 Library List in the SDT.CFG File

The PLIB line lists the path to the directory for OrCAD SDT libraries. If Capture cannot find this directory, it will attempt to substitute its own libraries. Each LIB line lists a library used in the OrCAD SDT design in the order that they are searched. The first library is .\Tutor8.lib. This is the archive library that contains all parts used in the design. The ".\" path indicates that this archive library is stored in the design directory. The archive library usually has the same name as the design.

Examination of the library list in the Sdt.cfg file will reveal possible difficulties that you may encounter during translation. If an archive library is listed in the Sdt.cfg file and this library is included in the design directory, you should not encounter any difficulties. If an archive library is not available and the Sdt.cfg file references specialized or custom libraries to which you do not have access, you cannot successfully translate the design.

Translation of SDT Part Fields into Capture Part Properties

OrCAD SDT allows up to eight part fields, which are roughly equivalent to Capture part properties. Part fields 1 through 7 are generally used for additional part description data, such as manufacturers' part numbers. Part field 8 is commonly used for PCB footprint values. Note that OrCAD SDT uses the term *module* in place of PCB footprint.

During the translation process, Capture reads the part field list in the OrCAD Sdt.cfg file to determine what part property names to use during the translation process. You can easily edit the part field names to match the desired Capture part property names. In the Tutor8 design example, additional part descriptions in the form of manufacturers' part numbers are in part field 2. These will be assigned the part property name VALUE_2. PCB footprint values are in part field 8. Use a text editor such as the Windows Notepad to edit the part field list as shown in Figure 11-2.

Lines labeled FN1 through FN8 correspond to the eight part fields. Note that the part field names must be enclosed in single quotation marks. Recall that Capture property names are case sensitive. The PCB footprint is a reserved property name and must be spelled and capitalized exactly as shown in Figure 11-2. The VALUE_2 name is arbitrary, and you could substitute a different name. To avoid confusion you should settle on consistent property names for all your designs.

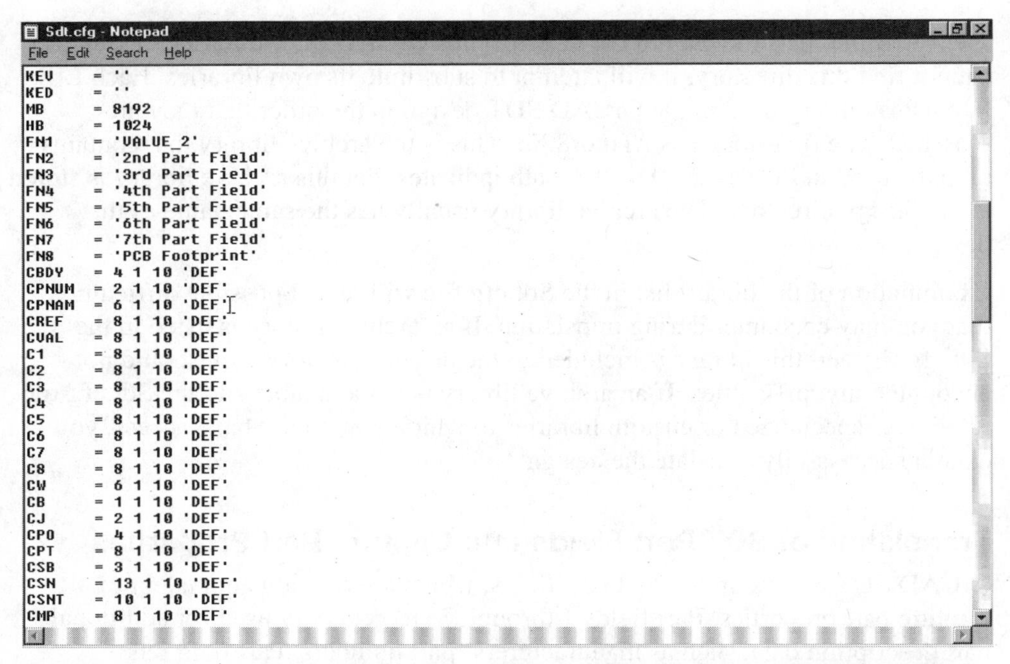

```
Sdt.cfg - Notepad
File  Edit  Search  Help
KEV     = ''
KED     = ''
MB      = 8192
HB      = 1024
FN1     = 'VALUE_2'
FN2     = '2nd Part Field'
FN3     = '3rd Part Field'
FN4     = '4th Part Field'
FN5     = '5th Part Field'
FN6     = '6th Part Field'
FN7     = '7th Part Field'
FN8     = 'PCB Footprint'
CBDY    = 4 1 10 'DEF'
CPNUM   = 2 1 10 'DEF'
CPNAM   = 6 1 10 'DEF'
CREF    = 8 1 10 'DEF'
CVAL    = 8 1 10 'DEF'
C1      = 8 1 10 'DEF'
C2      = 8 1 10 'DEF'
C3      = 8 1 10 'DEF'
C4      = 8 1 10 'DEF'
C5      = 8 1 10 'DEF'
C6      = 8 1 10 'DEF'
C7      = 8 1 10 'DEF'
C8      = 8 1 10 'DEF'
CW      = 6 1 10 'DEF'
CB      = 1 1 10 'DEF'
CJ      = 2 1 10 'DEF'
CPO     = 4 1 10 'DEF'
CPT     = 8 1 10 'DEF'
CSB     = 3 1 10 'DEF'
CSN     = 13 1 10 'DEF'
CSNT    = 10 1 10 'DEF'
CMP     = 8 1 10 'DEF'
```

Figure 11-2 Part Field List in the SDT.CFG File after Editing

Staring the Translation Process

Start Capture and open the Tutor8 design. You must select file type SDT schematic (*.sch) and then click on the root schematic, Tutor8.sch. Next, Capture prompts for the name used to save the translated design. Since Capture allows more descriptive names, use AVR_Board.dsn as the name for the translated design. The translation process takes 30-60 seconds. After the translation is complete, a message appears: "Warnings encountered – Please check Session Log." Click on Session Log from the Window menu. Messages in your session log will appear similar to those shown in Figure 11-3.

Most of the warning messages relate to OrCAD SDT part libraries that are not available on the system. Capture then informs you that it is using the Tutor8.lib archive library. Additional status messages appear as each schematic page is translated. You can ignore the warning messages about nonorthogonal wires and buses. The Tutor8 design has bus structures with 45-degree corners.

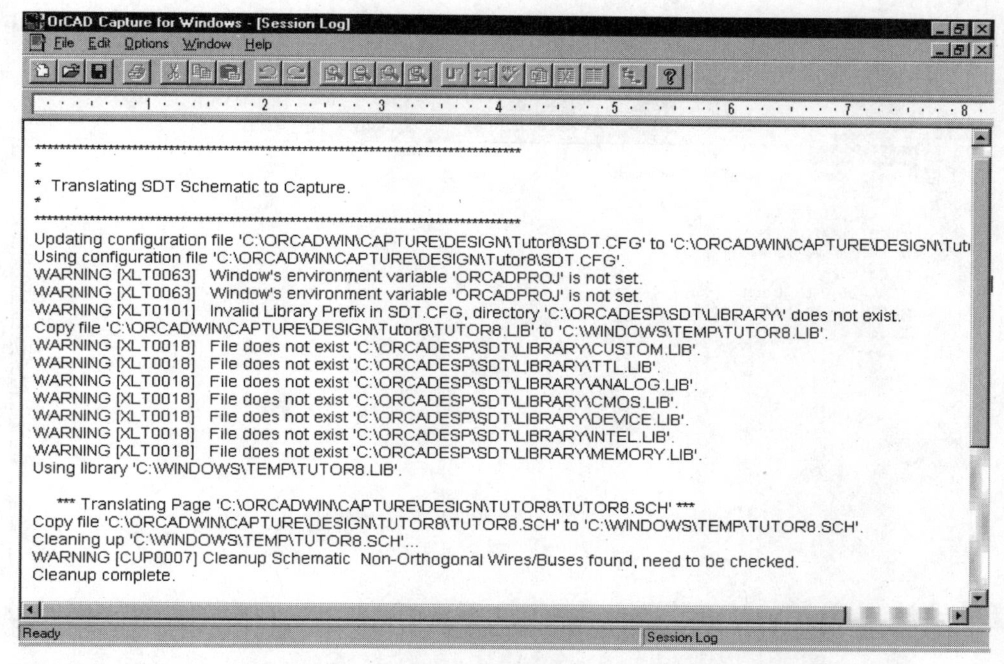

Figure 11-3 Session Log after Translation Process

Completing the Translation

Return to Project Manager by clicking on the design name on the Window menu. The design initially appears in Project Manger as shown in Figure 11-4. Edit the names of the schematic folders and pages as shown in Figure 11-5. You can easily do this by clicking on each object and then pressing the right mouse button to bring up a shortcut menu which includes the Rename command.

After renaming the schematic folders and pages, you must make corresponding changes to all hierarchical blocks in the design. In Project Manager, double click on the root schematic to open it. Then double click on the hierarchical block for the user interface to edit its properties as shown in Figure 11-6. Note that the implementation name for the hierarchical block must match the name for the corresponding schematic folder in Project Manager. Make the appropriate edits for both hierarchical blocks on the root schematic.

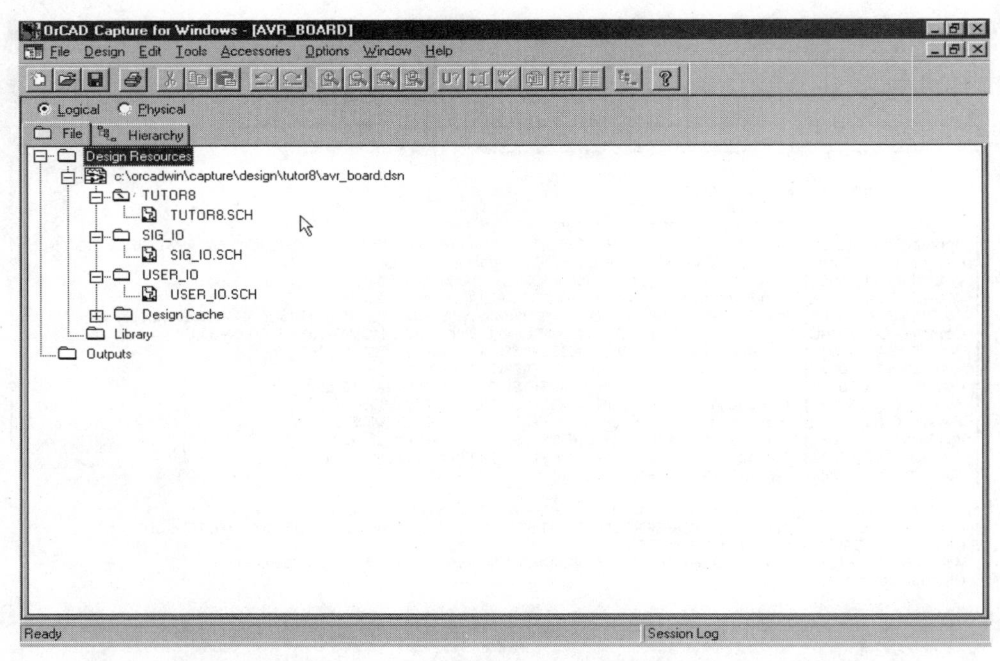

Figure 11-4 Initial Design Structure in Project Manager

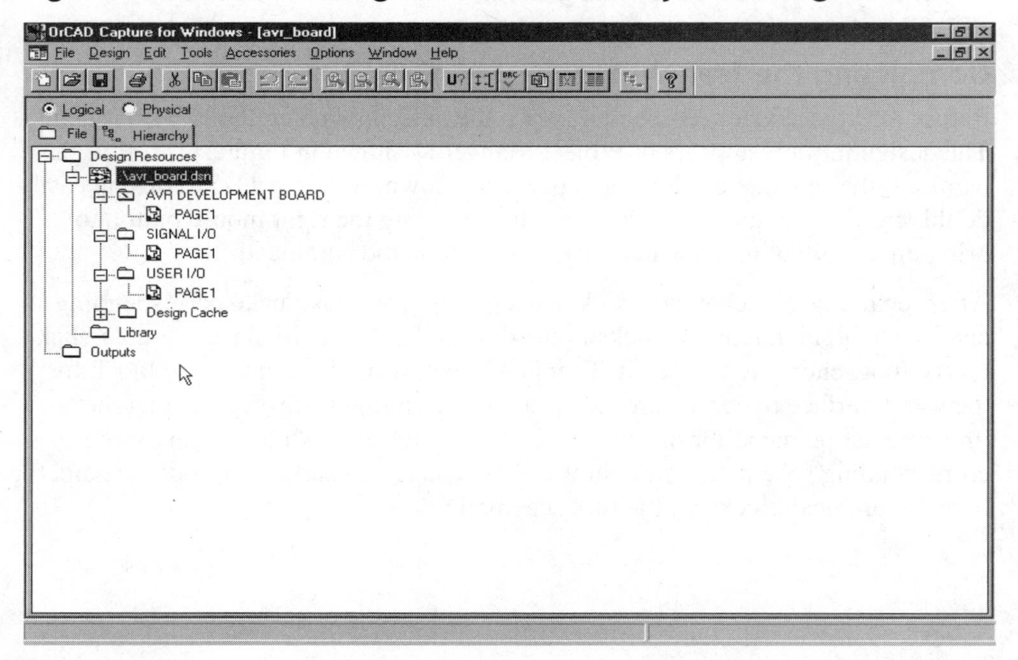

Figure 11-5 Design Structure after Editing Schematic Names

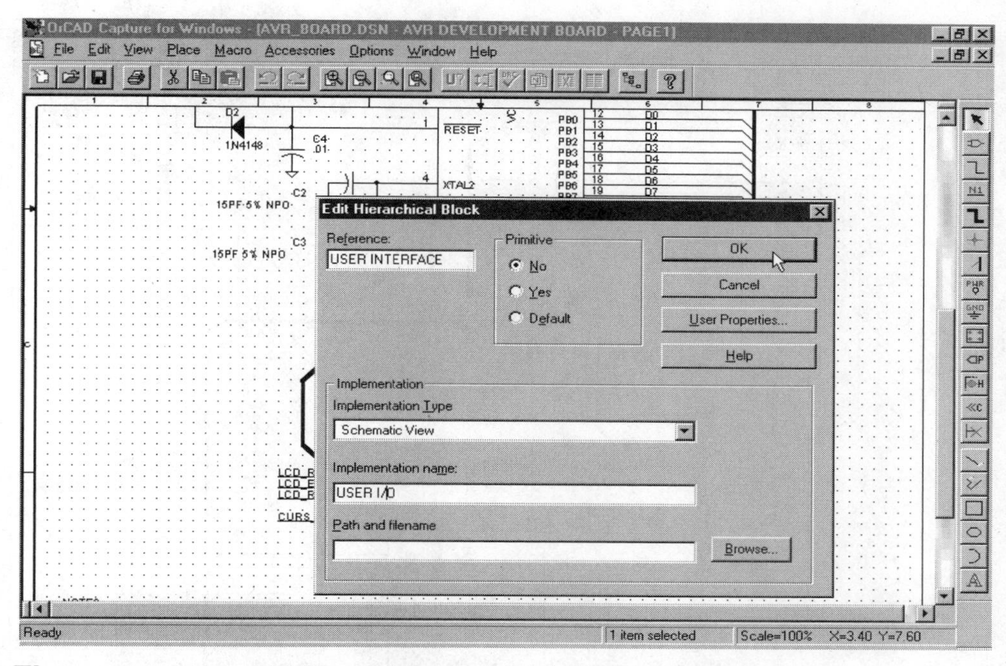

Figure 11-6 Editing Hierarchical Block Properties

The last step in the translation process is to do some cleanup editing on each schematic. You will have to realign company name and address text strings in the title block areas. You may have to realign some reference designators and part values with the part symbols.

The normal orientation for names is 90 degrees (reading from bottom to top). If you are translating a design that has many net aliases, port names, and off-page connectors with names oriented at 270 degrees (reading from top to bottom), you can select the affected objects and use the Rotate Aliases command from the Accessories menu to automatically reorient the names to 90 degrees.

The completed schematics should appear as shown in Figures 11-7 through 11-9. Save the design and print out a hard copy.

Part Properties in the Translated Design

Take a moment to examine the part properties in your translated design. In Project Manager, use the Browse tool from the Edit menu. If you correctly edited the Sdt.cfg file before starting the translation, you should see PCB footprint and VALUE_2 properties as shown in Figure 11-10.

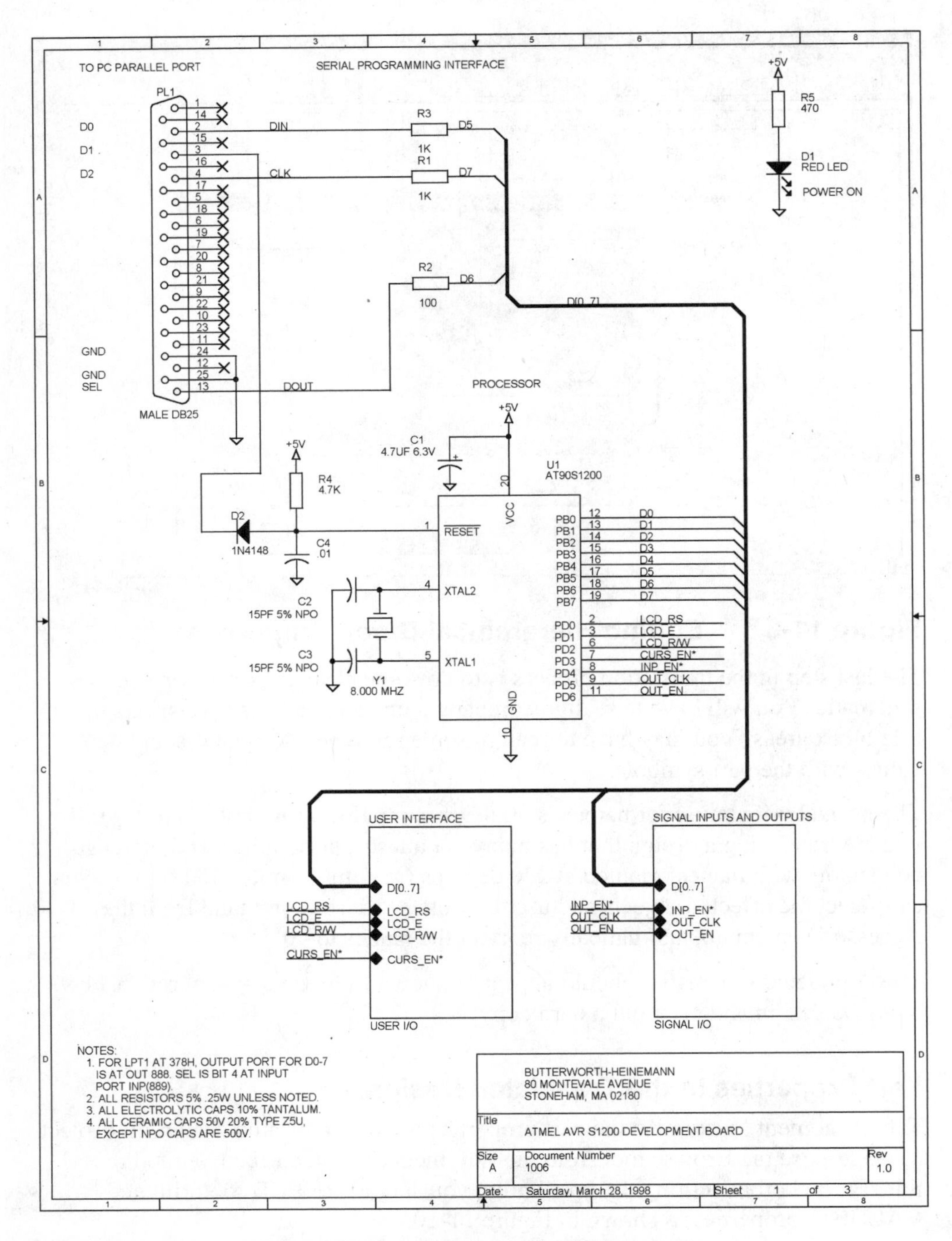

Figure 11-7 Translated Schematic (Sheet 1)

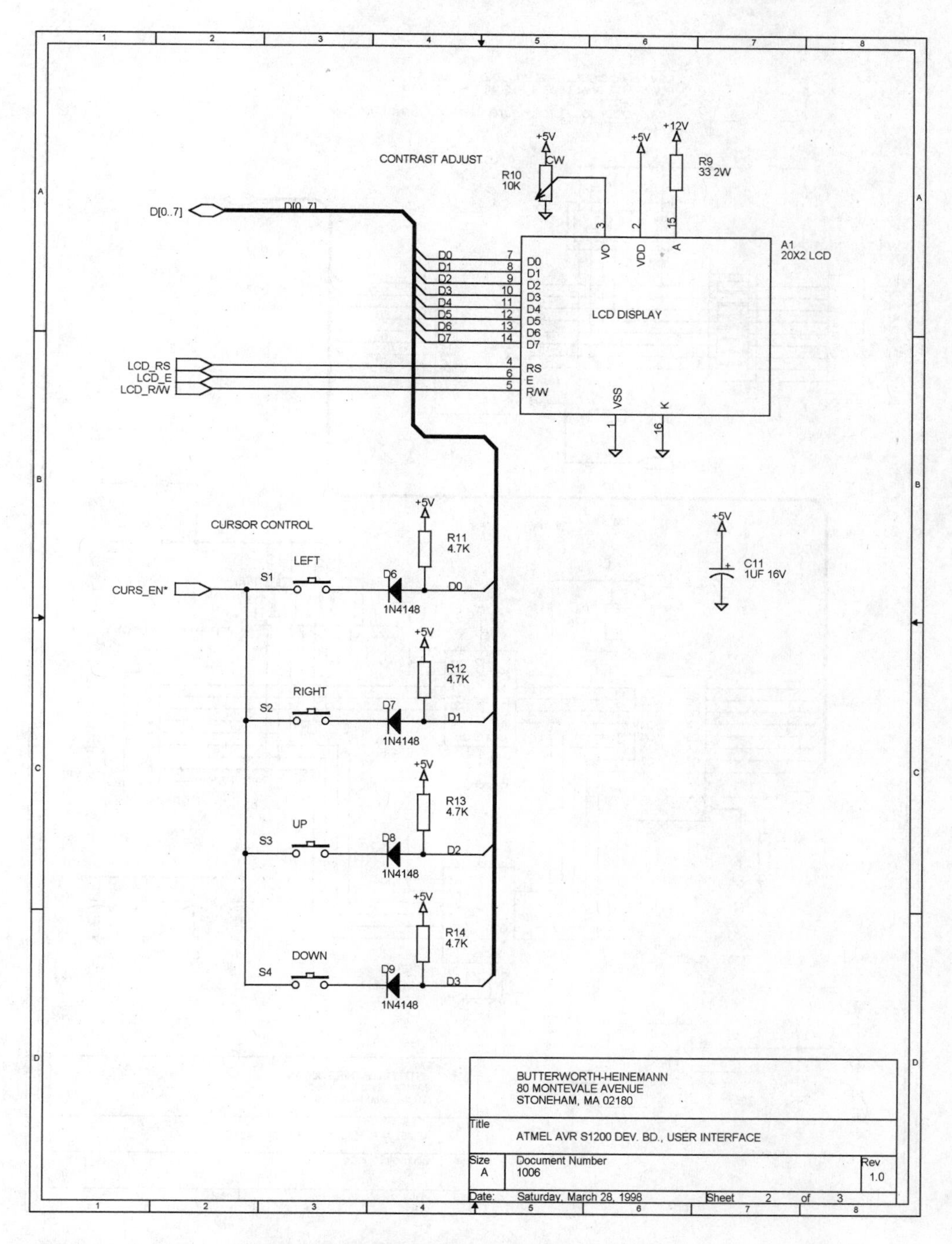

Figure 11-8 Translated Schematic (Sheet 2)

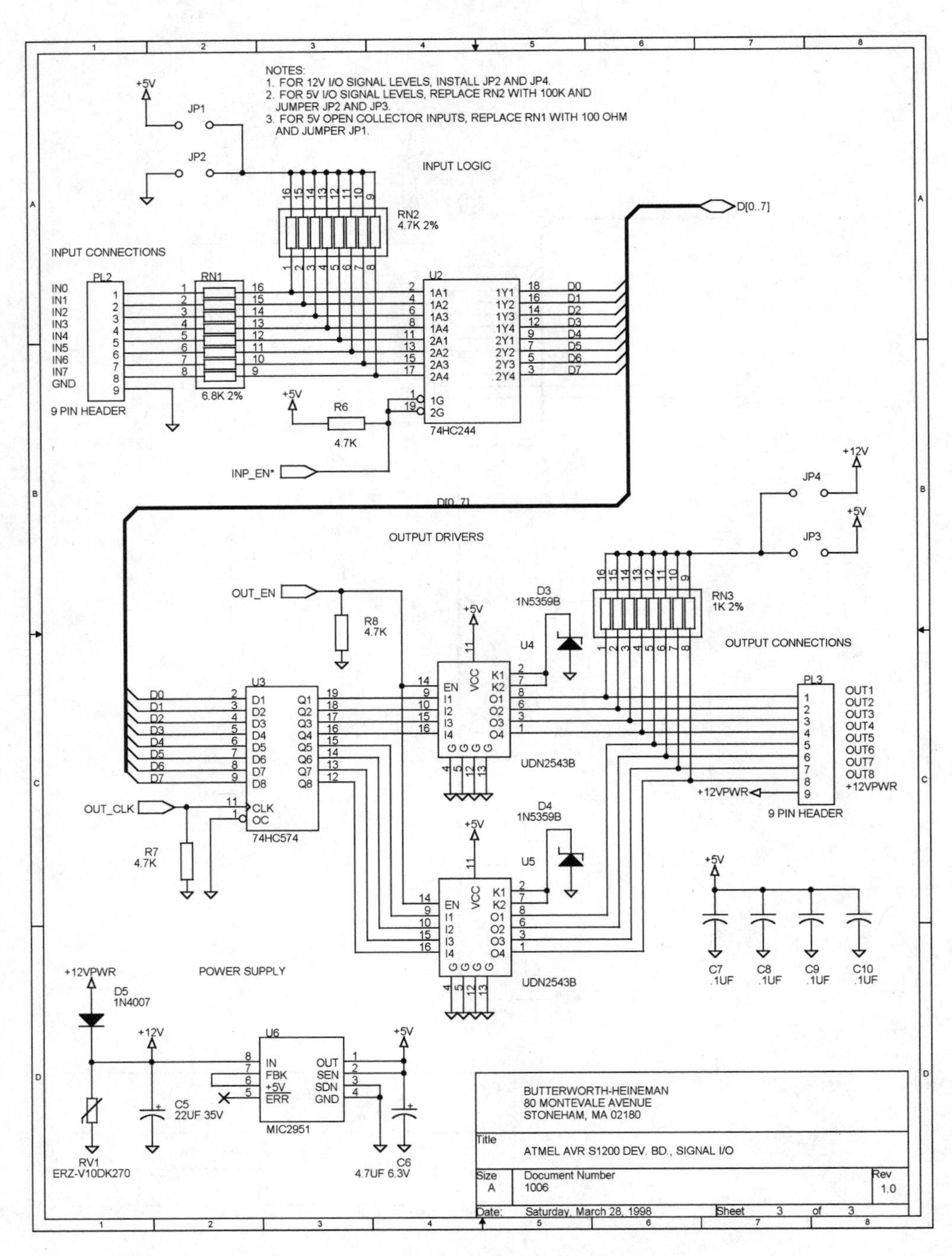

Figure 11-9 Translated Schematic (Sheet 3)

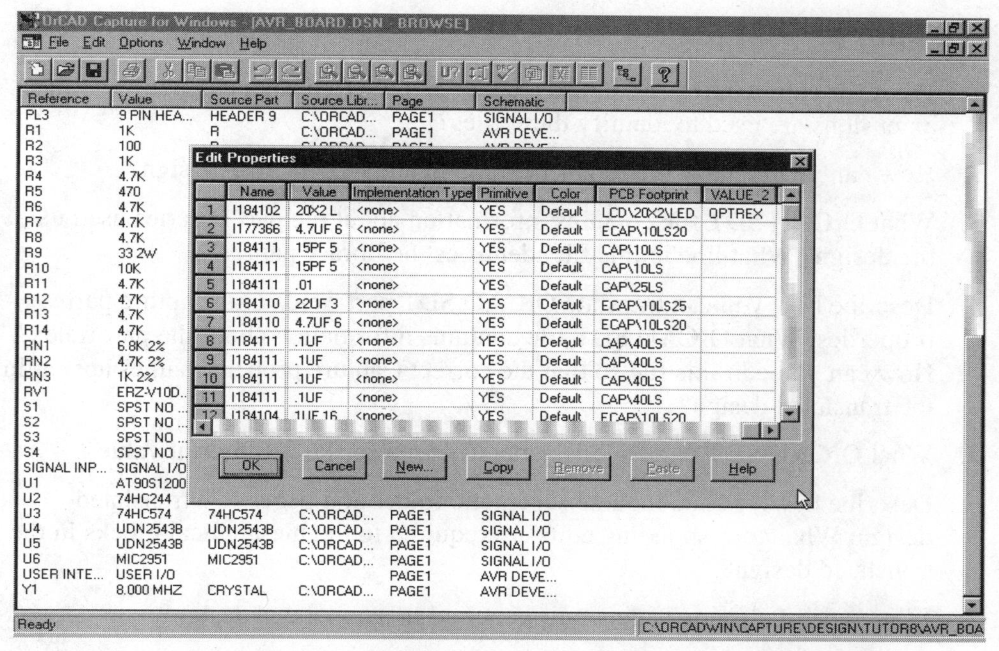

Figure 11-10 Part Properties in the Translated Design

Translating Library Files

You can easily translate OrCAD SDT part library files to Capture. Copy the OrCAD SDT part library files to your Capture library directory (typically C:\Orcadwin\Capture\Library). Start Capture and then open each OrCad SDT library file. Select file type SDT library (*.lib). When capture prompts for a name used to save the translated library, select file type Capture library (*.olb). The translation process may take several minutes for a large library.

You can also reverse the translation process from Capture back to SDT, if you need to create a part library in SDT format. If you receive OrCAD SDT designs without appropriate libraries, you can use this process to create SDT format libraries for use in the design translation process. Most of the Capture libraries are backward compatible with respect to part names. You can examine the Sdt.cfg file to determine which libraries are required. Note that you must change the library path line (starts with "PLIB") in the Sdt.cfg file to correspond to the directory that contains the translated libraries.

Review Exercises

1. What files are required to translate an OrCAD SDT design to Capture? What extensions are used to identify these files?
2. How can you identify the root schematic in an OrCAD SDT design?
3. What OrCAD SDT file contains information about the part libraries used in the design? What line in this file identifies the library path?
4. Describe how you can translate OrCAD SDT part fields into Capture part properties. What OrCAD SDT file contains information about the part fields? How can you edit this file so that the correct Capture property names appear in the translated design?
5. What OrCAD SDT part field is generally used for PCB footprint values?
6. Describe how you can rename schematic folder and pages in a translated design. What corresponding edits are required to the hierarchical blocks in the translated design?
7. What type of cleanup is generally required in a translated design?
8. Why is an archive part library so important when translating a design?
9. What can you do if no archive part library exists for a design that you must translate and you do not have access to OrCAD SDT libraries?

12

Image and Data Transfer

This chapter briefly covers techniques for image and data transfer between Capture and other Windows 95 programs. Preparation of technical manuals represents a typical application in which schematic images must be exported from Capture. Illustrating schematics with signal waveforms is an application in which graphic images are imported into Capture. Part and pin properties data can be exported to spreadsheet programs for editing. Data in EDIF (electronic design interchange format) also can be exported and imported. Refer to the Capture documentation and online help for more details about EDIF transfer.

Image data can be in either raster or vector form. Display devices, such as a CRT, and printers, including laser and inkjet varieties, use raster data. Raster devices generate an image from a pattern of small dots, typically 300 or 600 DPI for laser printers. Older mechanical pen plotters use vector data. Vector devices generate an image by drawing a series of straight lines, arcs, circles, and text characters.

Exporting Data in DXF Format

DXF (data exchange format) is a vector format originated by Autodesk (the makers of AutoCAD) during the 1980s. DXF is now widely supported throughout the EDA and CAD industry. You can export DXF data only for individual schematic pages. To export DXF data from Capture, open the desired schematic page with the schematic editor. Use the Export Design Command from the File menu and then select the DXF tab in the dialog box. You can enter a file name for the DXF file. The file path defaults to the current design directory, and the file name extension defaults to .DXF.

Once you have created a DXF file, you can import it into other Windows 95 programs that support DXF format. One common scenario in which DXF transfer proves useful involves annotating schematics with signal waveforms or details of specialized electromechanical devices using CAD software such as AutoCAD.

Like HPGL (Hewlett Packard Graphics Language) format, DXF format files contain ASCII data that you can examine with any text editor. Capture exports DXF files in AutoCAD Release 12 format which is backward compatible with most programs that accept DXF data. Note that the Microsoft Word 97 DXF import filter appears to be incompatible with DXF files created by Capture. Your EDA/CAD software documentation should include reference material that describes DXF format in detail.

Exporting Data in TIFF Format

TIFF (tagged image file format) data is a raster format originated by Adobe Systems during the 1980s. TIFF is now universally supported by desktop publishing, image editing, and even word processing programs such as Microsoft Word. Most programs that accept TIFF data offer some level of editing capability. A number of third-party TIFF print drivers are available for Windows 95. These programs allow you to write TIFF data to a file. A program that the author has used is Print-2-Image by Resource Partners, Inc. (Wakefield, NH). TIFF files in Huffman encoded CCITT (Comite Consultatif International Telephonique et Telegraphique) format generated by this program appear compatible with a wide range of applications that accept TIFF data.

Image transfer via TIFF format generally gives the best results for desktop publishing applications such as preparing technical manuals. Almost all the figures in this book are TIFF format images. Recommended practice is to use an image editing program such as Adobe Photoshop to read in the raw TIFF data, make any required edits, and then write out the data with LZW (Lempel-Zif-Welsh) compression enabled. LZW compression greatly reduces file size.

Before you can export TIFF data from Capture, you must install a TIFF print driver. In Capture, use the Print Setup command to select the TIFF printer. The file name extension .TIF is generally used for TIFF files.

Exporting Properties Data

Part and pin properties data can be exported from Capture by using the Export Properties command from the Tools menu in Project Manager. You can export the entire design or selected schematics. You can also select whether to export only part properties or part and pin properties. Properties are written to disk as tab-delimited ASCII data. The properties file path defaults to the current design directory, and the file name defaults to the design name with an .EXP extension.

You can load properties files into most spreadsheet programs. In theory you can edit the properties data and then use the Import Properties command on the Tools menu to load edited data back into Capture. However, you must be careful not to delete or add lines or change the order of lines in the file. Carefully study the online help material on this subject before attempting to edit property files. From a practical standpoint, using the Browse tool is a much safer and more efficient means of editing properties.

Importing Pictures into Capture

Capture provides a limited capability for importing raster images in the form of Windows 95 .bmp (bitmap) files. Most graphics-oriented Windows 95 programs support .bmp image files. You can use the Picture command from the Place menu to paste small images onto the schematic page. Unfortunately, this capability is still in a rather rudimentary stage. You cannot scale the bitmap image and it appears at screen resolution. You may find the Picture command useful for placing small icon-sized images representing signal waveforms. Expect to spend a considerable amount of time experimenting. It is hoped that future Capture releases will provide a more meaningful capability for import of raster images.

Review Exercises

1. Explain the difference between raster and vector graphics data.
2. What format is suggested for export of image data to EDA/CAD systems?
3. What format is suggested for export of image data to desktop publishing systems?
4. What limitations are placed on editing properties data if you plan to reimport the data into Capture? What alternative method is suggested for editing properties?

Appendix A

Information about the Disk Supplied with This Book

Disk Contents

This book includes a disk that contains sample files and a custom parts library for the tutorial exercises and the BOMSORT utility used for sorting OrCAD Capture bills of materials. The sample files for the tutorial exercises are in separate directories organized by chapter usage.

BOMSORT.EXE	Bill of materials sort utility used in Chapter 10.
CUSTOM.LIB	Custom library installed in Chapter 2.
TUTOR1	Sample files for exercises in Chapter 3.
TUTOR2	Sample files for exercises in Chapter 4.
TUTOR3	Sample files for exercises in Chapter 5.
TUTOR4	Sample files for exercises in Chapter 7.
TUTOR5	Sample files for exercises in Chapter 8.
TUTOR6	Sample files for exercises in Chapter 9.
TUTOR7	Sample files for exercises in Chapter 10.
TUTOR8	Sample files for exercises in Chapter 11.
README.TXT	Text file with any updated information not available when the book was printed.

The sample files are intended to help complete the exercises and to serve as examples of finished designs. In some cases, fine details on schematic figures within this book may be hard to make out because of the reduced size. You can use Capture to print out the sample schematic files for improved readability.

Requirements and Compatibility

The files are supplied on an MS-DOS formatted 1.4MB 3.5 inch floppy disk. The BOMSORT utility program should run on any MS-DOS/PC-DOS compatible computer capable of running OrCAD Capture schematic drafting software. BOMSORT will run in a Windows 95 environment but does not support long file names. At least 10MB of free hard disk space is recommended to allow for the files created during the tutorial exercises.

Index